Strategies to Enhance Environmental Security in Transition Countries

NATO Security through Science Series

This Series presents the results of scientific meetings supported under the NATO Programme for Security through Science (STS).

Meetings supported by the NATO STS Programme are in security-related priority areas of Defence Against Terrorism or Countering Other Threats to Security. The types of meeting supported are generally "Advanced Study Institutes" and "Advanced Research Workshops". The NATO STS Series collects together the results of these meetings. The meetings are co-organized by scientists from NATO countries and scientists from NATO's "Partner" or "Mediterranean Dialogue" countries. The observations and recommendations made at the meetings, as well as the contents of the volumes in the Series, reflect those of participants and contributors only; they should not necessarily be regarded as reflecting NATO views or policy.

Advanced Study Institutes (ASI) are high-level tutorial courses to convey the latest developments in a subject to an advanced-level audience

Advanced Research Workshops (ARW) are expert meetings where an intense but informal exchange of views at the frontiers of a subject aims at identifying directions for future action

Following a transformation of the programme in 2004 the Series has been re-named and re-organised. Recent volumes on topics not related to security, which result from meetings supported under the programme earlier, may be found in the NATO Science Series.

The Series is published by IOS Press, Amsterdam, and Springer, Dordrecht, in conjunction with the NATO Public Diplomacy Division.

Sub-Series

A. Chemistry and Biology	Springer
B. Physics and Biophysics	Springer
C. Environmental Security	Springer
D. Information and Communication Security	IOS Press
E. Human and Societal Dynamics	IOS Press

http://www.nato.int/science
http://www.springer.com
http://www.iospress.nl

Series C: Environmental Security

Strategies to Enhance Environmental Security in Transition Countries

Edited by

Ruth N. Hull
Cantox Environmental Inc.
Mississauga, Canada

Constantin-Horia Barbu
"Lucian Blaga" University of Sibiu
Sibiu, Romania

and

Nadezhda Goncharova
International Sakharov Environmental University
Minsk, Belarus

Published in cooperation with NATO Public Diplomacy Division

Proceedings of the NATO Advanced Research Workshop on
Strategies to Enhance Environmental Security in Transition Countries
Sibiu, Romania
6–9 September 2006

A C.I.P. Catalogue record for this book is available from the Library of Congress.

ISBN 978-1-4020-5995-7
ISBN 978-1-4020-5996-4 (eBook)

www.springer.com

Printed on acid-free paper

CONTENTS

PREFACE

Environmental security and sustainability are important considerations in NATO, Partner, and Mediterranean Dialogue countries. It is the goal of environmental security to protect people from the short- and long-term ravages of nature, man-made threats, and deterioration of the natural environment. Environmental security includes activities related to pollution prevention, environmental conservation, compliance with regulations, and environmental restoration. Transition countries have legacy issues (e.g., contamination issues associated with historical and continued industrial operations), and ongoing challenges (e.g., supplying safe, clean drinking water, dealing with solid waste and sewage), which are not unlike those in other countries. These issues can affect environmental security and sustainability within the transition countries, and may develop into transborder security and sustainability issues if left unresolved.

Transition countries (especially the Eastern and Central European ones) have the political and economical will to partner with or join international organizations (e.g., NATO, the European Union), although this requires adherence to many environmental laws and regulations. However, risk assessment and management frameworks, approaches, methods, priorities, and solutions have been developed primarily in the USA and other countries where funding is not a significant limiting factor when addressing environmental and health issues. Therefore, the environmental laws and regulations are not only new to many transition countries, but extremely expensive for them to meet.

Transition in Eastern and Central European countries, as well as in former Soviet Union countries, requires a practical approach to prioritizing, assessing and remediating significant environmental problems. Experience at similar sites and with similar issues around the world allows focus on the most critical issues and allows prioritization of actions and funds so that significant environmental security and sustainability improvement can be achieved, recognizing the constraints of limited resources, and different social, cultural, legal, political, and economic situations.

This volume presents case studies and other experience addressing a variety of environmental security and sustainability challenges. While the focus was to assist transition countries, the tools described herein are equally applicable for developed countries to ensure cost-effective methods are used and significant environmental issues are not overlooked.

The NATO Advanced Research Workshop (ARW) on "Strategies to Enhance Environmental Security in Transition Countries" took place from 6 to 9 September 2006 in historical Sibiu, Romania. As many as 35 participants from

14 countries participated in the ARW. Participants had diverse backgrounds and experiences, from current graduate students studying in former Soviet countries, to experienced professors from NATO, Partner, and Mediterranean Dialogue countries, to consultants and government researchers.

A field trip to a community close to Sibiu, impacted by historical and ongoing smelter emissions, promoted discussion of a real-world case. Copsa Mica is a site situated 40 km north of Sibiu where a smelter has been polluting the area with heavy metals for more than 60 years. Nowadays, the smelter is owned by a Greek holding company and is under pressure by Romanian authorities to comply with national (and European) regulations. The elevated levels of cadmium (50–100 mg/kg) and lead (500–1000 mg/kg) in soil, as well as in air, pose a great threat to the local people. Some people want the smelter to be closed others do not, due to the high unemployment rate in the area (60% of the active population). While at the site, the group discussed issues such as the most significant health and environmental concerns around the site, cost-effective approaches and tools for risk assessment, and risk management, risk communication challenges, and the balance between economic and environmental security priorities. Visiting the site provided the group with a better understanding of the environmental security challenges that such sites impose on transition countries.

Our hope is that academic researchers, private consulting firms, and government scientists will be encouraged to tackle the challenges related to environmental security using the ideas and tools that are presented in the various chapters of this book. Readers are welcome to contact the various authors of the chapters to continue the discussions.

Publication of this book would not have been possible without the assistance of Gary Williams and Victoria Putyrskaya who participated in the editing and formatting of the chapters. Our gratitude also to Dr. Deniz Beten, Programme Director of NATO's Public Diplomacy Division and her assistant Ms. Lynne Campbell-Nolan for their support for us holding this workshop, and their advice during workshop planning and completion. Thanks to NATO and to Mrs. Wil Bruins and Ms. Annelies Kersbergen of Springer for their support in bringing this book to completion. Finally, we thank all of the participants of this ARW for their willingness to share their experiences, their contributions to stimulating discussions, and the many wonderful personal and professional relationships that were built during our time in Sibiu.

Ruth N. Hull
Constantin-Horia Barbu
Nadezhda Goncharova

LIST OF CONTRIBUTORS

S. Allen-Gil
Environmental Studies Program, Ithaca College, Ithaca, NY 14850 USA

A. Amster
The Arava Institute for Environmental Studies, D.N. Hevel Eilot, 88840 Israel

D. Bairasheuskaya
Laboratory of Environmental Monitoring, International Sakharov Environmental University, Minsk, Belarus

D. Bănăduc
"Lucian Blaga" University of Sibiu, Faculty of Sciences, Ecology and Environmental Protection Department, Sibiu, Romania

H. Barbu
"Lucian Blaga" University of Sibiu, Sibiu, Romania

O. Blum
M.M. Gryshko National Botanical Garden, National Academy of Sciences of Ukraine, Timiryazevs′ka 1, 01014 Kyiv, Ukraine

O. Borysova
Kharkiv National Academy of Municipal Economy, 12 Revolutzii St., Kharkiv 61002, Ukraine

S. Brenner
The Arava Institute for Environmental Studies, D.N. Hevel Eilot, 88840 Israel

J. Bronders
Flemish Institute for Technological Research, Mol, Belgium

C. Bucşa
"Lucian Blaga" University of Sibiu, Faculty of Sciences, Ecology and Environmental Protection Department, Sibiu, Romania

D. Cadar
Technical University of Cluj-Napoca, Romania

A. Coman
Environmental Health Center, Cetatii 23 A, 400166 Cluj Napoca, Romania

A. Critto
Interdepartmental Centre IDEAS – University Ca' Foscari of Venice, Dorsoduro 2137, 30123 Venice, Italy

A. Curtean-Bănăduc
«Lucian Blaga» University of Sibiu, Faculty of Sciences, Ecology and Environmental Protection Department, Sibiu, Romania

W. Dejonghe
Flemish Institute for Technological Research, Mol, Belgium

K. Demnerova
Department of Biochemistry and Microbiology, Institute of Chemical Technology Prague; Technicka 5, 166 28 Prague 6, Czech Republic

L. Diels
Flemish Institute for Technological Research, Mol, Belgium

J. Dijk
Catholic University of Leuven, Heverlee, Belgium

V.G. Dikarev
Russian Institute of Agricultural Radiology and Agroecology, Kievskoe shosse, 109 km, 249020, Obninsk, Russia

N.S. Dikareva
Russian Institute of Agricultural Radiology and Agroecology, Kievskoe shosse, 109 km, 249020, Obninsk, Russia

T. Drozd
Laboratory of Molecular Markers of Environmental Factors, International Sakharov Environmental University, Minsk, Belarus

L. Eckardt
Eurofins-AUA GmbH Jena, Germany

T.I Evseeva
Institute of Biology, Komi Scientific Center, Ural Division RAS, Kommunisticheskaya 28, 167982, Syktyvkar, Russia

Z. Filip
Department of Biochemistry and Microbiology, Institute of Chemical Technology Prague; Technicka 5, 166 28 Prague 6, Czech Republic

L. Galatchi
Ovidius University, Department of Ecology and Environmental Protection, Mamaia Boulevard 124, RO-900527 Constanta-3, Romania

S. Geras'kin
Russian Institute of Agricultural Radiology and Agroecology, Kievskoe shosse, 109 km, 249020, Obninsk, Russia

I. Gishkeluk
International Sakharov Environmental University, 23 Dolgobrodskaya St., Minsk, 220009, Belarus

N. Goncharova
Laboratory of Environmental Monitoring, International Sakharov Environmental University, Minsk, Belarus

A. Gorobets
Sevastopol National Technical University, Management Department, Streletskaya Bay, Sevastopol 99053, Ukraine

P. Gramatikov
South-Western University "Neofit Rilski", Ivan Mihailov 66, 2700 Blagoevgrad, Republic of Bulgaria

I. Grdzelishvili
Cooperation for a Green Future, Tbilisi, Georgia

N. Grinchik
International Sakharov Environmental University, 23 Dolgobrodskaya St., Minsk, 220009, Belarus

A.E. Gurzau
Environmental Health Center, Cetatii 23 A, 400166 Cluj Napoca, Romania

E.S. Gurzau
Environmental Health Center, Cetatii 23 A, 400166 Cluj Napoca, Romania

K. Hamonts
Flemish Institute for Technological Research, Mol, Belgium

S.R. Hilts
Teck Cominco Metals Ltd, Trail, British Columbia, Canada

R.N. Hull
Cantox Environmental Inc. 1900 Minnesota Court, Suite 130, Mississauga, Ontario L5N 3C9, Canada

H. Kalka
Umwelt- und Ingenieurtechnik GmbH, Dresden, Germany

P. Kozubek
AQUATEST, Prague, Czech Republic

T. Kuhn
Forschungszentrum für Umwelt und Gesundheit, Neuherberg, Germany

J. Kuklik
AQUATEST, Prague, Czech Republic

S. Kundas
International Sakharov Environmental University, 23 Dolgobrodskaya St., Minsk, 220009, Belarus

T. Lange
Consulting und Engineering GmbH, Chemnitz, Germany

C. Lipchin
The Arava Institute for Environmental Studies, D.N. Hevel Eilot, 88840 Israel

A. Lopareva
Laboratory of Environmental Monitoring, International Sakharov Environmental University, Minsk, Belarus

M. Maesen
Flemish Institute for Technological Research, Mol, Belgium

G. Malina
Institute of Environmental Engineering, Częstochowa University of Technology, Brzeźnicka 60A, 42-200 Częstochowa, Poland

A. Marcomini
Interdepartmental Centre IDEAS – University Ca' Foscari of Venice, Dorsoduro 2137, 30123 Venice, Italy

P. Marozik
Laboratory of Molecular Markers of Environmental Factors, International Sakharov Environmental University, Minsk, Belarus

R. Meckenstock
Forschungszentrum für Umwelt und Gesundheit, Neuherberg, Germany

S. Melnov
Laboratory of Molecular Markers of Environmental Factors, International Sakharov Environmental University, Minsk, Belarus

T. Nawalaniec
Poznan University of Technology, Institute of Electronics and Telecommunications, ul. Piotrowo 3A, PL-60965 Poznan, Poland

W. Nawrocki
Poznan University of Technology, Institute of Electronics and Telecommunications, ul. Piotrowo 3A, PL-60965 Poznan, Poland

I. Neamtiu
Environmental Health Center, Cetatii 23 A, 400166 Cluj Napoca, Romania

L. Oprean
«Lucian Blaga» University of Sibiu, Sibiu, Romania

A.A. Oudalova
Russian Institute of Agricultural Radiology and Agroecology, Kievskoe shosse, 109 km, 249020, Obninsk, Russia

A. Ozunu
«Babes-Bolyai» University of Cluj-Napoca, Romania

S. Pazniak
International Sakharov Environmental University, 23 Dolgobrodskaya St., Minsk, 220009, Belarus

J. Perner
Eurofins-AUA GmbH Jena, Germany

N.H. Peters
Umwelt- und Ingenieurtechnik GmbH, Dresden, Germany

D. Petrescu
«Dimitrie Cantemir» Christian University, Cluj-Napoca, Romania

V. Putyrskaya
Laboratory of Environmental Monitoring, International Sakharov Environmental University, Minsk, Belarus

A. Rieger
Consulting und Engineering GmbH, Chemnitz, Germany

M. Rutgers
Laboratory for Ecological Risk Assessment, National Institute for Public Health and the Environment, RIVM. P.O. BOX 1, 3720 BA Bilthoven, The Netherlands

A. Ryngaert
Flemish Institute for Technological Research, Mol, Belgium

C. Sand
«Lucian Blaga» University of Sibiu, Sibiu, Romania

R. Sathre
Cooperation for a Green Future, Tbilisi, Georgia

E. Semenzin
Consorzio Venezia Ricerche, Via della Libertà 5-12, 30175 Marghera, Venice, Italy and Interdepartmental Centre IDEAS – University Ca' Foscari of Venice, Dorsoduro 2137, 30123 Venice, Italy

H. Smidt
Wageningen University, Wageningen, The Netherlands

D. Springael
Catholic University of Leuven, Heverlee, Belgium

M. Sturme
Wageningen University, Wageningen, The Netherlands

S.M. Swanson
Golder Associates Ltd., Calgary, AB

I. Tanasescu
University of Agricultural Sciences and Veterinary Medicine, Cluj-Napoca, Faculty of Animal Breeding and Biotechnology, Civil Engineering and Technical Design Department, Romania

V. Tarasenko
International Sakharov Environmental University, 23 Dolgobrodskaya St., Minsk, 220009, Belarus

V. Tasoti
Babes-Bolyai University, Faculty of Chemistry and Chemical Engineering, Str. Arany Janos 11, 400028, Cluj-Napoca, Romania

G. Torosyan
State Engineering University of Armenia 105 Teryan, 375009 Yerevan, Republic of Armenia

D.V. Vasiliev
Russian Institute of Agricultural Radiology and Agroecology, Kievskoe shosse, 109 km, 249020, Obninsk, Russia

J. Vos
Flemish Institute for Technological Research, Mol, Belgium

D. Wilzcek
Flemish Institute for Technological Research, Mol, Belgium

N. Zohdi
International Development and Environment Associates (IDEA), 37 Syria St., 4th Floor, Apt. #8, Mohandessin, Giza, Egypt

PART I

INTRODUCTION TO ENVIRONMENTAL SECURITY CHALLENGES

1. INTRODUCTION TO ENVIRONMENTAL SECURITY

CONSTANTIN-HORIA BARBU*, CAMELIA SAND,
LETITIA OPREAN
"Lucian Blaga" University of Sibiu, Sibiu, Romania

Abstract: The term "Environmental Security" refers to a range of concerns that can be organized into three general categories: (1) concerns about the adverse impact of human activities on the environment, seen as an asset to be inherited by the next generations; (2) concerns about the direct and indirect effects of various forms of environmental change (especially scarcity and degradation), which may be natural or human-generated on national and regional security; (3) concerns about the loss of security individuals and groups (from small communities to humankind) experience due to environmental change such as water scarcity, air pollution, global warming, and so on. In order to provide sustainability, environmental security represents the intersection between social and political systems with ecological systems. All individuals should have fair and reasonable access to environmental goods, and mechanisms to address disputes/crises/conflicts exist or should be created. The activities encompassed by environmental security may also be pollution prevention, environment conservation, compliance with existing regulations and norms, and environmental restoration. This advanced research workshop was developed due to recognition of the environmental security challenges faced by transition countries and in the hope that together we can identify practical approaches to prioritizing, assessing, and correcting significant environmental problems, which make the best use of limited financial and other resources.

Keywords: environmental security; sustainability

*To whom correspondence should be addressed. Dr. Constantin-Horia Barbu, "Lucian Blaga" University of Sibiu, Sibiu, Romania; e-mail: horiab@rdslink.ro

R. N. Hull et al. (eds.), Strategies to Enhance Environmental Security in Transition Countries, 3–12.

1. Historical and Conceptual Backgrounds

At the beginning of the 20th century, the general perception was that nothing new can be discovered in physics, and no war can occur. Because it took only 5 years until $E = mc^2$ opened new perspectives in physics, and another 9 years for the outbreak of the First World War, optimism was not the main characteristic in predicting how the 21st century would be. It was less than 2 years before 9/11 changed the world, proving that the conflicts and mistakes of the last century are more dangerous than anyone can imagine. Armed conflicts, terrorism on both large and small scales, floods, hurricanes, earthquakes, and tsunamis are topping the news.

In this context, environmental protection and the sustainable use of natural resources (which is part of sustainable development) is of paramount importance because neglecting them can cause either an adverse response from nature, or conflicts due to the scarcity of these resources.

Since the early 1990s, a great deal of research has tried to determine the relationship between environment and security. According to Oli Brown (2005), there are four main approaches concerning the links between security and environment:

1. The Toronto school approach (name given to the research groups led by Thomas Homer-Dixon). This approach focuses on resource scarcity as a cause for insecurity and conflict. The Toronto school argues that simple scarcity as a result of environmental change and population growth is only part of a much more complex picture. They focus on situations where elites extend their control over productive resources (in a process called "resource capture") and displace poorer communities ("ecological marginalization"). Resource capture and ecological marginalization, they argue, may lead to conflict (as people resist marginalization) and environmental damage (as displaced people move into fragile, marginal environments). In some cases, this process may be connected to state failure and political violence, especially in developing states where insurgencies are fueled by grievances related to injustice and inequity.

2. Another approach is proposed by the Swiss Environment and Conflicts Project (ENCOP), led by Günther Baechler. ENCOP research links environmental conflict more directly to a society's transition from a subsistence to a market economy. They argue that violence is most likely to occur in more remote areas, mountainous locations, and grasslands – places where environmental stresses coincide with political tensions and inequitable access to resources. In many cases, conflict occurs where communities resist the expropriation of resources and the environmental damage caused by large-scale development projects.

3. The approach proposed by the International Peace Research Institute in Oslo (PRIO) suggests that violence in many developing countries occurs when different groups attempt to gain control of abundant resources. World Bank studies indicate that countries heavily dependent for their income on the export of primary commodities are at a dramatically higher risk of conflict than other poor countries, particularly during periods of economic decline. Other studies suggest that many wars concern control over revenues from valuable resources – especially so if they are easy to transport and hard to trace.
4. A fourth approach argues that environmental degradation is one of the many "network threats" that face the world. Climate change, like epidemic disease or international terrorism, is an example of a network threat. People make decisions about their energy use based on their immediate social, economic, and ecological surroundings. These decisions constitute an informal, transnational web of individual behaviors that ultimately present a truly global security problem. Like epidemic disease, the threat is dispersed, and so is difficult to neutralize through negotiations or force. And although climate change could be extremely dangerous and costly, it is hard to identify an effective mitigation policy, since no single incentive structure can modify the behavior of all the actors.

Thus, the need to protect people from the short-term and long-term ravages of nature, man-made threats in nature, and deterioration of the natural environment has been defined as major aims of a new concept, named "Environmental Security".

If there is a consensus on the meaning of "environment", the concept of *security* has for too long been interpreted narrowly: as security of territory from external aggression, or as protection of national interests in foreign policy or as global security from the threat of a nuclear holocaust. It has been related more to nation–states than to people. The legitimate concerns of ordinary people were forgotten, disregarding that for many of them, security symbolized protection from the threat of disease, hunger, unemployment, crime, social conflict, political repression, and environmental hazards (UNDP, 1994).

Although sustainable development and environmental security are mutually reinforcing concepts and directions for policy, they are not the same thing. Sustainable development focuses on environmentally sound development that is economically, financially, socially, and environmentally sustainable.

Environmental security focuses more on preventing conflict and loss of state authority due to environmental factors, as well as the additional military needs to protect their forces from environmental hazards and repair military-related environmental damage. The UN and related commissions and conferences have popularized sustainable development as meeting the needs of the present without compromising the ability of future generations to meet their own needs. There is not yet a similar universal definition of environmental security (Glenn and Gordon, 2004).

2. Definitions and Aspects of Environmental Security

The term "Environmental Security" was officially introduced in 1994 by the United Nations Development Program (UNDP), which has stated in the Human Development Report that the scopes of global and human security should be (Henk, 2005):

- *Economic security*, assuring every individual a minimum requisite income.
- *Food security*, guaranteeing "physical and economic access to basic food".
- *Health security*, guaranteeing a minimum protection from disease and unhealthy lifestyles.
- *Environmental security*, protecting people from the short-term and long-term ravages of nature, man-made threats in nature, and deterioration of the natural environment.
- *Personal security*, protecting people from physical violence, whether from the state, from external states, from violent individuals and substate actors, from domestic abuse, from predatory adults, or even from the individual himself (as in protection from suicide).
- *Community security*, protecting people from loss of traditional relationships and values, and from sectarian and ethnic violence.
- *Political security*, assuring that people "live in a society that honors their basic human rights".

In the complex world in which we are living, besides or maybe due to the increase in the quality of life in most countries, there are a lot of major threats that can destroy the fragile equilibrium on earth. These major threats are:

- Human population growth and loss of biodiversity
- Climate change
- Water scarcity and pollution, including ground water contamination
- Food security
- Environmental refugees
- Deforestation
- Industrial contamination of air and oceans
- (Toxic) wastes

- Soil conservation/erosion
- Nuclear safety issues
- Ozone depletion
- Global warming

The extent of these threats and their causes are summarized in the table below (Glenn and Gordon, 2005).

	By ignorance and/or mismanagement	By intention	Mix of natural and human actions
Within a country	Aral Sea depletion in Russia Forest fires (Greece, Turkey) Groundwater contamination and freshwater scarcity Hazardous wastes Soil erosion Human settlement and development patterns	Sarin gas attack in Tokyo subway Chemical attacks and draining marshes in Iraq Poisoning or diversion, or misuse of water resources	Floods Famines Salinization Earthquakes Introduction of exotic species
Transborder	Rain forest depletion River usage (Jordan, Nile, Tigris, Euphrates) Chernobyl nuclear accident Diminishing biodiversity Ozone depletion Fisheries depletion Global climate change Acid rain and air pollution Poverty Radioactive waste	Burning oil fields in Kuwait Poisoning water Dam construction and water diversion Biological weapons	Solar radiation changes Global warming New, emerging, and drug resistant diseases such as AIDS and others affecting plants and animals Desertification Population growth Rich–poor gap

When analyzing these threats, one should consider two major aspects:

- Timescale in which an event/threat can occur (short-term: 0–5 years, mid-term: 5–25 years and long-term: >25 years).
- Responsibility of national and international bodies: international organizations, national governments, regional bodies, nongovernmental organizations (NGOs), and corporations.

Based on these considerations, from the multitude of definitions, we consider that the most significant one is: environmental security is the relative public safety from environmental dangers caused by natural or human processes due to ignorance, accident, mismanagement, or design and originating within or across national borders.

In this respect, the term "Environmental Security" refers to:

1. Concerns about the adverse impact of human activities on the environment, seen as an asset to be inherited by the next generations
2. Concerns about the direct and indirect effects of various forms of environmental change (especially scarcity and degradation), which may be natural or human-generated on national and regional security
3. Concerns about the insecurity individuals and groups (from small communities to humankind) experience due to environmental change such as water scarcity, air pollution, global warming, etc.

3. Global and Regional Actions to Enhance Environmental Security

Due to the complexity posed by environmental security problems and to the great efforts involved in enhancing it, this issue is part of the conjugated efforts of the United Nations (UN) (mainly of UNDP and United Nations Environment Program), of the Organization for Security and Cooperation in Europe (OSCE), and of the North Atlantic Treaty Organization (NATO).

Some major achievements in defining the actions and goals of environmental security are the elaboration of a long-scale program (Millennium Project) and the Resolution adopted by the UN General Assembly in September 2000 (United Nations Millennium Declaration), and the Environment and Security (ENVSEC) Initiative, focused mainly on Southeastern Europe (SEE) and Central Asian environmental problems.

One of the largest actions, the Millennium Project started in 2002, defines environmental security as environmental viability for life support, with three subelements:

- Preventing or repairing military damage to the environment
- Preventing or responding to environmentally caused conflicts
- Protecting the environment due to its inherent moral value

To make the world a safer place, ensure sustainability, and reverse the trend of environmental degradation, scholars and decision makers are examining the links between environment and security, the environmental risk factors that might trigger political tension, and the key players and their actions in order to achieve these goals. The Millennium Project conclusions are:

- Environmental security threats often involve transborder and/or global impacts that would require international cooperation.
- Nation–states acting alone can not provide environmental security.
- International organizations do not have the capacity to address the threats.
- The weight of decision power rests with national governments. As a result, national sovereignty can come in conflict with actions necessary to insure environmental security.

The OSCE, UNEP, and UNDP set up the ENVSEC Initiative in 2002 with the aim of identifying, together with regional stakeholders (such as governmental representatives and NGOs), environmental issues that are a threat to stability and peace, and using environmental cooperation as a confidence-building exercise in vulnerable regions. Since 2004, NATO has become associated with ENVSEC through the NATO Program for Security through Science and the activities of the Committee on the Challenges of Modern Society. NATO involvement coincides with focusing the activities on Eastern Europe countries (Belarus, Moldova and Ukraine), where the priorities are: rehabilitation of deprived areas, such as Chernobyl; pesticides destruction; transboundary river monitoring; underground pollution affecting wells; military dumps; and disposal of melange (a hazardous rocket fuel oxidizer).

In the future, NATO should continue to send information to the OSCE, UNEP, and UNDP on its newly awarded environmental projects for their screening against the ENVSEC criteria. These criteria relate to questions such as "Does the project have an impact on security? Does it integrate environment, security and policy? Is it focused on a vulnerable region?" In this way, NATO projects that meet all criteria are embedded into the ENVSEC Initiative. There are currently 16 Science for Peace projects embedded into ENVSEC, including, for example, "South Caucasus Co-operative River Monitoring" and "Study of Radioactive Waste Disposal in Turkmenistan". NATO will also look into the possibility of initiating more top-down projects on the basis of the ENVSEC assessments through which the countries identify their environmental security priorities.

The United Nations Millennium Declaration (the General Assembly Resolution no. 55/2 of 8 September 2000) considers certain fundamental values to be essential to international relations in the 21st century, among these being the respect for nature: "Prudence must be shown in the management of all living species and natural resources, in accordance with the precepts of sustainable development. Only in this way can the immeasurable riches provided to us by nature be preserved and passed on to our descendants. The current unsustainable patterns of production and consumption must be changed in the interest of our future welfare and that of our descendants".

Concerning *protection of our common environment*, the signatory head of states declared that:

- "We must spare no effort to free all of humanity, and above all our children and grandchildren, from the threat of living on a planet irredeemably spoilt by human activities, and whose resources would no longer be sufficient for their needs.
- We reaffirm our support for the principles of sustainable development, including those set out in Agenda 21, agreed upon at the UN Conference on Environment and Development.
- We resolve therefore to adopt in all our environmental actions a new ethic of conservation and stewardship and, as first steps, we resolve:
    - To make every effort to ensure the entry into force of the Kyoto Protocol, and to embark on the required reduction in emissions of greenhouse gases.
    - To intensify our collective efforts for the management, conservation and sustainable development of all types of forests.
    - To stop the unsustainable exploitation of water resources by developing water management strategies at the regional, national and local levels, which promote both equitable access and adequate supplies.
    - To intensify cooperation to reduce the number and effects of natural and man-made disasters".

The world's leaders, gathered at the UN headquarters in New York in September 2005, have recognized the serious challenge posed by climate change, being committed to take action through the UN Framework Convention on Climate Change and have agreed to create a worldwide early warning system for all natural hazards.

One of the most sensitive regions for environmental security concerns is SEE, i.e., Albania, Bosnia and Herzegovina, Bulgaria, Croatia, the former Yugoslav Republic of Macedonia, Romania, Serbia and Montenegro, and Kosovo, and the ENVSEC Initiative intends to solve the many environmental problems still existing here. For this, several meetings were or will be held:

- First Regional Meeting on ENVSEC in SEE, December 2002, Belgrade
- ENVSEC Consultations in SEE, September 2004, Skopje
- Reducing ENVSEC Risks from Mining in SEE and the TRB, May 2005, Cluj-Napoca
- Environment for Europe, 2007, Belgrade.

The programs are built around four interrelated "pillars": (1) in-depth vulnerability assessment, early warning and monitoring; (2) strengthening policies, institutions and awareness; (3) capacity building. In areas where specific risks are present ENVSEC also implements its "4th pillar" "Cleanup and remediation" by mobilizing financial support and technical expertise.

Besides this, many programs are under way, such as:

- Reducing Environment and Security Risks from Mining in SEE
- Management of Transboundary Natural Resources, focused on:
  (a) Reducing the Impact of Agriculture in Prespa Region
  (b) Reversal of Land and Water Degradation in the Tisza River Basin (TRB) Ecosystem
  (c) Network Development of Local Actors from the Sava River Basin on Water Resources Management
  (d) Enhancing Transboundary Biodiversity Management

As may be seen, the complexity of environmental security and the policies for its enhancement, enjoy worldwide recognition and efforts, in the hope that the future will be brighter and safer.

4. Conclusion

In order to provide sustainability, environmental security represents the intersection between social and political systems with ecological systems, where all individuals should have fair and reasonable access to environmental goods, and mechanisms to address disputes/crises/conflicts exist or should be created.

The activities included in environmental security may also be pollution prevention, environment conservation, compliance with existing regulations and norms, and environment restoration.

The mismanagement of natural resources lies at the heart of many conflicts and disaster vulnerability. Sustainable development is impossible without security. A better understanding of the links between environmental change and human security is vital for effective peace-building and postconflict reconstruction. Conflict-sensitive conservation could both protect biodiversity and contribute to peace-building. Better environmental management may help reduce community vulnerability to floods and droughts. Future work will seek to understand these links and provide practical advice on how careful management of our environment and our resources can contribute to conflict avoidance and disaster resilience.

References

Brown, O., 2005, *The Environment And Our Security – How Our Understanding Of The Links Has Changed*, International Institute for Sustainable Development (IISD), May 2005.

Glenn J.C., Gordon T.J., 2005, 2004, *State of the Future*, American Council for the United Nations, pp. 82–88.

Henk, D., 2005, *Human Security: Relevance and Implications, Parameters*, Summer 2005, pp. 91–106.

United Nations Development Program, 1994, United Nations Development Program, *Human Development Report, 1994*, available at: http://hdr.undp.org/reports/global/1994/en

2. ENVIRONMENTAL SECURITY: PRIORITY ISSUES AND CHALLENGES IN ROMANIA AND EUROPEAN TRANSITION COUNTRIES

ALEXANDRU OZUNU*
"Babes-Bolyai" University of Cluj-Napoca, Romania
DORIS CADAR
Technical University of Cluj-Napoca, Romania
DACINIA CRINA PETRESCU
"Dimitrie Cantemir" Christian University, Cluj-Napoca, Romania

Abstract: There is rationale in linking environment and security issues. The management of natural resources use and extraction can have very real and concrete impacts on human health and vulnerability, both at the regional and local levels. Often the impacts of environmental degradation are disproportionately borne by poor and marginalized groups. Southeastern Europe (SEE) is a complex area, strategically located. The crisis management environment is heavily conditioned by historical political and military events including numerous armed conflicts. Specifically, it is further complicated by the collapse of the Soviet system and resultant abandonment of its political and physical infrastructure, the major changes in land use, population migrations, and changes in political jurisdictions that created ethnic minorities within and among neighboring nations. This article offers a brief introduction to environmental security (ES) issues and the challenges faced by SEE.

Keywords: environment; security; environmental security; Southeastern Europe; Seveso II; Safety Report

---

* To whom correspondence should be addressed. Alexandru Ozunu, "Babes-Bolyai" University of Cluj-Napoca, Romania; e-mail: aozunu@enviro.ubbcluj.ro

R. N. Hull et al. (eds.), Strategies to Enhance Environmental Security in Transition Countries, 13–23.

1. Introduction

1.1. HISTORY

The term environmental security (ES) was primarily coined in 1970 regarding conflict situations caused by deforestation and soil erosion, and their related consequences – runaway population growth and poverty triggering famine in the highlands and mass migration to the Ogaden lowlands, located between Ethiopia and Somalia (Ogaden war; Myers, 2004). ES was then confirmed in the late 1970s, more related to the water quest and its preservation, being a continuous problem since 1950 (bold increase in world population and shift to modern life standard) – Jordan River (Israel threatened by Syria and Jordan; Blue Nile – Ethiopia overhanging Egypt).

The meta-problem of ES began to be widely recognized in the mid-1980s (Nermine Zohdi, Chapter 11, this volume; World Commission on Environment and Development, 1987): "The environmental problems of the poor will affect the rich as well in the not too distant future, transmitted through political instability and turmoil." And as Mikhail Gorbachev put it in 1989, "The threat from the skies today is not so much nuclear missiles as ozone-layer depletion and global warming." Since the 1980s the issue has steadily grown in importance, and especially in the second half of the 1990s, we have witnessed an outburst of interest from various ambits: research and science, industry and economy, international relations, etc.

1.2. CONCEPTUAL DETERMINATION

A central point in ES prevails over time: *there often rests a number of additional factors that undermine security* – faulty economic policies, inflexible political structures, oligarchic regimes, oppressive governments, and other adverse factors that have nothing directly to do with the environment. These deficiencies often aggravate environmental problems, and are aggravated by environmental problems in turn. Environmental deficiencies supply conditions, which render conflict all the more likely: serving to determine the source of conflict, or acting as multipliers that aggravate core causes of conflict, thus helping to shape the nature of conflict. Moreover they can not only contribute to conflict, but also stimulate the growing use of force to repress disaffection among those who suffer the consequences of environmental decline (Myers, 2004).

All these implications enforce national security's modern role, superannuating the basic need of fighting forces and weaponry alone (military prowess), but increasingly relating to watersheds, forests, soil cover, croplands, genetic resources, climate, and other factors rarely considered by military experts and political leaders. The new paradigm of national security is epitomized by the leader who no longer proclaims he will not permit 1 m^2 of national territory to be ceded to a foreign invader, but the one person who would not allow hundreds of square kilometers of topsoil to be eroded away each year.

The basic framework for understanding the relationship between environment and security is the Millennium Ecosystem Assessment (Figure 1), which looks at all the functions of ecosystems and the services they deliver to people and nature, depicting their strength linkages one to another (Institute of Environmental Security, 2005).

Broadly defined, ES affects humankind and its institutions and organizations anywhere and at anytime.

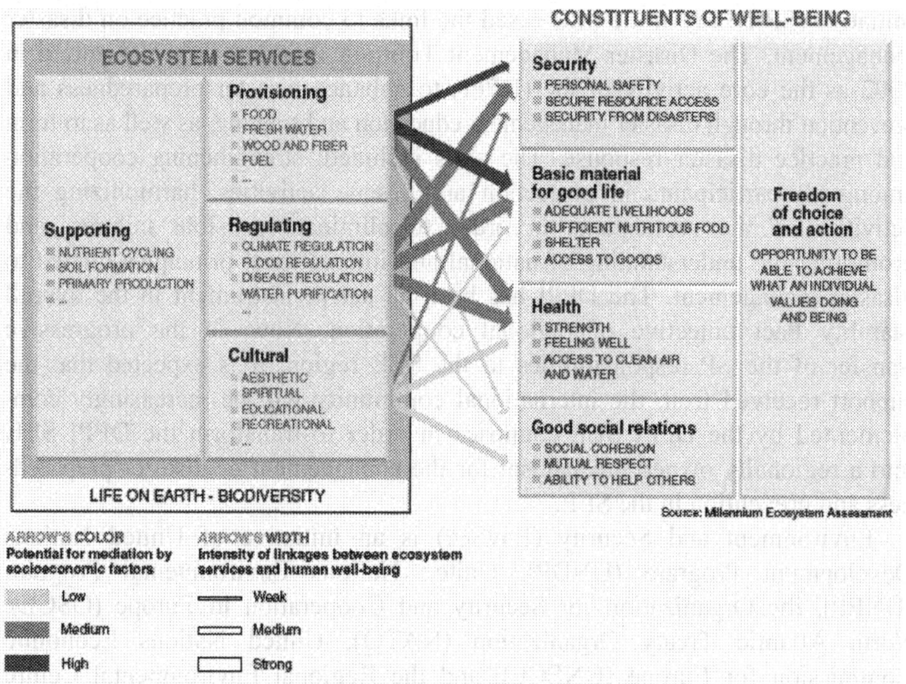

Figure 1. Conceptual determination – environment (consisting of ecosystems) and security (aiming to the well-being). (From Millennium Ecosystem Assessment, 2005, p. 50.)

2. SEE Perspective on Environmental Security

2.1. THE NEED FOR INNOVATIVE INSTITUTIONAL ARRANGEMENTS

Present global institutions governing worldwide well-being, were established in response to two world wars fought in the last century. A cause for great concern is that the world's political, legal, and economic institutions from the global to the local level operate on outdated concepts and trail behind in formulating and implementing preventive policies through improved and innovative institutional and financial arrangements. Global institutions need to be equipped for the 21st century and resolve ES challenges by peaceful means.

Notably, beginning with end of the 1990s, Southeastern Europe (SEE) began to operate programs and international institutions aiming to develop and govern upon ES. We consider in the following paragraphs just a few of them, to briefly point out their effect on ES within the SEE region.

The presence of the Stability Pact's Disaster Preparedness and Prevention Initiative (SP-DPPI) in the SEE eased the links to common practice on disaster management. The Disaster Management Training Program (DMTP) started in 2002 as the core activity of the (DPPI) to enhance disaster preparedness and prevention through disaster management education and training, as well as to train and practice disaster response. The aims included: strengthening cooperation among all participants in protection and rescue activities, harmonizing the activities of the participants in order to eliminate possible failures, and promoting the understanding of national and international principles related to disaster management. The DPPI has been an active instrument in the overall Stability Pact objective of regional cooperation. Now, in the progressive transfer of the SP responsibilities to the SEE region, it is expected that the support received from the international community will be increasingly complemented by the DPPI SEE countries in order to transform the DPPI SEE into a regionally owned mechanism for the enhancement of disaster preparedness and prevention in the SEE.

Environment and Security (EnvSec) is an initiative of United Nations Development Program (UNDP), United Nations Environmental Program (UNEP), the Organization for Security and Cooperation in Europe (OSCE), North Atlantic Treaty Organization (NATO), United Nations Economic Commission for Europe (UNECE), and the Regional Environmental Centre (REC) offer countries their combined pool of expertise and resources towards addressing the links between the natural environment and human security. It builds on the combined strengths and field presence of the leading organizations to perform three key functions: assessment of environment and security risks, capacity building and institutional development to strengthen environmental cooperation, as well as integration of environmental and security concerns and priorities in international and national policy making. Five demonstration projects will be finalized and launched in the period of 2006–2008 under the umbrella project "Environment and Security in South Eastern Europe: Improving regional cooperation for risk management from pollution hotspots as well as trans-boundary management of shared regional resources".

Southeastern Europe Simulation (SEESIM) was launched in October 2000, by the ten Defense Ministers of the SEE Defense Ministerial SEDM process that endorsed a US proposal for a SEESIM network, to promote cooperation, coordination, and interoperability among SEDM nations and the initiatives through effective use of computer modeling and simulation. SEESIM is designed to serve as a foundation for integrating several initiatives functioning within the SEDM framework and to be a prelude to regional live exercises and real world emergency response operations. Representatives from NATO and the Stability Pact for SEE have a standing invitation to attend SEESIM meetings and planning conferences.

The Regional Arms Control Verification and Implementation Assistance Centre (RACVIAC) was established within the framework of the Stability Pact, Working Table III on security issues. It was initiated by Germany in the middle of 1999 as a common project with the Republic of Croatia, but it started working in June 2000, in the city of Zagreb, as a multinational regional center for the assistance in implementing Arms Control agreements for all SEE states, which signed or are about to sign such agreements. RACVIAC is a multinational center with its twofold aim being: a forum for regional dialogue and cooperation in different arms control and confidence and security building measures, as well as to provide assistance in all matters of Arms Control and its implementation. RACVIAC contributes to the aims of the Stability Pact, supports and complements the activities of the OSCE, as well as other organizations working in similar fields in SEE. Since 2005, RACVIAC has expanded its mission to conducting seminars and workshops in the area of defense conversion. Seminars are organized not only for military, but also for civilian representatives (civilian experts, parliament members and academics) from SEE states. Topics for the seminars are as follows: military contacts and cooperation in SEE, regional dialogue, defense planning and military budget, democratic control of armed forces to improve stability in SEE, CSBMs in establishing stability and security in SEE, regional cooperation in combating illicit trade in small arms and light weapons, disaster assistance and procedures, border control and management, and defense conversion. Currently, there are 39 members from 16 European states working in RACVIAC. It is continuously improving its regional position, thus formulating its strategy in the beginning of April 2006.

In the period between 9 and 25 July 2006, DPPI organized an 11 SEE countries assessment, under the 3rd Field Work on Disaster Management preparedness in the SEE region, in order to establish a structure for disaster management preparedness for the period of 2007–2008 in the DPPI's member countries from the SEE Region to continue DMTP. Conclusions drawn from interviewing international experts and DMTP coordinators working closely in DM to enforce ES, pinpoint as the utmost objective to be achieved: set up innovative institutional arrangements in the SEE to bridge existing expertise and to benefit synergistically from it.

2.2. INSIGHT UPON ENVIRONMENTAL SECURITY IN THE SEE

SEE countries have to face challenges and problems of ever increasing dimension and complexity from a wide range of disasters and emergency arising from technological hazards. The environment (air, water, soil) is liable to technological aggression all the time and the progressive worsening of its quality determines social problems, having an impact over the quality of life.

We can safely make several broad statements that apply generally to the risks nations of SEE face from natural, man-made or technological disasters. In most cases national authorities were unable or unwilling to attempt to establish priorities for the risk. There are clearly a number of shared risks that can be readily recognized by nations in the SEE region and provide the impetus for mutual cooperation. In a number of cases vulnerabilities as well as the risks are shared by neighboring nations. Man-made or technological risks rival natural threats to the nations of the region. The potential problems caused by the introduction of nuclear power and research facilities and the growing volume of hazardous material created by economic activity or movement on major regional transportation arteries require significant planning and preparedness efforts. Safety standards and guidelines are not uniformly applied and insurance systems that could assist in managing the risk and providing compensation when there are casualties are in the early stages of development.

The consequence of recent military actions and the persistence of danger from munitions and land mines deployed in earlier conflicts have created a particular type of man-made technological risk requiring close cooperation between military and civilian authorities to reduce threat of injury and death.

Establishing transboundary environmental monitoring and early warning systems is key for obtaining relevant and timely information on the state of the environment for policy makers. Detailed synthesis and analysis of data, in accordance with predefined criteria, should form the basis for decision making and elaboration of measures for action. Initiatives of this kind, applying a regional approach to addressing environmental "hot spots", technically and financially supported by international institutions, deserve the support and efforts for their implementation.

In the primary EnvSec Initiative, strongly aimed at delivering ES in SEE, there has been one distinct cluster of priorities for the SEE to be dealt with, related to technological hazards: managing and reducing transboundary risks of hazardous activities, in particular related to mining and industrial "hot spots". Despite a number of chronic and acute contamination pathways from sites of mining and mineral processing activity, it is the release of mine tailings waste through the effects of wind and water, particularly in the form of catastrophic releases to waterways that remains the most pressing issue related to minerals activity. A large number of warning "vectors" are present in the region – ranging from waste types with high environmental toxicity, through an absence of adequate maintenance and monitoring, to the geographical and seismic conditions in the region. It was indicated that over 150 major sites of resource extraction and downstream processing had been cataloged for the region, and that roughly 30% of these had been identified as "candidate hot spots". The transboundary nature of risks required that such activities must involve transboundary cooperation.

3. Technological Hazards

3.1. SEE POPULATION AT RISK

Change of the political influence of areas, the development of democracy, and the change of certain frontiers all lead, from the point of view of industrial development, to the profound importance of good marketing positioning, to an increasing competition among producers, and to implementing environmental protection as a priority issue. All these problems continually occur to the detriment of poor and medium developed SEE countries.

All countries in SEE are exposed to various man-made disasters that menace both population and environment. Technological disasters in one country may imperil one or more neighboring countries in the region. It is therefore necessary that each country in SEE possesses a developed system of observing, informing, and warning the population, and, at the same time, a system for information sharing with other countries in the region.

At the moment, SEE countries lack centers for information sharing and networking systems in case of industrial accidents, local and regional experts, and the knowledge transfer flow among them. These countries' observation and information centers serve for local use only, i.e., exclusively for the needs of the country where the center is located. Given the condition that basic requirements for quick and effective information sharing hardly exist, the existing local institutions and agencies of SEE countries cannot react to man-made disasters with sufficient speed and means.

For the above reasons, countries of SEE should endeavor to follow the examples from European Community member countries and establish local centers with a joint operational system and standard operational procedures. This would result into fast and efficient information sharing to prevent all kinds of disasters in SEE, and preparation for networking and effective action.

3.2. ISSUES ON SAFETY REPORT

The requirement to have a major accident policy is a new duty, imposed on operators of establishments in SEE that comes within the scope of the Seveso II Directive. The necessity of implementing a Risk Management Program in accordance with a dedicated law is a priority for countries facing the EU-enlargement process, as in the case of Romania. Companies need to take inventory of their current operations to determine the existing risks on human health and the environment within their plants, with the scope of effectively communicating them to the stakeholders and the public.

SEE countries have to face challenges and problems arising from ever increasing dimension and complexity of a wide range of potential disasters and emergencies related to technological hazards. Neighborhoods with communities of industrial platforms/parks are dramatically influenced because of air and water pollution, and soil degradation. There are two major problems to be urgently addressed: risk assessment and recovery strategy for polluted sites, and risk analysis and management of technological hazards.

Answers to the two problems share most risk-related methodologies, IT tools, and data sources, so they can be accomplished in a synergistic coordinated manner. Companies need to take inventories of their current operations to determine exactly what health and environmental risks they present. It is difficult to argue before an Environmental Protection Agency (EPA) or the public that a particular plant poses no significant industrial hazard and risk when no one has ever assessed the level of risk from it. The social perceptions, especially in countries within Eastern European become critical when one tries to combine environmental and social values with economic issues. Which is more important in value judgment: the ecology or human health? This relationship is complementary. In real life we do not have a choice – if we want to maintain a high level of human health we must invest in the environment. What about a comparison between human health and ecological health against the economic risk? Social perceptions, especially among SEE countries, become critical when one tries to combine social values, land use, economic issues, property rights, and expectations.

Safety studies/reports and safety programs are a relatively new domain. Concerning these issues, there is no Romanian version of any methodological guide for environmental risk assessment in the case of an industrial incident or accident. International consultants' recommendations embrace guidelines provided by the work of the Major Accident Hazards Bureau, under the Joint Research Centre of the European Commission, from Ispra, Italy (Fabbri et al., 2005).

Industrial risk assessment (safety) and ecological risk assessment are complementary activities. The operators must undertake risk assessments according to the legislation. The necessity to implement a Risk Management Program (RMP) according to a specific law becomes a priority for SEE countries towards assuring their economic development. The RMP minimises the risks relating to the accidental release of hazardous substances, thus reducing potential risks to acceptable levels. Developing a RMP implies an analysis of the various possible hazardous scenarios and subsequent options. Facilities that produce, handle, or store toxic chemicals must be subject to the more rigorous constraints of RMP rules. The steps to be followed in developing a RMP should be:

- Identify hazardous chemical risks
- Estimate the on-site as well as the off-site consequences for the public in the worst case and alternative scenarios
- Design procedures for site facility maintenance
- Train employees in the execution of safe procedures
- Prevent releases from occurring
- Minimize the consequences of releases if they do occur
- Inform the public

The requirement to have a Major Accident Prevention Policy (MAPP) is a new duty imposed on industry that comes within the scope of Seveso II Directive. The MAPP developments are connected with effective arrangements to promote safety, health, and environmental (SHE) protection in place and auditing their performance. Based on the provisions of the Seveso II Directive and on the existing state of hazard prevention management, internal and external emergency plans must be prepared.

It is necessary to involve the public and their representatives in post-incident reviews for preventing further hazards at the incident ground and to learn for the future. The accidents databases are very important and relevant in emergency response plan preparing. The process industries facilities must ensure that they have an effective operational and communications strategy for addressing residual risks of old technologies. The main problems dealing with the occurrence of accidental water pollution is to estimate residual risk in the transport, handling and storage of toxic waste products and to develop an effective and adequate disaster management plan within european legislation.

4. Conclusions

The environment is the most transnational of transnational issues, and its security is an important dimension of peace, national security, and human rights that is just now being understood.

The SEE region, prior to its full integration into the EU, needs to create basics for evaluation, monitoring, preparedness, and intervention in any situation threatening its ES, either by a nation's own means or through regional cooperation.

ES is strongly linked to economic welfare as technological hazards are a priority issue to address, mainly for two reasons:

- The rapid adoption, enforcement, and implementation of EU Directives – mainly upon environmental aspects
- Marketplace acceptance and positive evolution of industrial operators in a European producer market

"Sustainable development is a compelling moral and humanitarian issue, but it is also a security imperative. Poverty, environmental degradation and despair are destroyers of people, of societies, of nations. This unholy trinity can destabilise countries, even entire regions." (Colin Powell, 1999).

References

Environment and Security Initiative, Map of Hazardous Industrial Sites, Water Pollution and Mines (2006), available at: http://www.envsec.org/see/maps/hazmines_cluj.jpg

Fabbri, L., Struckl, M., Woods, M. (eds.), 2005, European Commission EUR 22113 – Guidance on the Preparation of a Safety Report to Meet the Requirements of Directive 96/82/EC as Amended by Directive 2003/105/EC (Seveso II), Office for Official Publications of the European Communities, Luxembourg, 2005, ISBN 92-79-01301-7.

Institute of Environmental Security, Horizon 21 – Advancing Global Environmental Security, 2005, available at: www.envirosecurity.org

Millennium Ecosystem Assessment, 2005, Ecosystems and Human Well-Being: Synthesis, Island Press, Washington, DC. Available at: http://www.maweb.org//en/Products.Synthesis.aspx

Myers, N., 2004, Environmental Security: What's new and different? Background paper for The Hague Conference on Environment, Security and Sustainable Development, The Hague, The Netherlands 9–12 May 2004. Available at: http://www.envirosecurity.org/conference/working/newanddifferent.pdf

United Nations, World Commission on Environment and Development, 1987, *Our Common Future*, Report of the World Commission on Environment and Development, August 1987. Available at: www.are.admin.ch

PART II

STRATEGIES AND METHODS/APPROACHES

3. DEVELOPING HIGHLY EFFECTIVE ENVIRONMENTAL SECURITY POLICIES FOR A SUSTAINABLE DEVELOPMENT OF THE FUTURE. CASE STUDY: ROMANIA

LIVIU–DANIEL GALATCHI*
Ovidius University, Department of Ecology and Environmental Protection, Mamaia Boulevard 124, RO-900527 Constanta-3, Romania

Abstract: Seven objectives are identified that can increase the effectiveness of environmental security policy: (1) prevent environmental damage before it happens; (2) take countermeasures that concentrate on more basic causes; (3) implementing elements of the environmental framework in an integrated way; (4) combine environmental security policies with other policies; (5) make appropriate use of the market mechanism; (6) make appropriate use of information; (7) appropriately shape the mechanism for linkups and participation by concerned parties. Developing highly effective environmental security policies for a sustainable future, implies the need to analyze the environment and trade, environment and market, environment and information, new measures of local governments, consumers and business and their corporation, and the basic environment plan linked to a sustainable future.

Keywords: effectiveness of environmental policies; integrated ways to consider environment; sustainable future; Romania

1. Introduction

Modern civilization, which has developed by seeking new frontiers and expanding its range of activities, is approaching the limits of the environment on a global scale. This is an experience that mankind has never had before.

*To whom correspondence should be addressed. Liviu–Daniel Galatchi, Ovidius University, Department of Ecology and Environmental Protection, Mamaia Boulevard 124, RO-900527 Constanta-3, Romania; e-mails: galatchi@univ-ovidius.ro and liviugalatchi@yahoo.com

R. N. Hull et al. (eds.), Strategies to Enhance Environmental Security in Transition Countries, 27–34.

Developed countries, in particular, have used trade and other avenues to deepen their relations broadly with the nations of the world so as to achieve rapid economic growth. These developed countries have come to hold much influence in international society both economically and from the point of view of placing burdens on the environment. They also have a large role to play in international society toward resolution of environmental problems, together with the developing and underdeveloped countries.

All countries have to build a sustainable society. But while the basic direction of environmental policy has been set, the real task of developing an environmental policy that specifically changes the modern socioeconomic system still remains ahead. Environmental policy should be not merely the province of the government or of the local public authorities; businessmen, the people, and private sector groups also should be actively pursuing it autonomously, each working under a fair division of responsibility according to their positions. But the capital and labor available for these efforts are certainly not limitless. And if additional problems appear in the near future, such as global warming, the time frame within which these groups can work to promote their environmental policies is even more limited.

It is therefore extremely important for policy makers, whether they be governments, local public authorities, or the private sector, to select environmental policies that are as effective as possible in terms of the investment costs of capital, labor, and time.

2. Objectives that Boost the Effectiveness of Environmental Policies

The character of environmental security problems today has changed. As can be seen by such issues as global warming or waste disposal, the emphasis of environmental security today has shifted toward how to reduce the burden on the environment caused by business activity or daily life. Moreover, many environmental problems are global in scale, complexly intertwined in causing today's situations. It also is possible that such problems as climate change that are caused by global warming may appear and cause damage when it is too late for mankind to do anything about it.

It will be necessary in the future to incorporate environmental security policy into increasingly broad socioeconomic activities, from all sorts of government, local public authority, and private sector groups. Assuring a viewpoint that makes these policies as efficient and effective as possible will probably be an important point in the achievement of a sustainable economy and society that smoothly promotes this kind of integration between the environment and economy.

This is why we attempt to adjust specific objectives for increasing the effectiveness of environmental security policy. Methods of adjustment are not assured, while methods for quantitative analysis of the effectiveness of each viewpoint are not necessarily assured. These objectives are important for both the governments and local authorities, and also for the environmental policies of business and private sector groups, and future studies will be needed on assessment methods in terms of administration. However, we must remain fully aware of the fact that many of the costs and their effectiveness in relation to the environment may be difficult to evaluate in terms of money.

2.1. PREVENTION OF ENVIRONMENTAL DAMAGE BEFORE IT HAPPENS

The first objective is preventing environmental damage before it happens. It is more cost effective to develop a policy to promote prevention of damage before it happens than it is to develop a policy after environmental damage has begun to appear. Moreover, examples from all over the world are beginning to show that changing the production process itself as a method of preemptive prevention is more cost effective than countermeasures taken at the point of emissions. The objective of preventing damage before it happens is extremely important, especially as a policy for those environmental problems from which it is difficult to recover.

2.2. COUNTERMEASURES THAT CONCENTRATE ON MORE BASIC CAUSES

In studying countermeasures for environmental problems, it is important to go beyond the cause of primary pollution or the surface phenomena that arise to consider the underlying socioeconomic structural causes. So, regardless of likes or dislikes, when the socioeconomic structure leads to a burden on the environment, each individual and group must be aware of that basic cause and reach past the difficulties to incorporate countermeasures. When we can consider the basic cause and make systematic policy reforms, each group's environmental conservation activities can become more effective, or in other words, more cost effective. In waste disposal, for example, restrictions on waste generation, promotion of appropriate recycling, and promotion of appropriate disposal are important. These are related to the production side and are also related to the lifestyles of each and every citizen. This view also incorporates environmental education, environmental training, and other policies, and promoting it can be expected to boost its effectiveness.

2.3. IMPLEMENTING ELEMENTS OF THE ENVIRONMENTAL FRAMEWORK IN AN INTEGRATED WAY

In promoting countermeasures for chemical substances, for example, policies to date have mainly been framed according to the medium (e.g., soil, water, air) impacted. If, at the substance level, we consider that releases into a number of media and movement between media can lead to exposures from each medium, then framing countermeasures to account for a number of media should be more effective. Moreover, an integrated approach is important when taking countermeasures at the source of impact, or for geographical approaches. Compared with policies that deal with pollution in a single medium, the study of environmental conservation policies in the context of integrated support for elements of the environmental framework is much more effective.

2.4. COMBINING ENVIRONMENTAL POLICIES WITH OTHER POLICIES

Combining environmental policy with other policies can result in the promotion of more effective policies. For, as an example of this approach, this paper discusses the example, in such areas as traffic congestion policy combined with a policy to reduce automobile emission gases, or the reduction of fertilizer use combined with a soil conservation policy, studying policies that take both areas into account from the first stage is more effective than the relationship of environmental policy with trade policy in terms of the validity of mutual support of policies having differing objectives.

2.5. MAKING APPROPRIATE USE OF THE MARKET MECHANISM

Making use of the market mechanism can be expected to be a valid way to increase the effectiveness of environmental policy. The market uses price as a signal to distribute complex resources, and price movements are used to resolve surpluses or shortages of assets or services. One of the major causes of environmental problems is the lack of awareness of the costs to society of environmental impacts. These costs may be due to socioeconomic activities, and insufficient pollution countermeasures, so that the costs were not appropriately incorporated into the market mechanism.

For this reason, external environmental costs should be appropriately reflected in product and other prices, and an appropriate price assigned from the viewpoint of the environment. This process should result in the socioeconomic structure placing a smaller burden on the environment. Of course, while this environmental price cannot necessarily be completely calculated in monetary terms, it should open up fields that acquire marketability. Environment-related industries have been growing in scale in recent years, a trend that can in a sense be considered to be increasing activity of the market mechanism.

2.6. MAKING APPROPRIATE USE OF INFORMATION

Appropriate use of information concerning the environment is essential in promoting an appropriate understanding of the current state of and future forecasts for the environment, and in promoting effective efforts in the roles played by each of the major actors.

Information about the economy and society is advancing rapidly in scope. There are probably both pluses and minuses in determining what the effects on the environment will be from building the information infrastructure and from using it. It goes without saying that building the information society must be directed in such a way as to reduce the burden on the environment. Whatever happens, as information grows in importance, we should promote appropriate delivery of information concerning the environment, and the use by the major actors of such information, to make possible more effective environmental conservation.

2.7. APPROPRIATELY SHAPE THE MECHANISMS FOR LINKUPS AND PARTICIPATION BY CONCERNED PARTIES

Should, for example, the industrial sphere have a certain environmental conservation goal promoted by the industrial sphere only, or if consumers have a goal that is promoted by consumers only, their efforts will not be very effective. However, if governments and all other parties work toward a common goal, under an appropriate division of roles, then environmental conservation will be promoted more effectively. We have often seen spontaneous, active efforts by a number of major actors in recent years, by international linkups (Vadineanu, 2000).

3. Considering the Environment in Integrated Ways in Romania

Considering Romania as a case study, in the last 16 years, it has not attained as good environmental standards as was expected when compared with efforts and investment that went into its pollution policy. For example, in the improvement of energy efficiency or reduction of waste volumes, remarkable displays of efficiency may have been seen over the short term, yet expanding those results over the longer term would require making basic changes to socioeconomic activities (Guvernul Romaniei, 2001). This is why efforts are turning toward policies that consider the environment in integrated ways.

Some newly considered integrated security policies for the environment that can be seen in Romania include: assessment of environmental problems and ranking of their risks (Guvernul Romaniei, 2001); adoption and implementtation of environmental security plans; integrated strategies formed between policy departments; integrated environmental conservation through geographical approaches; study and implementation of life cycle assessments; and information supply and participation.

3.1. INTEGRATED FOCUS ON SUBSTANCE LIFE CYCLES

Government plans in Romania for the monitoring and control of chemical substances have been undergoing changes in recent years. Chemical substances are now beginning to be ranked by their degree of danger, with integrated consideration paid to their life cycles and the forms they take during use, and then putting the substances under control (Guvernul Romaniei, 2001). This kind of planning can be considered to be a shift from the conventional method of taking countermeasures at the point of emission toward methods that encompass the entire process and entire life cycle to control initial appearances of substances and to reduce their emissions. This kind of integrated viewpoint for studying policies at the substance level is now recognized as being very important.

3.2. INTEGRATED ENVIRONMENTAL VIEWPOINT FOR SOURCE OF OCCURRENCE POLICIES

Pollutants are known to traverse between such differing media as air, water, and soil. For example, 25% of the nitrogen burden in Chesapeake Bay in the USA is believed to be from air pollutants generated by automobiles and power plants. Moreover, pollutants discovered in the North Pole haze are said to be emissions from industrial districts transmitted through the air.

If waste water and exhaust gases are disposed of at their source, then the amount of sludge and other materials that must eventually be properly disposed of increases. There are cases where districts using tall smokestacks to reduce air pollution locally become the cause of damage to forests and lakes in quite distant regions (Guvernul Romaniei, 2001). These cases demonstrate that assessments taking time and space into account when considering the extent of the environment affected by policies implemented at the source of the occurrence will probably be more valid.

As a result, Romania has established (Guvernul Romaniei, 2001) that it is important to make objectives such as: (1) select the most advanced technology that can be acquired for each medium; (2) consider opportunities for reducing or activating waste materials; and (3) integrate and reduce the costs of environmental control systems and pollutant emissions.

3.3. GEOGRAPHICAL VIEWPOINT FOR INTEGRATED CONSERVATION OF THE ENVIRONMENT

The approach of taking a certain geographical form (distributed forests, lake fronts, etc.), declaring it a unified environment, and then moving to conserve it, is perhaps the most effective way to conduct integrated conservation of the environment. While many examples can be found, there have been some recent cases where this geographic approach has been used from the viewpoint of assuring sustainability and preventing pollution. In Canada's Fraser River Action Plan, for example, the federal government, province (British Columbia), local communities, and companies have joined together to develop a policy based on a sustainable development plan for the 2 million people living in the area, and for the river ecology (Gray, 1996). In addition, the United States Environmental Protection Agency (US EPA) has joined with seven states and Canada to promote a pilot program for the Great Lakes (US EPA, 2006). The goal of this program focuses particularly on the elimination of organic substances with high bioaccumulation, and using activities that aim at usage restrictions at the sources of occurrence, in order to reduce concentrations in the environment.

Thus, in proposing environmental policies in Romania, one basic objective is the consideration of geographical characteristics, in order to take a broad integration of the environment and so conserve the environment over an entire region.

Developing highly effective environmental security policies for a sustainable future requires the analyses of: environment and trade (controversies surrounding them, efforts toward mutual support between them), environment and market (environmental security policy and the market mechanism, eco-business trends – environment related industry), environment and information (the development of information-orientation and the environment, the preparation and utilization of environmental information), new measures of local governments, consumers, and business and their corporations, and a basic environmental plan linked to a sustainable future.

4. Towards a Sustainable Future in Romania

Today environmental problems have reached a global scale. Moreover, our mass production, mass consumption, and mass disposal type of economic and social system and lifestyle are continuing. In addition, when the problems of conservation of the environment of developing areas, and the situation of the accelerating increase in population are looked at, there is the danger that the global environment that is the foundation of human existence will be harmed during the next century if the situation continues in this way. In order to avoid this kind of situation, we must construct a sustainable society and achieve a greater environmental conservation effect at an early period by devising the necessary countermeasures comprehensively and systematically.

The official documents on the environment in Romania have the character of stipulating the way policy measures should be the roles of each constituency and the general principles of policies to promote more highly effective environmental policies that have aimed at the construction of a sustainable society that conserves the environment. In addition to cooperating with all related constituencies and following the direction of measures across a wide range of areas set forth in the plan, the government as one body must promote countermeasures that serve to heighten the environmental conservation effect. By these means we must construct a sustainable society before the danger of harming the global environment that is the foundation of human existence becomes an actuality.

By means of these countermeasures, we will be able to change our lives from lives that pursue only material abundance that does not take notice of the limited global environment that is the foundation of human existence, to lives that heighten true abundance and are supported by the great blessings of the global environment.

Acknowledgment

We thank Professor Stoica-Preda Godeanu for helpful suggestions during our research.

References

Gray, C., 1996, *Fraser River Action Plan*, Presentation at the Ecological Monitoring and Assessment Network, Second National Science Meeting, 17–20 January 1996, Halifax, Nova Scotia; http://eman-rese.ca/eman/reports/publications/SECOND/part23.html, accessed June 2006.

Guvernul Romaniei, 2001, *Strategia Nationala pentru Dezvoltare Durabila*, Bucharest, pp. 1–120.

US EPA, 2006, *Canada–United States Strategy for the Virtual Elimination of Persistent Toxic Substances in the Great Lakes*, United States Environmental Protection Agency; http://www.epa.gov/glnpo/p2/bns.html, accessed June 2006.

Vadineanu, A., 2000, *Dezvoltarea durabila*, University of Bucharest Press, Bucharest, pp. 1–224.

4. DSS-ERAMANIA: A DECISION SUPPORT SYSTEM FOR SITE-SPECIFIC ECOLOGICAL RISK ASSESSMENT OF CONTAMINATED SITES

ELENA SEMENZIN
Consorzio Venezia Ricerche, Via della Libertà 5-12, 30175 Marghera, Venice, Italy and Interdepartmental Centre IDEAS – University Ca' Foscari of Venice, Dorsoduro 2137, 30123 Venice, Italy

ANDREA CRITTO, ANTONIO MARCOMINI*
Interdepartmental Centre IDEAS – University Ca' Foscari of Venice, Dorsoduro 2137, 30123 Venice, Italy

MICHIEL RUTGERS
Laboratory for Ecological Risk Assessment, National Institute for Public Health and the Environment, RIVM. PO BOX 1, 3720 BA Bilthoven, The Netherlands

Abstract: The increasing demand for the assessment and management of contaminated land requires the development of new and effective tools for use in the site-specific Ecological Risk Assessment (ERA) procedure. For this purpose we propose the DSS-ERAMANIA, a decision support system (DSS) that aims at implementing the whole ERA procedure and supporting the expert/decision maker (DM) in its assessment. It was developed according to a Triad approach, where the results provided by a set of measurement endpoints are gathered and compared to support the assessment and evaluation of ecosystem impairment (i.e., assessment endpoint) caused by the stressor(s) of concern. In particular, in the Triad approach, the measurement endpoints refer to three Lines of Evidence (LoEs): environmental chemistry, ecotoxicology, and ecology. The rationale is that by integrating the three different LoEs the uncertainty in the risk estimation will be pragmatically reduced. For the DSS, a framework including three subsequent investigation levels (i.e., tiers) was proposed, that enables the completion of the risk assessment once the provided answer is unequivocal and characterized by a relatively small uncertainty,

*To whom correspondence should be addressed. Antonio Marcomini, Interdepartmental Centre IDEAS – University Ca' Foscari of Venice, San Giobbe 873, 30121 Venice, Italy; e-mail: marcom@unive.it

R. N. Hull et al. (eds.), Strategies to Enhance Environmental Security in Transition Countries, 35–46.

ensuring at the same time an adequate financial investment. The DSS consists of two multicriteria decision analysis (MCDA)-based modules: “Comparative Tables” and “Integrated Ecological Risk Indexes”. The former aims at comparing the different measurement endpoints belonging to each LoE (i.e., bioavailability tools, toxicity tests, and ecological observations) to guide the expert/DM in the choice of the suitable set of tests to be applied to the case study. The “Integrated Ecological Risk Indexes” module provides qualitative and quantitative tools allowing the assessment of terrestrial ecosystem impairment (i.e., the impairment occurring on biodiversity and functional diversity of the terrestrial ecosystem) by integrating complementary information obtained by the application of the measurement endpoints selected in the first module.

Keywords: ecological risk assessment; decision support system; riad; megasite; integrated risk indexes; multicriteria decision analysis; soil ecology

1. Introduction

Criteria and methodologies for the assessment and rehabilitation of contaminated sites are urgently needed worldwide because of the huge number of contaminated sites and the financial implications representing significant constraints for site redevelopment.

In order to address the rehabilitation of contaminated ecosystems, ecological risk assessment (ERA) is the appropriate process for identifying environmental quality objectives and the ecological aspects of major concern (US EPA, 1998; Suter et al., 2000). At the international level, the principles and procedures that have been established (UK–EA, 2003; CLARINET, 2001; INIA, 2000; ECOFRAME, 1999; CARACAS, 1998), suggest an ERA framework based on a tiered approach, including: (1) a screening phase that allows the definition of land use–based soil screening values (SSVs), and (2) a site-specific phase in order to accomplish a more comprehensive risk characterization. Within the risk characterization, the application of weight of evidence (WoE) methods (Burton et al., 2002; Chapman et al., 2002) were proposed to determine possible ecological impacts based on multiple LoEs (US EPA, 1998). Recently, in the Netherlands an update of the Soil Protection Act is foreseen (VROM, 2003), clarifying the way for a site-specific risk assessment based on one of the WoE methods: the Triad approach (Rutgers and Den Besten, 2005). The Triad approach requires three major LoEs for an accurate assessment of contamination: chemical contaminant characterization, laboratory-based toxicity to surrogate organisms, and indigenous biota community characterization (Long and Chapman, 1985). In accordance with international suggestions, the new Dutch approach, to be launched in the new Soil Policy Act of 2006, will be a tier-based approach and will include: a comparison with standard values in the first step; an estimation and evaluation of effects due to a mixture of contaminants, in the second step, by means of multisubstances potentially affected fraction (msPAF) calculation (Posthuma et al., 2002); and a simple Triad in the third step. The proposed Triad approach will be based on the calculation of specific msPAFs for metals and organic compounds for the chemical LoE, the application of Microtox (or comparable tests on, e.g., algae, plants, earthworms) for the toxicological LoE, and a site visit (possibly observing vegetation or nematode communities) for the ecological LoE.

The results obtained in each tier have to support the decision-making process, whose main challenge is to find sustainable solutions by integrating different disciplinary knowledge, expertise, and views (Siller et al., 2004; Kiker et al., 2005). Technical tools are therefore needed in order to integrate the wide range of decisions related to contaminated land management and reuse (CLARINET, 2002a). Instruments, proved to be an effective support for any kind of decision-making process including contaminated sites management, are decision support systems (DSSs) (CLARINET, 2002b). DSSs can be defined generally as tools that can be used by decision makers (DMs) in order to have a more structured analysis of a problem at hand and define possible options of intervention to solve the problem (Jensen et al., 2002; Georgakakos, 2004; Salewicz and Nakayama, 2003). To the purpose of contaminated sites management, the currently available systems address in different ways the several involved issues: supporting the human health risk assessment procedures, providing selection of the most suitable technology to deal with the contamination of concern, or facilitating the identification of the best management solution for the site requalification (CLARINET, 2002b). Some of them can facilitate the achievement of a shared vision for the redevelopment of the site of interest between both experts and stakeholders (Pollard et al., 2004); for example, by including multicriteria decision analysis (MCDA) in order to allow consideration of different stakeholder priorities and objectives (Linkov et al., 2004; Giove et al., 2006; Kiker et al., 2005) and also integration of different issues.

Although some specific tools for human health risk assessment are included within the existing DSSs, relevant development should be made in order to implement specific support systems for site-specific ESA (CLARINET, 2002b).

For this reason, within the "Ecological Risk Assessment: development of a Methodology and Application to the contaminated site of National Interest Acna di Cengio (Italy)" (ERA-MANIA) project, the DSS-ERAMANIA was developed, as an MCDA-based system capable to support experts/DMs in the site-specific phase of ERA of contaminated sites and in particular of megasites. A megasite is defined as a large (km^2 scale) contaminated area or impacted area (current or potential), characterized by multiple site owners, multiple stakeholders, multiple end users, unacceptable costs for complete cleanup (due to political, economical, social, or technical constraints) within currently used regulatory time frames, and by need for an integrated risk-based approach at a regional scale (WELCOME project, 2005). Due to its complexity, the remediation process may be required to address different issues (e.g., environmental, economic), to involve different stakeholders and to take into consideration many different data sets (Agostini et al., 2006).

As far as the ERA-MANIA project is concerned, the Acna di Cengio contaminated megasite was selected as the case study for the DSS design and application. Extending over ~550,000 m^2, it is located in Valbormida (Savona province) and is bordered by the Bormida river. From 1882 to 1999, several industrial activities were carried out at this site, including: production of explosives, dyes, chemical intermediates for pharmaceutical industries, agrochemical compounds, and additives for rubber. Since December 1998, the site is included in the list of Contaminated Sites of National Interest to be reclaimed. The complexity of the site contamination was especially useful for the DSS design that resulted from a close collaboration among different experts within a multidisciplinary team (including risk assessors, ecotoxicologists, ecologists, environmental chemists, mathematicians, and informatics) and thanks to the valuable support of an international steering committee, including also DMs and stakeholders.

This paper is intended to provide an overview of the DSS-ERAMANIA, with a major emphasis on its structure and functionalities. The detailed presentation of the MCDA-based procedures implemented in the DSS-ERAMANIA and of the DSS application to the Acna megasite is beyond the scope of this paper. They are the subject of four manuscripts either submitted or in preparation (Critto et al., 2006; Semenzin et al., 2006).

2. The DSS-ERAMANIA

In order to support experts and DMs in the application of site-specific ERA for contaminated soil, a specific framework was developed and implemented into the DSS-ERAMANIA. The proposed framework is mainly based on the Triad approach (Rutgers and Den Besten, 2005). Moreover, it includes three subsequent investigation levels, named tiers, each one allowing the assessment to stop depending on the estimated risk and related uncertainty.

The first one (i.e., Tier 1) can be regarded as a preliminary investigation level, in which the analysis can stop only if the estimated risk is considered to be acceptable and affected by a minimum level of uncertainty. The unacceptable risks found in Tier 1 can be investigated in the second tier (i.e., Tier 2), to reduce the uncertainty in the risk estimation and to achieve a more accurate risk assessment. Because of economic and time reasons, it is preferable to stop the analysis at Tier 2, when a satisfactory estimation of the investigated risks is achieved. Nevertheless, site-specific aspects of particular interest, for which the risk uncertainty could not be reduced in Tier 2, can be analyzed in the third level (i.e., Tier 3) that represents the deepest degree of investigation. However, this effort has to be justified by adequate saving of money for the site remedial actions.

The developed framework identifies two different phases in the site-specific ERA process. The first phase concerns a comparison of tests (i.e., measurement endpoints) within each Triad LoE and the selection of a suitable set of tests to be applied to the case study at each Triad tier. The second one concerns the estimation and evaluation of impairments occurring in the ecosystem due to soil contamination.

For each phase, specific MCDA-based tools were developed and implemented in the two modules of the DSS-ERAMANIA: "Comparative Tables" (i.e., Module 1) and "Integrated Ecological Risk Indexes" (i.e., Module 2) (see Figure 1).

In the following paragraphs each module of the DSS-ERAMANIA is presented.

2.1. MODULE 1: COMPARATIVE TABLES

The DSS-ERAMANIA Module 1 aims at comparing the different tests belonging to each of the Triad LoE (i.e., chemistry/bioavailability, ecology, and ecotoxicology) to guide the expert/DM in the choice of the most suitable set of tests to be applied to the case study at each investigation tier.

For this purpose three specific tables were developed to compare the measurement endpoints belonging to the three LoEs: bioavailability assessment tools, ecological observations, and ecotoxicological tests. These comparative tables were respectively named: Comparative Tools Table for bioavailability (i.e., BAV Table), Comparative Observations Table for ecology (i.e., ECO Table), and Comparative Tests Table for ecotoxicology (i.e., ETX Table) (Figure 1).

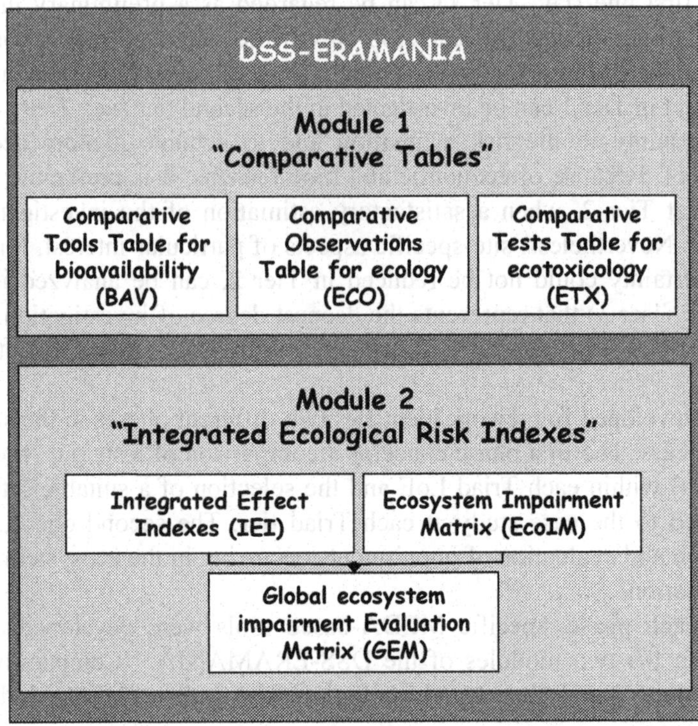

Figure 1. Modular structure of the DSS-ERAMANIA.

They can first of all be regarded as databases for the collection of information related to the specific measurement endpoints. In addition, the tables support experts in: excluding those measurement endpoints, which cannot be applied to the case study; comparing the tests belonging to each Triad LoE in order to define their suitability to be applied at the first, second, or third Triad tier; selecting the best set of measurement endpoints for the case study. As shown in the general flowchart reported in Figure 2, the multiple measurement endpoints (i.e., bioavailability tools, ecotoxicological tests, and ecological observations) included in the comparative tables are compared by means of an MCDA-based procedure where multiple criteria are used to support the tests comparison performed by different experts.

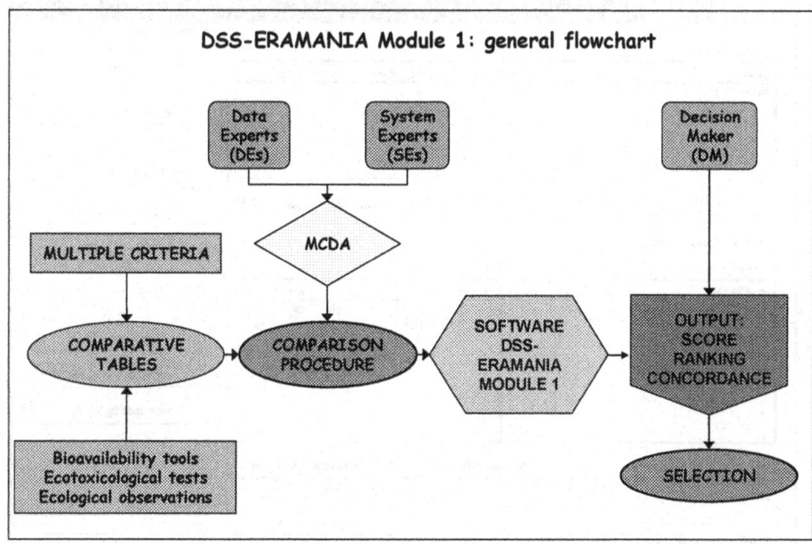

Figure 2. General flowchart of the DSS-ERAMANIA Module1.

The selected criteria are specific for each table (i.e., BAV, ECO, and ETX Tables) but generally they refer to three main aspects: specific information required by the investigation tier (e.g., ecological relevance of the test species), evaluation of economic and temporal efforts (e.g., cost, analysis time), and test reliability (e.g., repeatability, test protocol). The involved experts are called data expert (DE) and system expert (SE), according to their specific role in the procedure. In fact, the DE should be an expert on the measurement endpoints to be compared, while the SE should know the tiered Triad-based ERA procedure. Their judgments are collected in specific tables and elaborated by the software "DSS-ERAMANIA Module 1" (see Figure 3).

Three outputs are then available: score, ranking, and concordance. For each tier a score, resulting from the mathematical combination of expert judgments by means of simple additive weights (SAW) method (Vincke, 1992), is assigned to each test. Then, for each tier and each LoE (i.e., chemistry/bioavailability, ecology, ecotoxicology), a ranking of the measurement endpoints is provided, according to the calculated scores. Finally, an estimation of the judgments' concordance among DEs and SEs is performed by means of specific indices (i.e., mean expert concordance and global mean concordance). On the basis of this information, theDMs, who should represent the responsible people for the risk management of the contaminated site, can proceed with the selection of the suitable set of tests to be applied to the megasite of concern.

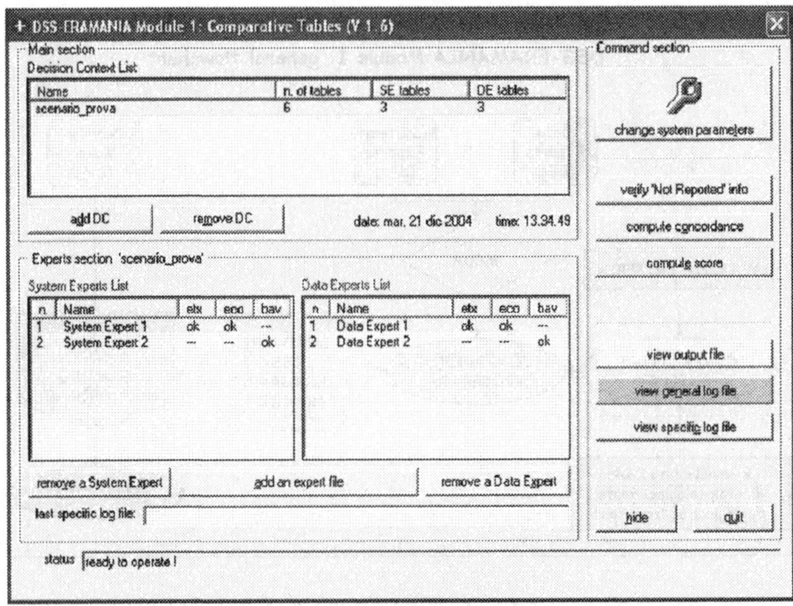

Figure 3. Software interface of the DSS-ERAMANIA Module 1.

2.2. MODULE 2: INTEGRATED ECOLOGICAL RISK INDEXES

The DSS-ERAMANIA Module 2 provides qualitative and quantitative tools that allow the assessment of terrestrial ecosystem impairment (i.e., the impairment occurring on biodiversity and functional diversity of the terrestrial ecosystem) by integrating the complementary information obtained by the measurement endpoints' application. As shown in Figure 1, it includes the following tools: the integrated effect index (IEI), the ecosystem impairment matrix (EcoIM), and the global ecosystem impairment evaluation matrix (GEM).

The general flowchart of Module 2 is presented in Figure 4.

The core of the procedure is the impairment analysis, where the measurement endpoint results, obtained from both contaminated and reference sites, are evaluated in terms of impairment for the tested endpoints and normalized on a 0–1 scale, according to test-specific thresholds (i.e., negligible impairment threshold and relevant impairment threshold) defined by the involved experts. This way, for each LoE, the scaled test results can be aggregated into a single index (i.e., IEI). The adopted MCDA aggregation operator is a weighted average, where the weights are criteria that take into account the relevance of the information provided by each measurement endpoint to the assessment endpoint (i.e., the terrestrial ecosystem).

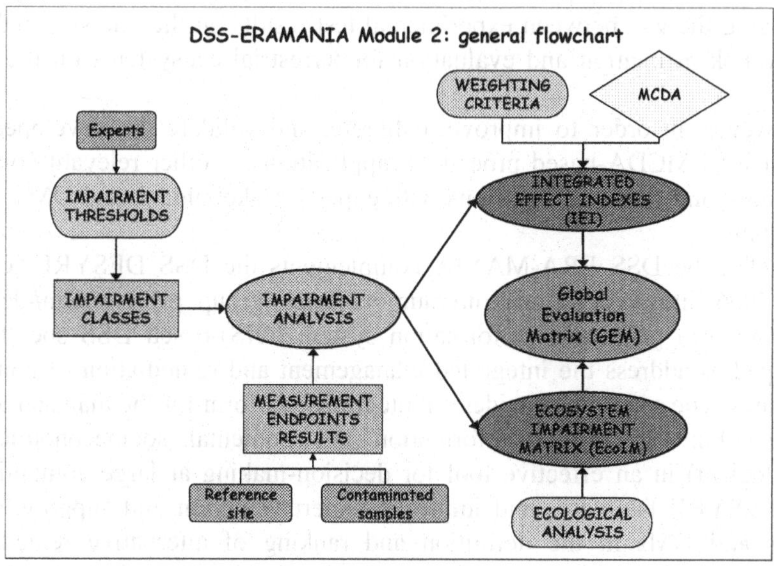

Figure 4. General flowchart of the DSS-ERAMANIA Module 2.

The comprehensive understanding of the quantitative estimation provided by the calculated IEIs is supported by the EcoIM. In fact, using the same scaled results obtained from the impairment analysis, the EcoIM provides a qualitative assessment of the impairment at the ecosystem level evaluating both biodiversity and functional diversity of the terrestrial ecosystem, and highlighting the ecosystem factors stressed by soil contamination. MCDA methods and soil ecology have been used to develop this tool for supporting risk management. Finally, the outputs of both IEI and EcoIM tools, obtained for the analyzed contaminated and reference sites, can be reported and summarized by the experts in the GEM, allowing a comprehensive evaluation of the impairment occurring on the terrestrial ecosystem of concern.

3. Conclusions

The preliminary application of DSS-ERAMANIA to the Acna megasite led to a transparent selection of measurement endpoints for each Triad LoE (i.e., Module 1 application), and to a comprehensive (quantitative and qualitative) evaluation of the impairment occurring on the site of concern (i.e., Module 2 application). Moreover, Module 1 turned out to be a useful tool for promoting discussions among experts to reach a common judgment on advantages and drawbacks of existing experimental tests; while Module 2 confirmed its capability of bridging the gap between experimental test results on the one side and site-specific risk estimation and evaluation for terrestrial ecosystems on the other side.

However, in order to improve, calibrate, and validate the developed and implemented MCDA-based procedure, applications to other relevant contaminated sites, and further discussions with experts, stakeholders, and DMs would be desirable.

Finally, the DSS ERA-MANIA complements the DSS DESYRE (Carlon et al., 2006; http://venus.unive.it/eraunit/research_group_projects.htm#desyre). The latter is a geographic information system (GIS)-based DSS specifically developed to address the integrated management and remediation of contaminated sites. The software provides an integrated platform for the management of a large volume of different information (environmental, socioeconomic, and technological) in an effective tool for decision-making at large contaminated sites. DESYRE facilitates and improves expert valuation and supports stakeholders and DMs in the definition and ranking of alternative remediation scenarios, which can be described on the basis of technological options, human health risk reduction, cost, time, environmental impacts, and socioeconomic benefits. The simultaneous application of DSS-ERAMANIA and DESYRE will guarantee a comprehensive risk assessment (i.e., human health and ERA), according to international guidelines and national regulatory requirements.

Acknowledgments

We specially thank Miranda Mesman and Ton Schouten for their contribution to the ERA-MANIA project. We also thank Professor Silvio Giove for developing the MCDA procedures of the DSS-ERAMANIA and Stefano Silvoni for the implementation of the software "DSS-ERAMANIA Module 1: Comparative Tables". This work and the ERA-MANIA project have been funded by the Italian Government Commissary for the rehabilitation of the Bormida Valley and were developed by University of Venice, Ca' Foscari, Interdepartmental Centre for Sustainable Development (IDEAS) in collaboration with Italian National Environmental Protection Agency (APAT), Dutch National Institute for public Health and Environment (RIVM), and Consorzio Venezia Ricerche (CVR).

We are grateful to the Steering Committee members (Trudie Crommentujin, Bona Piera Griselli, John Jensen, Giuseppe Marella, Cristina Menta, Nick Pacini, Gianniantonio Petruzzelli, Eugenio Piovano, Milagros Vega, and Aldo Viarengo) for the helpful discussions during the testing phase of both Module 1 and 2 development.

References

Agostini, P., Critto, A., Semenzin, E., Marcomini, A., 2006, Decision support systems for contaminated sites: a review, *Integr. Environ. Assess. Manag.* (accepted).

Burton, G.A. Jr., Chapman, P.M., Smith, E.P., 2002, Weight-of-evidence approaches for assessing ecosystem impairment, *Hum. Ecol. Risk Assess.* 8: 1657–1673.

CARACAS, 1998, *Risk Assessment for Contaminated Sites in Europe*, Final publication, Vol. 1 – Scientific basis. LQM Press, Nottingham.

Carlon, C., Critto, C., Ramieri, E., Marcomini, A., 2006, DESYRE: decision support system for the rehabilitation of contaminated mega-sites, *Integr. Envion. Assess. Manag*, (in press).

Chapman, P.M., McDonald, B.G., Lawrence, G.S., 2002, Weight of evidence frameworks for sediment quality (and other) assessment, *Hum. Ecol. Risk Assess.* 8: 1489–1516.

CLARINET, 2001, *Proceedings of the CLARINET Workshop on Ecological Risk Assessment for Contaminated Sites*, 17–19 April 2001, Nunspeet, NL.

CLARINET, 2002a, *Sustainable Management of Contaminated Land: An Overview*, A report from the Contaminated Land Rehabilitation Network for Environmental Technologies, pp. 128.

CLARINET, 2002b, *Review of Decision Support Tools for Contaminated Land Management, and Their Use in Europe*. A report from the Contaminated Land Rehabilitation Network for Environmental Technologies, pp. 180.

Critto, A., Torresan, S., Semenzin, E., Giove, S., Mesman, M., Schouten, A.J., Rutgers, M., Marcomini, A., 2006, *A Multi-Criteria Based System for the Selection of Ecotoxicological Tests and Ecological Observations in a Site-specific ERA* (submitted).

ECOFRAME, 1999, *Ecological Committee on FIFRA Risk Assessment Methods*, ECOFRAM terrestrial draft report.

Georgakakos, A., 2004, Decision support systems for integrated water resources management with an application to the Nile basin, in: *Proceedings of the IFAC Workshop Modelling and Control for Participatory Planning and Managing Water Systems*, Venice, 29 September–1 October 2004.

Giove, S., Critto, A., Agostini, P., Semenzin, E., Marcomini, A., 2006, Decision support system for the management of contaminated sites: a multi-criteria approach, in: *Environmental Security and Environmental Management: Decision Analytical and Risk-Based Approaches*, Kiker, G., Linkov, I. (eds.), Springer, Amsterdam (in press).

INIA – Instituto Nacional de Investigacion y tecnologia Agraria y alimentaria, 2000, *Desarrollo de criterios para la caracterizacion de suelos contaminados y para determinar umbrales para la proteccion de los ecosistemas*, Madrid, Espana.

Jensen, R.A., Krejcik, J., Malmgren-Hansen, A., Vanecek, S., Havnoe, K., Knudsen, J., 2002, River basin modelling in the Czech Republic to optmise interventions necessary to meet the EU environmental standards, in: *Proceedings of the Internation Conference of Baisn Organizations*, Madrid 4–6 November 2002, DH ref 39/02.

Kiker, G.A., Bridges, T.S., Varghese, A., Seager, T.P., Linkov, I., 2005, Application of multicriteria decision analysis in environmental decision making, *Integr. Environ. Assess. Manag.* 1(2): 95–108.

Linkov, I., Sahay, S., Kiker, G., Bridges, T., Seager, T.P., 2004, Multi-criteria decision analysis: a framework for managing contaminated sites, in: *Strategic Management of Marine Ecosystems,* Levner, E., Linkov, I., Proth, J.M. (eds.), Kluwer, Amsterdam.

Long, E.R., Chapman, P.M., 1985, A sediment quality triad: measures of sediment contamination, toxicity and infaunal community composition in Puget Sound, *Mar. Poll. Bull.* 16: 405–415.

Pollard, S.J.T., Brookes, A., Earl, N., Lowe, J., Kearny, T., Nathanail, C.P., 2004, Integrating decision tools for the sustainable management of land contamination, *Sci. Tot. Environ.* 325: 15–28.

Posthuma, L., Suter II, G.W., Traas, T.P., 2002, *Species Sensitivity Distribution in Ecotoxicology*, Lewis, Boca Raton, FL.

Rutgers, M., Den Besten P., 2005, Approach to legislation in a global context, B. The Netherlands perspective – soils and sediments, in: *Environmental Toxicity Testing*, Thompson K.C., Wadhia K., Loibner A.P. (eds.), Blackwell Publishing/CRC Press, Oxford, pp. 269–289.

Salewicz, K.A., Nakayama, M., 2003, Development of a web-based decision support system (DSS) for managing large international rivers, *Glob. Environ. Chang.* 14: 25–37.

Semenzin, E., Carlon, C., Critto, A., Giove, S., Rutgers, M., Marcomini, A., 2006, Selection of bioavailability assessment tools in a site-specific Ecological Risk Assessment (submitted).

Siller, D., Blodgett, C., Cziganyik, N., Omeroglu, G., 2004, Factors involved in urban regeneration: remediation and redevelopment of contaminated urban sites – a comparison of France and the Netherlands, in: *Proceedings of CABERNET 2005: The International Conference on Managing Urban Land*, Compiled by Oliver, L, Millar, K, Grimski, D, Ferber, U, Nathanail, C.P., Land Quality Press, Nottingham, ISBN 0-9547474-1-0.

Suter II, G.W., Efroymson R.A., Sample B.E., Jones D.S., 2000, *Ecological Risk Assessment for Contaminated Sites*, Lewis, Boca Raton, FL.

UK–EA, 2003, Ecological Risk Assessment: a public consultation on a framework and methods for assessing harm to ecosystems from contaminants in soil; www.environment-agency.gov.uk

US EPA – US Environmental Protection Agency, 1998, *Guidelines for Ecological Risk Assessment*, EPA/630/R-95/002F, Risk Assessment Forum, Washington, DC.

Vincke, P., 1992, *Multicriteria Decision-Aid*, Wiley, Chichester.

VROM, 2003, Soil Policy Letter. Circular BWL/2003 096 250, Ministry of Housing, Spatial Planning and the Environment, The Hague, The Netherlands; http://international.vrom.nl/ docs/ internationaal/7178menarev_1.pdf

WELCOME project, March 2005; http://euwelcome.nl/kims/index.php

5. ENHANCEMENT OF ENVIRONMENTAL SECURITY BY MATHEMATICAL MODELLING AND SIMULATIONS OF PROCESSES

VALENTINA TASOTI*
Babes-Bolyai University, Faculty of Chemistry and Chemical Engineering, Str. Arany Janos 11, 400028, Cluj-Napoca, Romania

Abstract: The objective of this article is to review benefits of mathematical modeling and simulations of different processes that are happening and which describe a system. A classification of different types of mathematical models is done as well as the steps in mathematical modeling. Depending on the system and the case study, an analytic mathematical model could be described by mass balance, energy, momentum and/or voltage equations, which are the building blocks in mathematical modeling of environmental systems. The mathematical models are used for steady state simulation, dynamic simulation, design, or process control. Packages for model development are identified. Computer implementation of the mathematical models for a few examples is presented. Numerical simulations aid in the prediction of future environmental concentrations of pollutants under various waste loading and/or management alternatives; in environmental impact assessment; to design, operate, and optimize the reactors; and, to offer pollution control alternatives. All of these uses contribute to the enhancement of environmental security.

Keywords: environmental security; mathematical modeling; numerical simulation balance equations

*To whom correspondence should be addressed: Valentina Tasoti, Babes-Bolyai University, Faculty of Chemistry and Chemical Engineering, Str. Arany Janos 11, 400028, Cluj-Napoca, Romania, e-mail: vtasoti@yahoo.com

R. N. Hull et al. (eds.), Strategies to Enhance Environmental Security in Transition Countries, 47–56.

1. Introduction

Environmental security is an important issue not only at the national level, but also at the international level, as environmental problems produced in one state, potentially have spillover effects on neighboring countries. Prevention of conflicts and peace keeping can be accomplished by protection of the environment across national borders and sustainable use of natural resources. Major forms of environmental degradation, including air and water pollution, nuclear waste, acid rain, deforestation, soil erosion, depletion of marine resources, and climate change are the concerns of environmental security. Sustainable development must be a guiding principle to evaluate measures in the field of environmental security. Mathematical modeling could be a very good instrument to enhance environmental security.

2. Mathematical Modeling

A mathematical model is a system of equations that defines the intrinsic interdependencies of a system (Hangos and Cameron, 2001). Mathematical modeling in the environmental field has a multidisciplinary character, dealing with different systems (chemical, biological, ecological), which must be described. Analytical; statistical (empirical), or mixed models can express relationships between variables of these systems.

Analytical models are based on conservation equations (mass, energy, momentum) and laws of physical and chemical processes that are taking place in the system. Analytical mathematical models have the following advantages: extended validity domain, increased flexibility, and use in modeling similar processes. Besides these important advantages, they also have some disadvantages:

- They necessitate a good knowledge of the phenomena and processes that are taking place inside the modeled system
- Users need special training to develop the specific equations of the modeled system
- They need an additional stage to validate/verify the model, which requires obtaining experimental measurements in the modeled system

A statistical model is based exclusively on observation dates and measurements performed on the modeled system. Statistical mathematical models are advantageous because: they are simple from a mathematical point of view; they do not need extended knowledge about the system, or phenomena and processes that take place within the system; they need only minimal mathematical knowledge, regardless of the system modeled. The main disadvantages of statistical models are: they necessitate an extended set of experimental data obtained by measurements performed on the system modeled; the mathematical models obtained in this way are not valid outside the domain of experimental data used initially. To use these mathematical models on other systems, even similar systems, requires making new measurements on those systems.

The mixed models are based on conservation equations (mass, energy, momentum), laws of the processes that are taking place in the system, together with equations obtained by measurements data correlation based on experimental data obtained in the modeled system.

Equations of any mathematical model could be algebraic linear or nonlinear, in the case of a steady state model (when time is invariant), or, a system of differential equations and algebraic equations, in the case of a dynamic model, when the equations of the system express the variation with time of the system variables.

To obtain a mathematical model of a process we need to:

- Write mass, energy, moment, and voltage (in the case of electrochemical processes) conservation equations; write equations to describe physical and chemical processes that are taking place in the system; this results in an analytic mathematical model.
- Make experimental measurements of all variables of the process and identify the equations that correlate the input and output variables by using statistical techniques; this results in a statistical mathematical model.
- Combine the previously presented modalities to obtain a mixed mathematical and experimental model.

A short review of the uses of mathematical models is presented below (Himmelblau and Bischoff, 1968; Linnhoff and Akinradewo, 1999):

- System design (process optimization; dynamic analysis of processes; optimization of process parameters; analysis of the factors, which influence a process; economic analysis of a process; waste minimization; control analysis of a process; retrofit analysis)
- Process control (examination of different control strategies; optimal control; modeling predictive control; expert systems)
- Identification of possible failures (identification of perturbative factors; identification of factors that cause the decline of plant performance)
- Training of working personnel (start-up and shutdown operation; current operation of the plant; operation in limit conditions)

- Safety operation and environmental impact of a process (detection of dangerous operation regimes; estimation of waste products; estimation and prevention of accidents; characterization of economic and social impact; dispersion prediction of waste products)

Mathematical models could typically be written for different environmental media (atmosphere, surface water, groundwater, subsurface, ocean) in order to solve different issues. These issues include: hazardous air pollutants, air emissions, acid rain, global warming, wastewater treatment plant discharges, industrial discharges, potable water source, food chain, leaking underground storage tanks, leakages from landfills, agriculture, solid and hazardous wastes, sludge disposal, spills.

2.1. SOFTWARE FOR SIMULATION OF PROCESSES

There are three approaches for the development of mathematical models (Gani and O'Connell, 1999): visual modeling using process flow diagrams (Flowsheet Modeling Environment – FME); modeling using a general programming language (Generic Modeling Language – GML); modeling using knowledge-based systems (Knowledge-based Systems – KBS). These approaches make possible and easier the solving of a mathematical model that contains many equations, which describes a system or phenomena. Commercial software, which could be used in order to solve mathematical models or to simulate various processes, are as follows.

2.1.1. MATLAB/SIMULINK

MATLAB (MATrix LABoratory) is an interactive environment for numeric calculation using matrix (that describe linear equations, keeping track of the coefficients of linear transformations, and recording data), developed by The MathWorks Inc., USA. The MATLAB software package can be extended with over 500 external functions named toolboxes. These external functions are characteristic for specific domains. MATLAB software is able to: perform model fitting and analysis (curve fitting); acquire and send out data from plug-in data acquisition boards (data acquisition); design and simulate fuzzy logic systems (fuzzy logic); control large, multivariable processes in the presence of constraints (model predictive control); solve standard and large-scale optimization problems (optimization solve); apply statistical algorithms and probability models (statistics).

SIMULINK is an extension of the MATLAB software package created as an independent modeling and simulation environment of dynamic systems (The MathWorks Inc., 1999).

SIMULINK is provided with its own graphic language that uses blocks for model construction.

2.1.2. FEMLAB

FEMLAB software package is an interactive environment for modeling and solving scientific and engineering problems based on partial differential equations. In order to use it, you must define variables and expressions that can be used within a model; customize the material library; link equations on multiple geometries; fine-tune the model mesh; and solve and postprocess the solution. Also it has flexibility giving the possibility of users to customize predefined equations (PDE), or to write their own PDEs (Comsol Multiphysics, 2001a). It contains different modules (Comsol Multiphysics. 2001b): Chemical Engineering Module, Earth Science Module, Electromagnetics Module, Heat Transfer Module, MEMS Module, and Structural Mechanics Module.

2.1.3. ChemCAD

ChemCAD is a software package for mathematical modelling and simulation of chemical processes developed by CHEMSTATIONS Inc., USA.

The program has the following modules (Nor-Par, 1999a):

- ChemCAD – used for steady state simulation of chemical processes, technological design, cost estimation etc.
- CC-Batch – used for simulation of batch distillation processes
- CC-Reacs – used for dynamic simulation of chemical processes and batch reactors (Nor-Par, 1999b)
- CC-Therm – used for design and simulation of shell and tube heat exchangers
- CC-DColumn – used for dynamic simulation of distillation columns
- CC-Props – used for calculation of physical and thermodynamic properties of components mixtures
- CC-LANPS – used for management of ChemCAD products in a Local Area Network (LAN)

2.1.4. HYSYS

HYSYS Plant is a software package for mathematical modeling and simulation of chemical processes developed by HYPROTECH Ltd., Canada (recently merged with ASPEN TECHNOLOGY Inc., USA). Using HYSYS software package, steady state and dynamic simulation of chemical processes can be accomplished (Hyprotech, 1995). The main advantages of HYSYS software package for the mathematical modeling and the simulation of chemical processes are (Hyprotech, 1995, 1996):

- Integrated system
- Intuitive and interactive system
- Open and extensible system

HYSYS Plant provides the user with four interfaces to allow information exchange between the program and the user. Any modification of one property value within these four interfaces leads to the property value change in the whole application.

2.1.5. ASPEN PLUS

Aspen Technology software packages are designed for a large diversity of chemical processes (petrochemical processes, general chemical processes, pharmaceutical products manufacture, electrolytic processes, polymerization processes, metallurgical processes, etc.) (Aspen Technology, 1999a, b).

Aspen Technology software packages are provided with an interactive graphic interface that makes the programs friendly to use. Also, apart from the mathematical modeling and the simulation, Aspen Plus software package allows the accomplishment of broad specific chemical engineering tasks: estimation and regression of physical property, complex system analysis, process optimization, cost estimation etc. (Aspen Technology, 1999c).

2.1.6. PRO II

PRO/II is a software package for mathematical modelling and simulation of chemical processes developed by SIMULATION SCIENCES Inc., USA.

This software package is designed for a large diversity of chemical processes. It can be used with success for a large range of chemical processes from petrochemical processes to pharmaceutical processes (Simulation Sciences, 1999a, b).

The program can be used in a wide variety of purposes: new plant design; modernization of existing plant; evaluation of possible plant designs; chemical processes optimization.

2.2. COMPUTER IMPLEMENTATION OF THE MODELS

Different research groups from around the world are trying to develop and implement mathematical models in order to better define a certain system, and to understand how this reacts and how it could be controlled (Turner, 1994; Schnelle and Dey, 1999; Bumble, 2000; Datta, 2002; Niessen, 2002; McCuen, 2003; Chien et al., 2004). Examples of these activities include:

- There are remarkable activities in fuel cell research and development worldwide, located in the EU, Japan, and the USA, and not at least in Australia, Brasilia, China, and Korea. Within the EU, CO_2 reduction is the key target and climate change is mentioned as the main motivation (Argyropoulos et al., 2002; Lunghi and Boye, 2002; Yoon et al., 2004; Jurado, 2005).
- Magnesium and its alloys are widely used as structural materials for automobiles and aircraft, due to their high strength/weight ratio. Their corrosion resistance is relatively poor, especially in a brine environment. Using classical anodization a protecting oxide film is formed on the surface. The disadvantage of this procedure is that the electrolyte used (NH_4HF_2, Na_2CrO_7, H_3PO_4) contains chromium and is not an environment-friendly substance and that is why we have not chosen this procedure. Chromate-based conversion coatings are cheaper, but the hexavalent chromium involved is both a carcinogenic and a hazardous air pollutant. The Directive 2000/53/EC on end-of-life vehicles makes explicit reference to the reduction or elimination of its use in cars. An alternative by developing clean and environment-friendly coatings could be microarc oxidation. In order to gain a better understanding of the electrochemical behavior of microarc oxidation (an electrochemical oxidation with a high voltage spark treatment in an alkaline electrolyte) was developed and implemented into Femlab, a model which describes the system (Tasoti et al., 2003).
- Other researchers are orientated to the study of sludge systems for nitrification and denitrification. The mathematical model, based on long-term experimental results, could be useful for the design of a new treatment system and operation of an existing one (Argaman et al., 1999). The model consists of 22 equations with 54 parameters.
- A steady state mathematical model is developed for describing the completely mixed biofilm–activated sludge reactor (hybrid reactor) (Fouad and Bhargava, 2005).
- Mathematical models can also serve as a tool to formulate operational and control strategies for high-rate anaerobic treatment systems. High-rate anaerobic treatment systems are used for treatment of industrial wastewater containing large amounts of organic matter in the form of carbohydrates or proteins. Good strategies will reduce operating costs, improve process

stability, and enhance treatment efficiency (Ramsay and Pullammanappallil, 2005).

- Different types of pollutants (pesticides, fertilizers, and herbicides) are applied on the soil surface. Massoudieh et al. (2005) developed a mathematical model in order to simulate transport of herbicides applied to roadsides into adjacent surface water. They have considered herbicide transport by overland flow, infiltrating water, and subsurface flow (Massoudieh et al., 2005).
- The rate of sediment transport is needed for various purposes such as the control and management of watersheds, river channels, reservoir sedimentation, and pollutant transport. Chiu et al. (2000) developed a mixed model, which combined application of the deterministic and probabilistic concepts.

Sustainable development is a challenge that gives also opportunities for new business with new and better products. This fits very well into modeling. Mathematical model could be used to optimal process design for minimization/maximization of quantitative/qualitative parameters or in optimal process control, to determine the optimal parameter of control systems.

3. Conclusions

Modeling of large-scale environmental systems is often a complex and challenging task. Mathematical modeling is used to help better understand the transport of pollutants in the environment, to determine short-term and long-term chemical concentrations in different compartments of the ecosphere, or to implement the design, operation, and optimization of reactor processes. At the same time, modeling of environmental systems uses simulation of a complex system because it is more feasible to work with a substitute than with the real, often complex system. Also, generated data could be postprocessed (such as statistical analysis or visualization) for better understanding and dissemination of information. To conclude, it could be said that mathematical modeling is a useful instrument for the enhancement of environmental security with real benefits for research and decision support.

References

Argaman, Y., Papkov, G., Ostfeld, A., Rubin, D., July, 1999, Single-sludge nitrogen removal model calibration and verification, *J. Environ. Eng.* 125(7): 608–617.

Argyropoulos, P., Scott, K., Shukla, A.K., Jackson, C., 2002, Empirical model equations for the direct methanol fuel cell, *Fuel Cells* 2: 78–82.

Aspen Technology, 1999a, Aspen Plus – User Guide, Vol. 1–3, Aspen Technology Inc., Cambridge, MA.

Aspen Technology, 1999b, Aspen Plus – Physical property, Methods and Models, Aspen Technology Inc., Cambridge, MA.

Aspen Technology, 1999c, Aspen Plus – Building and Running a Process Model, Aspen Technology Inc., Cambridge, MA.

Bumble, S., 2000, *Computer Simulated Plant Design for Waste Minimization/Pollution Prevention,* Lewis Publishers/CRC Press, Boca Raton, FL.

Chien, C.C., Medina, M.A. Jr., Pinder, G.F., Reible, D.D., Sleep, B.E., Zheng, C., 2004, *Contaminated Ground Water and Sediment – Modeling for Management and Remediation,* Lewis Publishers/CRC Press, Boca Raton, FL.

Chiu, C.-L., Jin, W., Chen, Y.-C., 2000, Mathematical models of distribution of sediment concentration, *J. Hydraul. Eng.*, ASCE 126(1): 16–23.

Comsol Multiphysics, 2001a, *FEMLAB – Reference Manual,* Comsol Multiphysics Inc., Wilmington, MA.

Comsol Multiphysics, 2001b, *FEMLAB – Model Library*, Comsol Multiphysics Inc., Wilmington, MA.

Datta, A.K., 2002, *Biological and Bioenvironmental Heat and Mass Transfer*, Marcel Dekker, New York.

Fouad, M., Bhargava, R., 2005, Mathematical Model for the Biofilm–Activated Sludge Reactor, *J. Environ. Eng.* 131(4): 557–562.

Gani, R., O'Connell, J.P., 31 May–2 June, 1999, *Properties and CAPE: From Present Uses to Future Challenges*, ESCAPE 9, Budapest, Hungary.

Hangos, K., Cameron, I., 2001, *Process Modeling and Model Analysis*, Academic Press, London.

Himmelblau, D.M., Bischoff, K.B., 1968, *Process Analysis and Simulation*, Wiley, New York.

Hyprotech, 1995, HYSYS Reference, Hyprotech Ltd., Calgary, Alberta, Canada.

Hyprotech, 1996, HYSYS and HYSIM, Process Engineering and Solutions, Hyprotech Ltd., Calgary, Alberta, Canada.

Jurado, F., 2005, Experience with non-linear model identification for fuel cell plants, *Fuel Cells* 1: 105–114.

Linnhoff, B., Akinradewo, C.G., 31 May–2 June, 1999, *Linking Process Simulation and Process Integration*, ESCAPE 9, Budapest, Hungary.

Lunghi, P., Bove, R., 2002, Reliable fuel cell simulation using an experimentally driven numerical model, *Fuel Cells* 2: 83–91.

Massoudieh, A., Huang, X., Young, T.M., Mariño, M.A., 2005, Modeling Fate and Transport of Roadside-Applied Herbicides, *J. Environ. Eng.*, ASCE (July): 1057–1067.

McCuen, R.H., 2003, *Modeling Hydrologic Change – Statistical Methods*, Lewis Publishers/CRC Press, Boca Raton, FL.

Niessen, W.R., 2002, *Combustion and Incineration Processes*, 3rd edn., Marcel Dekker, New York.

Nor-Par, 1999a, ChemCAD, ChemCAD for Windows, Nor-Par a.s.

Nor-Par, 1999b, ChemCAD, CC-Reacs, User's guide, Nor-Par a.s.

Ramsay, I.R., Pullammanappallil, P.C., 2005, Full-Scale Application of a Dynamic Model for High-Rate Anaerobic Wastewater Treatment Systems, *J. Environ. Eng.* ASCE 131(7): 1030–1036.

Schnelle, K.B. Jr., Dey, P.R., 1999, *Atmospheric Dispersion Modeling Compliance Guide*, McGraw-Hill, New York.

Simulation Sciences, 1999a, *PRO/II – A Product Demonstration*, Simulation Sciences Inc., Brea, CA.

Simulation Sciences, 1999b, *Process Engineering Suite – Tutorial Guide*, Simulation Sciences Inc., Brea, CA.

Tasoti, V., Agachi, S., Caire, J.P., 2003, Modeling and simulation of the electrochemical process of micro-arc oxidation, 30th International Conference of Slovak Society of Chemical Engineering, P146, pp. 1–6.

The Mathworks Inc., 1999, SIMULINK, Dynamic System Simulation for MATLAB – Using Simulink, The MathWorks Inc., Natick, MA.

Turner, D.B., 1994, *A Workbook of Atmospheric Dispersion Estimates*, 2nd edn., Lewis Publishers/CRC Press, Boca Raton, FL.

Yoon, S.P., Nam, S.P., Han, J., Tae-Hoon Lim, Seong-Ahn Hong, Sang-Hoon Hyun, 2004, Effect of electrode microstructure on gas-phase diffusion in solid oxide fuel cells, *Solid State Ionic.* 166: 1–11.

6. APPLICATION OF COMPUTER MODELING FOR THE ANALYSIS AND PREDICTION OF CONTAMINANT BEHAVIOR IN GROUNDWATER SYSTEMS

SEMJON P. KUNDAS*, IGOR GISHKELUK,
NIKOLAY GRINCHIK
International Sakharov Environmental University,
23 Dolgobrodskaya St., Minsk, 220009, Belarus

Abstract: The application of computer modeling for analysis and prediction of contaminant behavior in groundwater systems is discussed. A new mathematical model of impurity migration in soils and its application for the analysis of accessibility of toxicants for plants, and prediction of groundwater pollution, is studied.

Keywords: mathematical modeling; contaminant transport; groundwater system

1. Introduction

Owing to current intensive development of industry and agriculture, the increase in domestic wastes and the influence of other factors, soil and groundwater depletion, and pollution have achieved threatening levels. For the Republic of Belarus, soils contaminated by radionuclides became the most acute problem after the Chernobyl Accident. Therefore, the study of migration conditions and contaminant reduction (radionuclides, heavy metals, and organic compounds) in soil and groundwater is the most acute problem to ensure environmental safety and protection. Solving of this important problem is impossible without scientific knowledge about the dominant processes influencing both migration and physical–chemical changes of the pollutants in different geological and environmental conditions. In our opinion, the leading role in this research should belong to the computer modeling methods that will help to understand this complicated and dynamic process more precisely as an

*To whom correspondence should be addressed. Semjon P. Kundas, International Sakharov Environmental University, 23 Dolgobrodskaya str., Minsk, 220009, Belarus; e-mail: kundas@iseu.by

R. N. Hull et al. (eds.), Strategies to Enhance Environmental Security in Transition Countries, 57–68.

influence of pollution on human health. Application of modeling methods will help us analyze different migration mechanisms and the reduction of toxicants in soils and groundwaters, determine the influence of contaminated soil on the biosphere, reduce the number of toxicants in crops produced, and prevent their accumulation in humans and animals (Serebryannyi and Zhemzhurov, 2003).

The main purpose of computer modeling methods is to formulate a mathematical model of the contaminant migration as an algorithm. In the future, this model will be used as a computer program that will allow us to carry out numerical experiments and gain insight into the processes involved (Kudryashov, 1998). Thus, research on the influence of factors such as weather conditions, physical–chemical characteristics of toxicants, etc. on contaminant migration in soils is carried out using the mathematical model and determining the dependence of these factors on any parameter. Such an experiment helps to investigate questions more deeply in the environment, and reduce the number of the expensive measurement experiments.

Thus, the main objective of the work outlined in this paper was to research the possibilities of applying computer modeling to the analysis and prediction of contaminant behavior in groundwater systems, in particular, radionuclides in soil.

2. Mathematical Model of Contaminant Behavior in Groundwater Systems

For the analysis and prediction of the migration and concentration of a contaminant in soils, it is necessary to create mathematical models providing the description of the main processes that influence the transfer and redistribution of these substances (Figure 1), while understanding that the level of influence of each process can differ between various geological environments.

The main physical–chemical processes determining contaminant transport and redistribution in soils include: (Bulgakov, 1992):

- Convective-dispersion transport of contaminants in solute form
- Movement of groundwater in the liquid and gas phases under the influence of moisture and temperature gradients
- Redistribution between contaminants dissolved in water and absorbed by the matrix of solid material
- Diffusion of sorbed contaminants in soil by the solid material
- Chemical and radioactive transformation of the contaminants
- Aerosol fallout of contaminants on the soil surface
- Infiltration of water from the soil surface

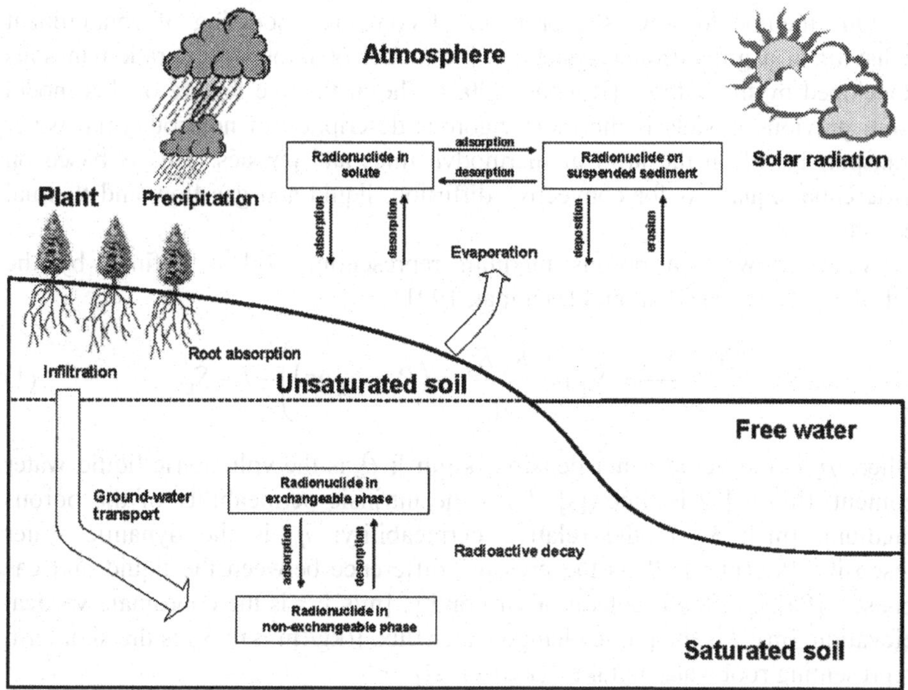

Figure 1. Basic processes determining the contaminants redistribution in groundwaters systems.

- Deduction of contaminants from the soil during the evaporation of water
- Absorption of toxicants by plants
- Evaporation of moisture from the soil surface
- Loss of moisture due to the process of evapotranspiration by plants
- Contaminant deposition on the soil surface and ingress to the atmosphere due to wind erosion

Analysis of the foregoing processes shows that the dominant mechanism of contaminant transport in soils is dispersion and convection in the soil water (Kudryashov, 1998). Here the soil moisture exists both in the liquid and gas phases, and its distribution is influenced by moisture and temperature gradients. In this application, the modeling problem of contaminant distribution in soil has two subproblems: modeling of the nonisothermal water transport and contaminant migration.

In our opinion, progress in the application of computer modeling for the analysis and prediction of contaminant behavior in groundwater systems is connected to the substitution of purely empirical descriptions for the informal descriptions based on the physical mechanisms in nature. This makes the model more applicable to the processes taking place in nature.

One method to solve the problem of computer modeling of contaminant behavior in a groundwater system is the model of impurity migration in soils developed by the authors (Kundas, 2005). The distinctive feature of this model from previous models is the more rigorous description of nonisothermal water transport based on the laws of thermodynamics and physics. This is based on differential equations for convective diffusion, liquid and gas flow and thermal conductivity.

Water flow in a porous medium representing soil is defined by the following equation (Bear and Bachmat, 1991):

$$\frac{\partial(\rho_L\theta_L)}{\partial t}=\nabla\left(\rho_L\frac{K_o K_L}{\eta_L}\nabla(P_L-\rho_L gz)\right)-I-S_W \tag{1}$$

where ρ_L is the liquid water density, [kg/m^3]; θ_L is the volumetric liquid water content, [m^3/m^3]; t is time, [s]; K_0 is the intrinsic permeability of the porous medium, [m^2]; K_L is the relative permeability; η_L is the dynamic water viscosity, [kg/(m·s)]; P_L is the pressure difference between the liquid and gas phases, [Pa]; g is the acceleration of gravity, [m/s^2]; z is the coordinate vertical elevation, [m]; I is the phase change water flux, [(kg/(m^3·s)]; S_W is the sink term representing root water uptake, [(kg/(m^3·s)].

For modeling of water and vapor transport, knowledge of the phase change water flux, the dependence of the relative permeability from the water content and the dependence of pressure from the water content and temperature are necessary. The last relationship can be determined using the moisture sorption isotherm, which is the experimentally obtained dependence of the mass water content (u) from relative humidity (φ) at different temperatures.

Relative humidity and water vapor pressure (P_V) are related by the equation:

$$\varphi=\frac{P_V}{P_S} \tag{2}$$

where P_S is the saturation vapor pressure as a function of temperature.

The mass water content u is related to the volumetric liquid water content θ_L by the equation:

$$\theta_L=\frac{\rho_S}{\rho_L}u \tag{3}$$

where ρ_s is the soil density, [kg/m^3].

Then, using the sorption isotherm, the functional dependence of moisture content on water vapor pressure and temperature can be determined as $\theta_L=\psi(P_V,T)$.

From this, the function defining the change of water vapor pressure from the mass moisture content and temperature is derived: $P_V = f(\theta_L, T)$.

The relationship between pore pressure, saturation vapor pressure, and relative humidity is given by Kelvin's equation:

$$P_L = P_S + \frac{RT\rho_L}{M_L}\ln\varphi \tag{4}$$

where R is the universal gas constant, [8.31 J/(mol·K)] and M is the molar weight of water, [0.018 kg/mol].

This equation, using the relationship of water vapor pressure to the water content and temperature is then written in the following form (Grinchik, 2003):

$$P_L(\theta_L, T) = P_S + \frac{RT\rho_L}{M_L}\ln\frac{f(\theta_L, T)}{P_S} \tag{5}$$

Equation (5) relates the physical and chemical properties of liquid water to the structural characteristics of the soil, and is necessary for the solution of the equation of water transport as the dependence of pore pressure on water content and temperature.

In accordance with the hypothesis of local thermodynamic balance widely used in the theory of sorption, the phase change water flux is defined by the equation (Grinchik, 2003):

$$I = \rho_S\frac{\partial u}{\partial t} = \rho_S\left[\left(\frac{\partial u}{\partial P_V}\right)\frac{\partial P_V}{\partial t} + \left(\frac{\partial u}{\partial T}\right)\frac{\partial T}{\partial t}\right] \tag{6}$$

The dependence of relative permeability on water content is defined by the following empirical equation (Genuchten, 1980):

$$K_L = \left(\frac{\theta_L - \theta_r}{\theta_s - \theta_r}\right)^b \left[1 - \left(1 - \left[\frac{\theta_L - \theta_r}{\theta_s - \theta_r}\right]^{\frac{1}{m}}\right)^m\right]^2 \tag{7}$$

where θ_r and θ_s denote the residual and saturated water contents, respectively, [m^3/m^3]; b and n are the parameters that denote soil characteristics.

Infiltration of atmospheric precipitations can be modeled using Equation (1) by defining specific boundary conditions. In the beginning of the precipitation event, when the soil is not saturated with water, water flux from the soil surface is defined by the rainfall rate (f):

$$\frac{K_o K_L}{\eta_L} \nabla(P_L - \rho_L gz) = f \text{ for } \theta_L < \theta_s \text{ and } \varphi < 1 \tag{8}$$

When the near-surface soil layer approaches saturation, i.e., pores in the soil are fully filled with water, boundary conditions on the surface of the soil are written in the form:

$$P_{L\,sat}(t) = P_S + \frac{RT\rho_L}{M_L} \ln \frac{f(\theta_s, T)}{P_S} \text{ for } \theta_L = \theta_s \text{ and } \varphi = 1 \tag{9}$$

where P_{Lsat} is the liquid pressure on the surface of the soil at full saturation of the soil with water [Pa].

Water vapor flow in the soil, accounting that the content of air in pores contributes no significant influence on the movement of the water vapor, can be described by the following equation: (Bear and Bachmat, 1991):

$$\frac{\partial(\rho_V \theta_V)}{\partial t} = \nabla\left(\rho_V \frac{K_o K_V}{\eta_V} \nabla(P_V - \rho_V gz)\right) + I \tag{10}$$

where ρ_V is the water vapor density, [kg/m^3]; θ_V is the volumetric water vapor content, [m^3/m^3]; K_V is the relative permeability; η_V is the dynamic water vapor viscosity, [kg/(m·s)]; P_V is the vapor pressure, [Pa].

The relative permeability of the water vapor is defined similarly to the relative permeability of water:

$$K_L = \left(\frac{\theta_L - \theta_r}{\theta_s - \theta_r}\right)^b \left[1 - \left(\frac{\theta_L - \theta_r}{\theta_s - \theta_r}\right)^{\frac{1}{m}}\right]^{2m} \tag{11}$$

The general gas law relates the vapor density with the vapor pressure:

$$\rho_V = \frac{MP_V}{RT} \tag{12}$$

And the relation between the mass water content and the volumetric water vapor content is:

$$\theta_V = m - \frac{\rho_S}{\rho_L} u \tag{13}$$

where m is the total porosity.

Using Equations (12) and (13), the left part of Equation (10) is written as:

$$\frac{\partial(\rho_V \theta_V)}{\partial t} \approx \left(m - \frac{\rho_S}{\rho_L} u\right) \frac{M}{RT} \frac{\partial P_V}{\partial t} \tag{14}$$

Evaporation from the soil surface in the proposed model can be taken into account by defining the following boundary conditions:

$$P_V(t) = \varphi_{surf}(t) P_S(T) \tag{15}$$

where φ_{surf} is the soil surface relative humidity.

The equation for temperature distribution in the soil considering latent heat of the phase change is given by (Grifoll, 2005):

$$C_{eq} \frac{\partial T}{\partial t} = \nabla\left(K_{eq} \nabla T\right) + qI \tag{16}$$

where C_{eq} is the volumetric heat capacity, [J/(m^3·K)]; K_{eq} is the soil thermal conductivity, [W/(m·K)]; q is the latent heat of the phase change, [J/kg].

Volumetric heat capacity of the soil (C_{eq}) is the sum of the volumetric heat capacities of the soil constituents:

$$C_{eq} = \frac{\sum \theta_i \rho_i C_i}{\sum \theta_i} \tag{17}$$

where ρ_i, C_i, and θ_i are the density, specific heat capacity and volumetric fraction of the ith soil constituent, i.e., soil-solids, water, water vapor, and air.

The thermal conductivity of the soil can be calculated in the following way:

$$K_{eq} = \frac{\sum \theta_i k_i}{\sum \theta_i} \tag{18}$$

where k_i and θ_i are the thermal conductivity and volumetric fraction of the ith soil constituent.

Latent heat of the phase change is computed with the use of the Clausius–Claperon equation:

$$q = \frac{RT^2}{M}\left(\frac{\partial \ln P_V}{\partial \ln T}\right) \tag{19}$$

The heat exchange of the soil with the atmosphere is defined by the following boundary conditions:

$$T(t) = T_{surf}(t) \tag{20}$$

where T_{surf} is the temperature on the soil surface.

Contaminant transport, taking into account sorption, is described by the equation (Serebryannyi and Zhemzhurov, 2003)

$$\frac{\partial(\theta_L C)}{\partial t} + \frac{\partial(\rho_b N)}{\partial t} = \nabla(\theta_L D_{LS} \nabla C - V_L C) - R - S_S \tag{21}$$

where C is the dissolved concentration, [kg/m^3]; N is the mass of the adsorbed contaminant per dry unit weight of the soil, [kg/kg]; ρ_b is the bulk density [kg/m^3]; D_{LS} denotes the combination of the hydrodynamic dispersion tensor for water and diffusion in soil-solid [m^2/s]; V_L is the water velocity, [m/c]; R is the term representing the change of the concentration due to radioactive disintegration or chemical reactions [kg/(m^3·s)]; S_S is the sink term representing the root solute uptake, [kg/(m^3·s)].

Solute spreading includes mechanical dispersion in the water plus molecular diffusion for the water and the soil-solid. These three processes appear in the liquid–solid dispersion tensor:

$$D_{LS} = \alpha \frac{V_L}{\theta_L} + \tau D_m + \rho_b K_d \frac{D_S}{\theta_L} \tag{22}$$

where α is the soil dependent constant for the dispersivity, [m]; τ is the soil dependent constant for the tortuosity; D_m is the molecular diffusion coefficient, [m^2/s], D_S is the diffusion constant in the soil-solid [m^2/s].

Distribution of the impurity between soil-solid and liquid phases is described by using the distribution coefficient K_d:

$$N = K_d C \tag{23}$$

The value of K_d can be determined experimentally or calculated by using the available information about the sorption properties of the system.

The movement of contaminants in and from the soil to the atmosphere is defined by the boundary conditions of the second type:

$$\theta_L D_{LS} \nabla C = q \tag{24}$$

where q is the flow of the contaminants between the soil and the atmosphere. The value of the flow is calculated according to appropriate expressions depending on the movement of the contaminant in and from the soil (aerosol fallout of the contaminant, fallout on the surface of the soil and the movement of contaminants into the atmosphere due to wind erosion etc.). If contaminant transport from the soil into the atmosphere and vice versa is insignificant, the flow at the boundary is equal to zero: $q=0$.

The above mentioned mathematical model in the one-dimensional formulation is solved with the use of MATLAB and FEMLAB, and allows computer modeling of contaminant migration in the groundwater system.

3. Results of the Computer Modeling and Discussion

Prediction of the radionuclide migration depth in the soil, and analysis of the contaminated groundwater used as a drinking water source, is of practical interest to the Republic of Belarus after the Chernobyl accident. It is also interested in predicting the change in the amount of radionuclides available for absorption by plants. It is in this application that the models outlined above will be discussed further.

It is assumed that as a result of the nuclear fallout, the surface of the soil is contaminated by ^{90}Sr that, under the influence of infiltrating rainfall, migrates deep into the soil. It is necessary to determine the quantity of radionuclides that remain in the root layer versus the quantity impacting the groundwater after a specified period of time. Experimental measurements would allow us to determine the concentration of radionuclides in the soil at given depths, and their concentration in groundwater and crops, but would not allow prediction of the concentrations in the future.

Physical–chemical soil properties which determine the process of migration were determined for the podzolic soil. The influence of climatic factors was also taken into account by defining the temperature and rainfall intensity changes with time.

As a result of the modeling, changes in ^{90}Sr content with depth for different total amounts of deposited radionuclides were determined (Figure 2A).

Comparing the radionuclide distribution in soil with the root system distribution of the crop (Figure 2B) it is obvious that during the analyzed period of time (20 years) most of the nondecayed radionuclides remain in the soil layer from 0 to 0.60 m, making it available for plant uptake by root absorbtion. At the same time, it can be determined that there is no danger of the radionuclide entering the groundwater situated deeper (0.8–2.0 m).

Analysis of the modeling results (Figure 3) shows that the contribution of vertical migration to the purification of the root zone (0–0.6 m) from radionuclides during the first stage (20 years) is not significant in comparison to radioactive decay. But in the next period of time (>20 years) a sharp decrease of ^{90}Sr in the root zone is observed due to the vertical migration of radionuclides to the lower soil layer.

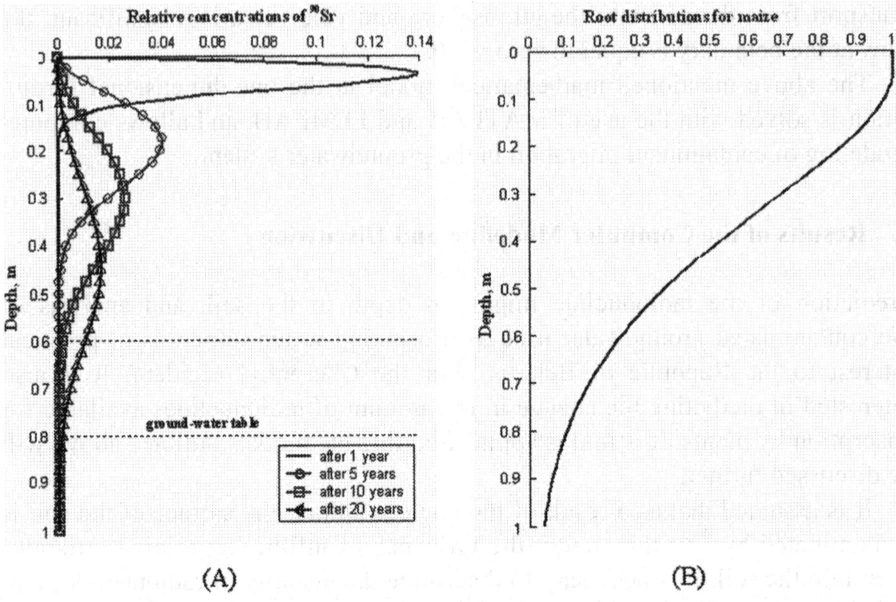

Figure 2. The distribution of ^{90}Sr (A) and root system for the maize (B) in the soil.

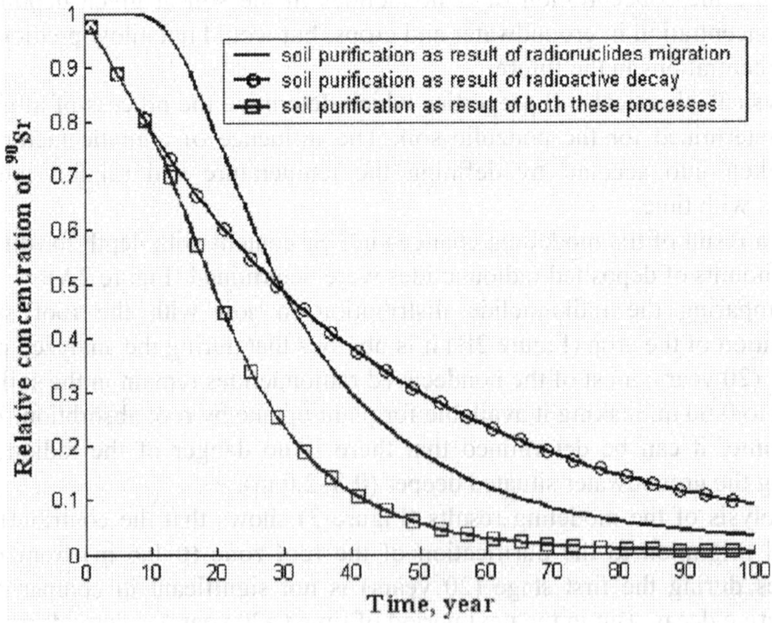

Figure 3. The role of the vertical migration and radioactive decay in the removal of ^{90}Sr from the soil level: 0–0.6 m.

The decrease of radionuclide content in the root zone twice in comparison with total amount of fallout on the soil surface, due to the processes of radioactive decay and vertical migration can be expected in 19 years and 100 times in 78 years (Figure 4A).

Radionuclides that have fallen on the soil surface will reach the level of groundwaters (1.4 m depth) in 40 years. The maximum accumulation of ^{90}Sr by groundwaters will be observed 110 years after nuclear fall-out, and will be 25% of the total amount of radionuclides that have fallen on the soil surface (Figure 4B).

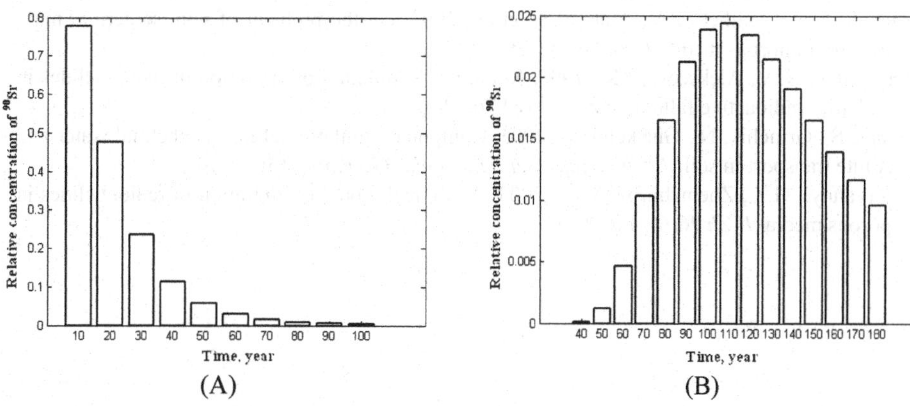

Figure 4. The dynamics of the relative concentration of ^{90}Sr in root-inhabited layer (A) and ^{90}Sr pollution density of groundwaters (B).

4. Conclusion

It is shown that the application of computer modeling for analysis and prediction of contaminant behavior in groundwater systems is an effective instrument to investigate the problem of environmental safety.

The contaminant migration model was discussed and analyzed in the article. The model was based on strict mathematical description, taking into account physical processes and environmental factors essential to the examined process.

Results of modeling the vertical migration of radionuclides in the soil have been obtained with the use of the proposed mathematical model, which illustrates its effective application for the prediction of long-term change of the radiological situation in a defined area.

Progress in the use of computer modeling methods to research radionuclide migration in the environment can be achieved with the use of universal software integrating mathematical models, databases with the information about different soil types, and long-term hydrometeorological conditions for the defined area and geographic information systems.

References

Bear, J., Bachmat, Y., 1991, *Introduction to Modeling of Transport Phenomena in Porous Media*, Kluwer, Dordrecht, 584 p.

Bulgakov, A.A., Konoplev, A.V., Popov, V.E., 1992, Prediction of ^{90}Sr and ^{137}Cs behavior in soil-water system after the Chernobyl accident, in: *Ecological and Geophysical Aspects of Nuclear Accidents*, Gidrometeoizdat, Moscow, pp. 21–42.

Genuchten, M.Th., 1980, A closed-form equation for predicting the hydraulic conductivity of unsaturated soils, *Soil Sci. Soc. Am. J.* 44: 892–898.

Grifoll, J., Gastó, J.M., Cohen, Y., 2005, Non-isothermal soil water transport and evaporation, *Adv. Water Res.* 28: 1254–1266.

Grinchik, N.N., Akulich, P.V., Kuts, P.S., et al., 2003, On the problem of non-isothermal mass transfer in porous media, *IFZh* 76(6): 129–141.

Kudryashov, N.A., Alekseeva, I.K., 1998, Numerical simulation of migration of radionuclides in soil after radioactive fallout, *IFZh* 71(6): 976–982.

Kundas, S., Grinchik, N., Gishkeluk, I., 2005, Computer simulation of non-isothermal water and solute transport in soil, *J. Univ. Appl. Sci. Mittweida (Germany)* 13: 27–30.

Serebryannyi, G.Z., Zhemzhurov, M.L., 2003, Analytical model of migration of radionuclides in porous media, *IFZh* 76(6): 146–150.

7. INTEGRATED MANAGEMENT STRATEGY FOR ENVIRONMENTAL RISK REDUCTION AT POSTINDUSTRIAL AREAS

GRZEGORZ MALINA*
Institute of Environmental Engineering, Częstochowa University of Technology, Brzeźnicka 60A, 42-200 Częstochowa, Poland

Abstract: An approach is presented on how to deal with postindustrial areas in Poland. First, an overview is given on the status of postindustrial areas, and the current legal framework is briefly discussed. Then, the National Program for Postindustrial Areas recently approved by the Polish government is presented. The rationale for application of integrated management systems for risk reduction is provided. Activities related to preparing, adjusting, and verifying the integrated management strategy, are based on the classical structure of Deming's cycle. Information systems and criteria, and sets of activities required for running such systems are discussed. Examples of integrated management for environmental risk reduction applied at postindustrial areas in Poland are presented, including the cases of a chemical plant (Tarnowskie Góry), a zinc smelter (Piekary Śląskie), and a lignite opencast mine (Bogatynia) in the "Black Triangle" area (Southwest Poland).

Keywords: postindustrial areas; legal framework; integrated management systems; National Program for Postindustrial Areas; case studies

1. Introduction

Long-term diverse activities in industrial areas of Poland resulted in direct emissions of waste products into environmental compartments (atmosphere, surface water, and soil/groundwater), which consequently led to major environmental problems (GIOŚ, 2003; GUS, 2003; Siuta and Kucharska, 1997). Of these, soil and groundwater systems respond much slower to remediation

*To whom correspondence should be addressed. Grzegorz Malina, Institute of Environmental Engineering, Częstochowa University of Technology, Brzeźnicka 60A, 42-200 Częstochowa, Poland; e-mail: gmalina@is.pcz.czest.pl

R. N. Hull et al. (eds.), Strategies to Enhance Environmental Security in Transition Countries, 69–94.

measures and need more time to improve their quality. Moreover, soils and groundwater reservoirs have been contaminated by numerous spills and landfills. This has resulted in diffuse pollution on a mega scale (e.g., brownfields, post-industrial areas, military bases, etc.), which makes the active conventional remediation within an intermediate time frame (25 years) not feasible, or even impossible for technical and economical reasons (Malina et al., 2006). Such areas are, therefore, continuous and long-term potential and actual sources of regional contamination of soils, surface water, groundwater, and/or sediments. Shifts in environmental policy have been observed since the mid-1990s from multifunctionality into risk reduction for humans and ecosystems to acceptable levels (Malina, 2005). Consequently, remediation goals changed from target-oriented to risk-oriented. The currently applied risk evaluation procedures are commonly based on the source–pathway–receptor sequence.

Postindustrial areas are defined as degraded, abandoned, or not fully used areas previously designated for industrial/business activities that have been terminated (Program Rządowy, 2004). These areas (excluding those having military or agricultural functions) are degraded to a degree that limits the possibilities of development and/or returning to their previous economic functions. Risk related to humans, and correlation between degraded area and economy are taken into consideration. These criteria can be applied for areas where no activity is carried out, and for areas still having industrial functions, thus causing further degradation.

The goal of this paper is to indicate the need of applying integrated management for postindustrial areas. In particular, such management should take into account soil, surface water, and groundwater quality protection, including information systems (environmental monitoring, statistical reporting, official information, and research and development – R&D), as well as criteria and activities required for implementing it. As examples, integrated management for environmental risk reduction is applied at three postindustrial areas in Poland, including the cases of: a chemical plant (Tarnowskie Góry), a zinc smelter (Piekary Śląskie), and a lignite opencast mine (Bogatynia) in the “Black Triangle” area.

2. Current Status of Postindustrial Areas in Poland

Postindustrial areas may have economic, social, and spatial functions. Economic functions may include: (i) exploitation and processing of natural resources, (ii) generation of gross national product (GNP), and (iii) induction of other economic sector development. Social functions are related to generation of jobs, and improvement of living conditions and education levels. Spatial functions are responsible for transformation of the environment (e.g., coal mines, factories, urbanization).

Based on the degrading factor, the Organization for Economic Cooperation and Development (OECD) distinguishes the following types of postindustrial areas (Założenia Programu Rządowego, 2003): (i) degraded chemically – cleanup required, (ii) degraded physically – reclamation and redevelopment (natural landscape) required, (iii) not degraded physically or chemically, but having lost their previous economic functions. Postindustrial areas are usually degraded physically and/or chemically, thus requiring remediation. During their migration from the sources at the surface towards the receptors (e.g., aquifers), chemical compounds contaminate surface water, soil, unsaturated zones, and aquifers. As a result, degradation of the area is observed, at a scale dependent on the contamination source, migration rates, and persistence of the contaminants.

Within postindustrial areas one can observe: (i) changes in hydrogeological conditions, (ii) changes in consolidation (degree of noncompact soils due to varying loads), and (iii) penetration of contaminants (especially chemical compounds) to the bedrock from wastewater and solid waste.

Depending on the area, the following types of degradation can be distinguished (Bolingier, 2004):

- Point degradation – an effect of one or several industrial objects that form the concentrated source of degrading factors (e.g., depots of industrial, mine and steel wastes, chemical plants, steelworks, etc.)
- Surface degradation – a result of more than one source of degradation factors (e.g., industrial complexes, recently often referred to as mega sites)
- Regional degradation – caused by big cities and towns, the so-called agglomerations, with significant contribution of industry hazardous to the environment

Three basic types of postindustrial areas can be distinguished:

1. Areas used for production: exploitation of natural resources, energy production, production, and distribution of goods, as well as waste management/neutralization/disposal. They can be: (i) part of the production process, e.g., quarries, open pits, shafts, dumps, landfills, water reservoirs; (ii) places of the production process (repository of resources, engine rooms, workshops, etc.; and (iii) places, where the infrastructure (building sites, motorways, and highways, railways, lawns, etc.) is located.
2. Areas associated with production, such as administration and research centers, design offices, centers of culture and sports, medical and education, logistic ad telecommunication centers, water supplies, and wastewater treatment plants, etc.
3. Areas affected by production: (i) physically/chemically (contamination, degradation), and (ii) economically (generation of GNP).

Degradation is often a result of industrial waste generated mainly by black coal (and other natural resources) mining, the power engineering industry, and metallurgy. Due to their origin, chemical and biological composition, and other properties and circumstances, hazardous waste is particularly important in terms of the risk it poses to humans and the environment. In 1996, hazardous waste constituted 3.5% of the total amount of industrial waste. This dropped to 1.1% in 2001 (Ciesielski and Motak, 2001). The maximum amount of hazardous waste in 2001 was generated in Małopolska (~55%), Lower Silesia (~10%) and Silesia (~8%) provinces.

In Poland, no full inventory of postindustrial areas has been made to date. It is estimated that 8.5×10^3 km^2 has been affected by degradation processes, including 1.5×10^3 km^2 degraded to a "high" and "very high" degree (Założenia Programu Rządowego, 2003). Moreover, an additional 40×10^3 km^2 is potentially threatened.

2.1. AREAS OF ECOLOGICAL HAZARD

Areas of ecological hazard have a high density of contamination sources and intensive degradation of the environment, often connected with bad sanitary status and health risk, which have led to disturbance or collapse of ecological balance. The basic criteria used to separate these areas include: (i) exceeding contamination standards, or serious degradation of at least two environmental compartments, and (ii) multiple or particularly heavy exceeding of acceptable levels in the case of a particular toxic substance, or serious degradation of one environmental compartment.

In 1982, 27 areas of ecological hazard were identified, which comprised ~11% of the total area of Poland and supported approximately one-third of the total population (~13 million) (Założenia Programu Rządowego, 2003). There are three types of areas of ecological hazard (Program Rządowy, 2004):

- Of particularly strong hazards also known as areas of ecological disaster (regions of Upper Silesia, Krakow and Rybnik, the Legnica-Głogów copper district, the Gdańsk Bay, including the Puck Bay)
- Of strong hazards (regions of Opole, Konin, Szczecin, Turoszów, Bydgoszcz-Toruń, Tarnów, Wałbrzych, Kielce, Wrocław, Częstochowa, Inowrocław, Łódź, Poznań, Myszków-Zawiercie, Chełm, Bełchatów, Tomaszów, Włocławek, Puławy, and Płock)
- Of strong contamination of the atmosphere (regions of Tarnobrzeg, Włocławek, Puławy, and Płock)

Currently, the situation is systematically improving, and more than half of the areas of ecological hazard no longer exist, while the quality of others is continuously improving (Bolingier, 2004).

2.2. AREAS DEGRADED BY INDUSTRIAL ACTIVITIES

Since the 1990s, significant environmental improvement in general has been observed in Poland. New principles of using and managing degraded areas have been introduced, including strict injunctions not to build or develop industrial plants that are particularly hazardous to the environment, resulting in an improvement in environmental quality (Założenia Programu Rządowego, 2003). It is assumed that currently the highest load of multifactor degradation of the environment is in the provinces described below (Program Rządowy, 2004).

2.2.1. The Silesia Province

The degraded areas here originate from extractive industry, mainly black coal and nonferrous metal mining. Mining activities require continuous drainage, and pumped saline waters are often discharged to rivers. Mining is also responsible for the biggest surface transformations in Poland, including depressions and cavities, as well as dumps (~300 dumps covering ~3,000 ha). Toxic substances, including heavy metals, chlorides, and sulfates leached into subsoil due to long-term infiltration of precipitation. The atmosphere is impacted by dump fires, due to easy access of air into dumps, as well as high contents of flammable fractions and sulfides in dumped materials. Other dumps consist of energy-related waste materials from electric power plants, thermal–electric power stations, and the steel industry.

Degradation is also the result of processing industries:

- Fuel and power – electric power plants: “Rybnik”, “Jaworzno III”, “Łaziska”, “Siersza”, and refineries in Czechowice-Dziedzice and Trzebinia
- Steel – big plants “Katowice” and “Częstochowa”, smaller and older steelworks in w Dąbrowa Górnicza, Chorzów, Ruda Śląska, Katowice, Świętochłowice, Sosnowiec, Gliwice, Będzin, a steel processing plant in Zawiercie, and zinc and lead plants in Miasteczko Śląskie, Piekary Śląskie, and Szopienice
- Chemical – the chemical plant in Tarnowskie Góry, the cement plant in Rudniki near Częstochowa
- Wood and cellulose – plants in Klucze, Żywiec, Krapkowice, Myszków, Kalety, and Koniecpol

2.2.2. *The Lower Silesia Province*

This province in Poland has the second highest number of degraded (postindustrial) areas. It is the region where one of the most contaminated areas in Europe, known as "the Black Triangle", is located. Near Wałbrzych, there are also many postindustrial areas (dumps, embankments, settlers, and quarries) related to the terminated coal mines, as well as coke, porcelain, and textile plants. These facilities caused heavy air and soil pollution, and resulted in a number of mining waste dumps. They are also responsible for surface water contamination (mainly with phenols) due to the direct discharge of industrial wastewaters (Wójcik, 1997).

In 1996, the Legnica-Głogów copper district was designated as an area of ecological disaster. Existing copper ores deposits (one of the biggest in the world) are exploited in numerous mines, and processed afterwards in copper plants in Głogów and Legnica. Substantial amounts of industrial waste are main sources of degradation at this area (Berkel, 1996).

2.2.3. *The Małopolska Province*

Environmental degradation in this province is mainly due to industry located in the neighboring Silesia Province. Long-term emissions of contamination from electric power, steel, and chemical plants resulted in soil acidification and contamination with heavy metals.

The local industry of the Province responsible for environmental problems includes (GIOŚ, 2003): the Sendzimir steel plant; the aluminum plant in Skawina; the mining and steel plant "Bolesław"; the electric power plant in Skawina; two thermal–electric power stations: "Kraków" and "Skawina"; the chemical plants in Alwernia, Tarnów, and Oświęcim; the pharmaceutical plant "Polfa"; and the sodium plant "Solvay" in Krakow.

2.2.4. *The Podkarpackie Province*

Environmental problems here are related to sulfur and sulfuric acid industry based on the sulfur deposits (some of the biggest in the world) near Tarnobrzeg and Lubaczów. The exploitation of sulfur deposits by strip mining resulted in strong geo-mechanical transformations of the land surface, similar to that originating from the lignite mining, with the formation of internal and external dumping grounds and final excavations. On the other hand, the application of hot water pumping into deposits and subsequent extracting of hot solutions, resulted in increased areas (~3000 ha) of geo-mechanical, hydrological, and chemical transformations.

2.2.5. The Province of Łódź

The biggest lignite open pit in Poland is operating in Bełchatów with a dumping site for waste (ashes and slag) from the electric power plant in Bełchatów. Due to emissions of SO_2 above the standards and deposition of waste products, this power plant was in 1990 included in the "List of 80" most hazardous plants in Poland.

2.2.6. The Province of Lublin

Emissions of industrial dusts from cement plants located near Chełm and Rejowiec are responsible not only for air pollution but also for soil erosion. In the mine "Bogdanka" near Łęczyca, black coal is exploited. In Puławy, the biggest plant of nitrogenous fertilizers in Poland is operating since 1996. The neighboring forest complex has been degraded due to the emissions of gases and dusts.

2.3. LEGAL FRAMEWORK

According to the Constitution: *Poland protects national heritage and assures protection of environment by following the sustainable development principle.* This principle is achieved through a number of acts of the Polish Parliament. Principles of land protection, including soil, surface water, and groundwater, are given by: the Environmental Protection Law (Dz.U.2001.62.627 with further changes), the Act on the Waste (Dz.U.2001.62.628 with further changes), the Law on Nature Protections (Dz.U.2001.99.1079 with further changes), and the Water Law (Dz.U.2005.239.2019 with further changes) (Figure 1). In addition, some issues of soil and land protection are included in a number of other legal documents (Malina, 2006).

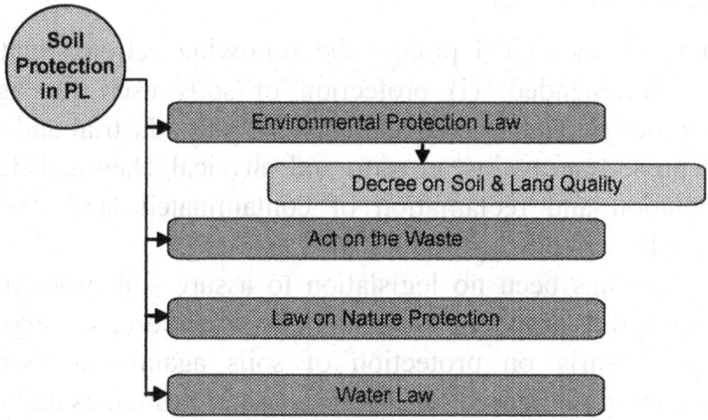

Figure 1. Polish soil and land protection related acts and laws.

Based on the Environmental Protection Law, the Decree on Soil and Land Quality Standards (Dz.U.2002.165.1359) identifies when soil is considered as contaminated. It is based on soil standards for the actual and planned land functions, for the following groups (Malina and Kwiatkowska, 2006):

(A)
- (i) Land included in the areas protected under the Water Law
- (ii) Areas protected under the Law on Nature Protection, if the actual contamination is not considered hazardous for humans and environment

(B) – Croplands and grasslands, forest lands, barren and urban lands
(C) – Industrial, mining, and communication areas

To achieve sustainable soil/land protection, the New National Ecological Policy was established (M.P.2003.33.433), with the strategic goals including: effective management of natural resources – sustainable development principle, counteraction/prevention of impacts to the environment causing its degradation, and restoration of the environment. This policy is based on five main principles:

1. The prevention principle – (i) prevention of contamination and other arduousness, (ii) recycling (closing materials and resources cycle, reuse of energy, water, and resources from wastewater and solid waste), and (iii) integrated neutralization of contaminants and hazards
2. The foresight principle – solving problems already at established reasonable probability of occurrence
3. The Best Available Technology (BAT) principle
4. The subsidiary principle – decentralized decisions related to environmental protection ("closer to citizens")
5. The efficiency principle – costs minimization and maximization of effects of environmental protection

Within the frame of ecological policy, the following actions related to soil protection are recommended: (i) protection of soils used for agricultural/horticultural purposes (including redevelopment of postindustrial and abandoned areas), (ii) soil protection against erosion and physical–chemical degradation, (iii) soil remediation and reclamation of contaminated land, (iv) holding/returning high quality of soil required for food growing.

Until now, there has been no legislation to assure soil protection against overuse of natural soil productivity potentials. Therefore, ecological policy should focus particularly on protection of soils against intensification of agricultural production (increased fertilization, use of pesticides and herbicides, concentration of pigs and stock-breeding, areas of monoculture cultivation, and transportation for agricultural production), remediation of degraded soils to include them again for natural use (forestation, plant growing, tillage), or redevelopment in the case of postindustrial areas.

The goals of ecological policy from the soil protection viewpoint in Poland until 2010 are as follows (Malina and Kwiatkowska, 2006):

- Education of soil and land users on soil overuse issues and irreversibility of soil degradation processes
- Organization of structures for protection and efficient soil/land use
- Agricultural production according to the principles of ecological farming
- Monitoring of soils, including physical–chemical and biological changes due to types, intensity of use and negative effects (erosion, urbanization, industry, emissions, waste, wastewater, etc.)
- Introducing principles and procedures to limit soil and land overuse (e.g., through changes of the direction of use), and to determine the minimum action required (e.g., risk analysis procedures)
- Identification of risks and intensification of remediation and reclamation actions at contaminated areas
- Redevelopment of postindustrial areas to include them again in the economic cycle (revitalization)

Based on the document: "Towards Thematic Soil Protection Strategy" (COM, 2002), all EU members, including Poland, are obliged to prepare a number of regulations and economic instruments to reduce soil degradation and contamination related to main hazards (Figure 2).

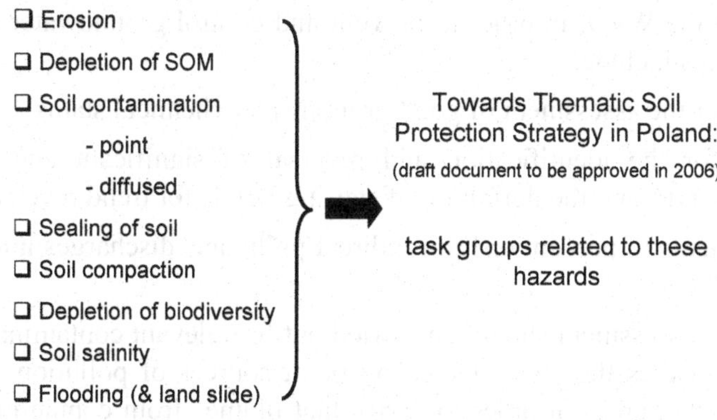

Figure 2. Main soil protection related hazards.

The industrial activities having a significant environmental impact are indicated in the Integrated Pollution Prevention Control (IPPC) Directive

(96/61/EC), which is one of the most important legal acts of the EU related to environmental protection (Kozłowska, 2004). Its strategic goal is to systematically adjust methods of planning and production to the requirements of permanent sustainable development and to assure environmental protection by prevention of pollutants emissions (or, if not possible, by maximal reduction of emissions). The most important problems are related to a full control of process production, modernization of existing plants, and technological processes (known as the "cleaner production" principle). The IPPC Directive requires that appropriate authorities force upon the installation operators defined duties for integrated environmental protection, in particular through applying the BAT, and avoiding environmental pollution and waste generation. After terminating the operation of installations, all necessary cautions should be applied to avoid hazards related to the contaminated area where the activity was carried out, and to return this area to the proper status.

Directive 82/501/EC addresses the problems of risk related to some industrial activities, and Directive 96/82/EC addresses risk control from severe accidents involving toxic substances.

Environmental management systems are addressed by one decree and two decisions. Decree EC/1836/93 comprises voluntary participation of an enterprise in the environmental management system and ecological audits within the EU. Decision 97/264/EC is related to the procedures of certification approval, while Decision 97/265/EC to the approval of ISO 14001, and the European Standard EN ISO 14001, as consistent with the system of voluntary participation.

Finally, management of postindustrial areas is important with regard to the Water Framework Directive (WFD) (2000/60EC), and Groundwater Directive (GWD) – under preparation – that establishes specific measures as set out in Article 17 of the WFD, in order to prevent and control groundwater pollution. These measures include:

- Criteria for the assessment of good groundwater chemical status
- Criteria for the identification and reversal of significant and sustained upward trends, and the definition of starting points for trend reversal
- Requirements to prevent or limit indirect pollutants discharges into groundwater

Specific trend assessment should be carried out for relevant contaminants in the groundwater bodies that are affected by point sources of pollution, including historical point sources in order to verify that plumes from contaminated sites do not expand over a defined area and do not deteriorate the chemical status of the groundwater body. The integrated management for postindustrial areas should take these EU requirements into account, as well.

2.4. NATIONAL PROGRAM FOR POSTINDUSTRIAL AREAS

The National Program for Postindustrial Areas approved by the Polish government in 2004 provides the framework for the integrated management to be developed to reclaim and redevelop contaminated/degraded abandoned areas (Program Rządowy, 2004). The organizational structure of the program is presented in Figure 3. The strategic goal is to create conditions and mechanisms for revitalization of postindustrial areas according to the sustainability principle. The direct goals include:

- Development of the management system to revitalize postindustrial areas, and consequently reduce use of nondegraded areas (i.e., greenfields) for industrial purposes,
- Development of the sector engaged in remediation and revitalization of degraded areas, thus generating new jobs

The Minister of Environment is responsible for the realization of the program. He is coordinating via the program coordinator, and with the assistance of the Steering Committee, all activities including:

- Verification of the approved task plan
- Determination of the order of works planned for a given year
- Giving opinions on annual breakdown of costs and tasks, and implementation reports
- Monitoring the performed works and expenditures
- Essential and organizational support in realization of works
- Application for funding from the national budget, and the National Fund for Environmental Protection and Water Management (NFEP)
- Assistance for potential beneficiaries in preparing applications for remediation and revitalization project financing
- Selection of pilot areas for revitalization
- Invitation tenders for program tasks and calls for research projects

The program tasks introduced in parallel should be complementary and realized with different strengths under the supervision of Field Coordinators at three fields: (i) pilot revitalization projects, (ii) developing the management system for postindustrial areas, and (iii) R&D and education.

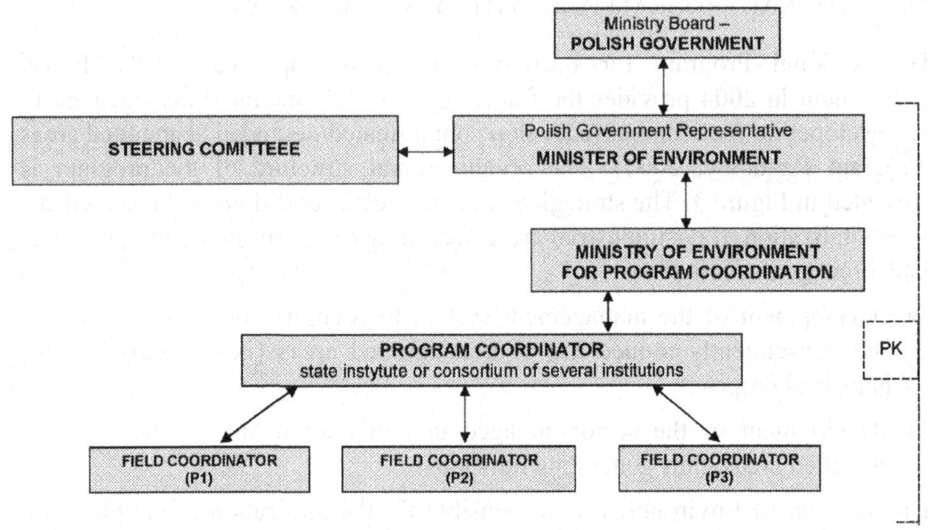

Figure 3. Organization of the National Program for Postindustrial Areas.

The Act on Spatial Site Planning (Dz.U.1994.15.139 with further changes) defines goals and principles of development plans in communes, districts, and provinces, based on the sustainability principle. Transferring many rights to local governments creates possibilities of undertaking decisions important for given areas with respect to postindustrial areas.

Registers carried out by local governments according to the Environmental Protection Law are done exclusively based on contaminated land criteria, for both industrial and agricultural areas which have an owner or ruler. They do not take into account the specificity of postindustrial areas and, therefore, only to a certain degree can be used as a management tool. It is then necessary to make an inventory of postindustrial areas at the national level, upon which a comprehensive information system can be developed with the help of geographic information system (GIS) databases and associated mathematical models. Furthermore, priorities for revitalization can be set up based on previously defined uniform criteria and, consequently effective policy of land protection can be planned and implemented. The management of postindustrial areas should be in accordance with development plans of the communities where they are located.

Effective management of postindustrial areas requires also introducing certain legal changes related to: (i) range and ways of carrying out tests on soil and land quality; (ii) principles and rules of carrying out registers of post-industrial areas, taking into consideration liquidation of industrial plants, mines, hazardous waste deposits, etc., which are recognized as activities having significant impacts on the environment; (iii) technical projects of revitalization, negotiations, and implementation; and (iv) creation of financial benefits for potential investors.

3. Integrated Management for Postindustrial Areas in Poland

3.1. INTEGRATED MANAGEMENT SYSTEMS

Community, economy, and environment create a macro system with strong interactions, thus requiring integrated management. This management should be based on conscious and rational formation of relations between individual elements of a system, with regard to the following: organizational structure, planning, sources of financing, procedures, processes, and resources. Management is defined as an efficient use of human, capital, and material resources (Matuszak-Flejszman, 2002).

Management of any system is always based on the same principles and requirements, and is represented by the Deming's circle (Hamrol and Mantura, 1998). Planning lies in determining the goals and intentions, preparing procedures of actions and required documents, implementing planned actions, and controlling the process, verification, i.e., comparing the effects with the assumptions and intentions, and correction of inconsistencies and indication of possible revisions. Important issues of the management process are: programming (i.e., designing the most desired events from a prospective situation viewpoint) and planning (i.e., transcription of goals into tasks and working out the organizational and financial ways of execution). These principles were first introduced in the ISO standards (series 9000). Integrated management systems allow for complex examination of the problem, an integrated working plan that includes elements required by different standards, simultaneous inspection of the system and certification process, open access and flexibility for introducing further requirements, as well as getting the synergistic effect that rises during cooperation of different elements of management systems.

The most popular way to create an integrated management system is to build it for one component and, subsequently, gradually integrate the next systems into the existing one(s). Integration is possible by connecting different ways of acting at a given area with the management consistent with the approved goals. Such connection should be dynamic, i.e., currently verified and updated, and experiences gained during creation of the first system can be used to build the next one(s). Integration comes gradually and in a natural way.

Creation of separate management systems for individual elements can be applied if, due to complexity of a process to be managed or organizational changes (e.g., restructuring), it is preferred to separate the systems during the implementation process. In this case, integration takes place gradually after some years of functioning of individual systems. The final stage after implementing individual management systems is the certification according to the standards: ISO 9000, ISO 14001, and/or PN-N/OHSAS 18001 (Matuszak-Flejszman, 2002; Podskrobko, 1998).

The sources of specific information for integrated management systems for postindustrial areas are environmental monitoring, statistical reporting, official information, and R&D (Matuszczak-Flejszman, 2002; Kozłowski, 2002). Environmental monitoring is a system of measuring, collecting, processing, transferring, and providing data, as well as evaluating and diagnosing environmental quality status based on uniform and standardized methods, and calibrated analytical equipment. National, regional, and local monitoring networks provide information comprising air, soil/land, surface and groundwater quality, waste generation, and management, as well as predicted results of operational use of the environment on national, regional, and local scales. Statistical reporting on activities that impact environmental quality, e.g., reports on hazardous waste, is the responsibility of enterprises that use the environment directly, and national and local governmental offices. Official information should be included in registers of land's evidence and real estate, cadastres of natural resources, maps of soil classification, etc. R&D provides supplemental and verification of data from environmental monitoring and statistical reporting. In Poland, data are collected by the Central Statistical Office (GUS) and distributed via statistical yearbooks and special reports.

3.2. CONCEPT OF MANAGEMENT SYSTEMS AT POSTINDUSTRIAL AREAS

In Poland, the integrated management strategy for postindustrial areas is not yet fully defined. The integrated management system for any enterprise that combines environmental, quality and occupational hygiene managements, and integrated permits and licenses, should avoid creation of new degraded areas. For the past few years, studies have been conducted and redevelopment actions have been undertaken for postindustrial areas.

At postindustrial areas one has to address degradation, which encompasses: (i) environmental compartments (atmosphere, hydrosphere, and lithosphere), (ii) societies (illness, unemployment, social degradation), and (iii) economy (bankruptcy and/or, restructuring of enterprises, plants, etc.). Therefore, the integrated strategy should rely on conscious and intentional merging activities aimed at rebuilding and returning industrially degraded areas for reuse, as well as restructuring economy and managing human resources. These activities must be in agreement with local development plans (communes, districts, etc.), therefore, require cooperation with local authorities. This is also important for obtaining funds.

Returning useful properties of postindustrial areas may comprise (Malina, 2006):

- Remediation (sanitation) – soil cleanup using physical, chemical, and biological methods
- Reclamation (recultivation) – actions directed at returning the natural earth surface to reach a quality of soil according to standards, to create/return useful/natural values for deteriorated land
- Revitalization (redevelopment) – land reclamation followed by remediation, change, modernization of the earth, etc., to return the status of land that allows for fulfilling its useful functions

The management system for postindustrial areas should include: (i) qualitative evaluation of an environmental status, (ii) diagnosis and assessment of health (HRA) and environmental (ERA) risk, (iii) selection of effective technologies, (iv) evaluation of the effects of undertaken decisions on the economy and societies, (v) use of research achievements in environmental engineering/ technology and ecology, and (vi) national, regional and local policies.

The system management concept for postindustrial areas that should assist in implementing the National Program for Postindustrial Areas is presented in Figure 4.

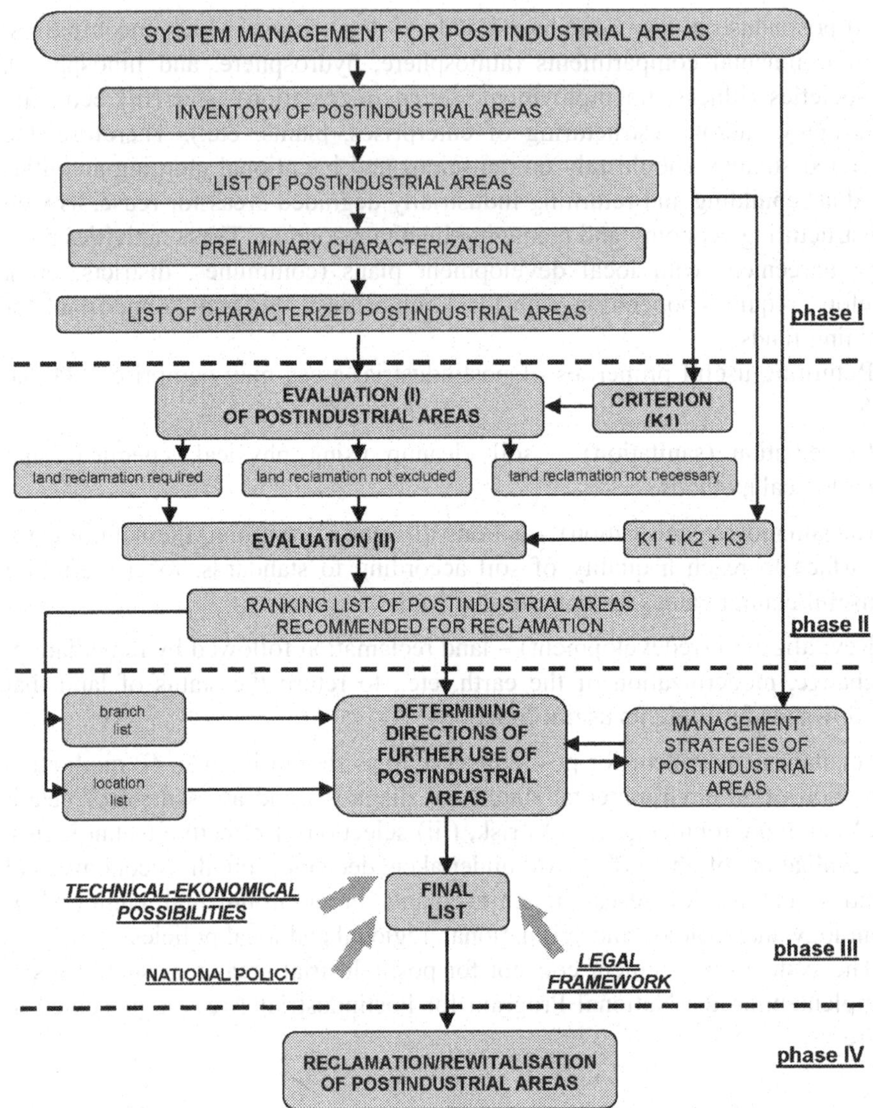

Figure 4. Concept of system management for postindustrial areas.

3.3. INTEGRATED MANAGEMENT STRATEGY MODEL

Based on the author's experiences with land contamination and redevelopment, and on the extensive description of the situation presented in this paper, the model of integrated management strategy for postindustrial areas is recommended that is presented in Figure 5, and described more in detail afterwards.

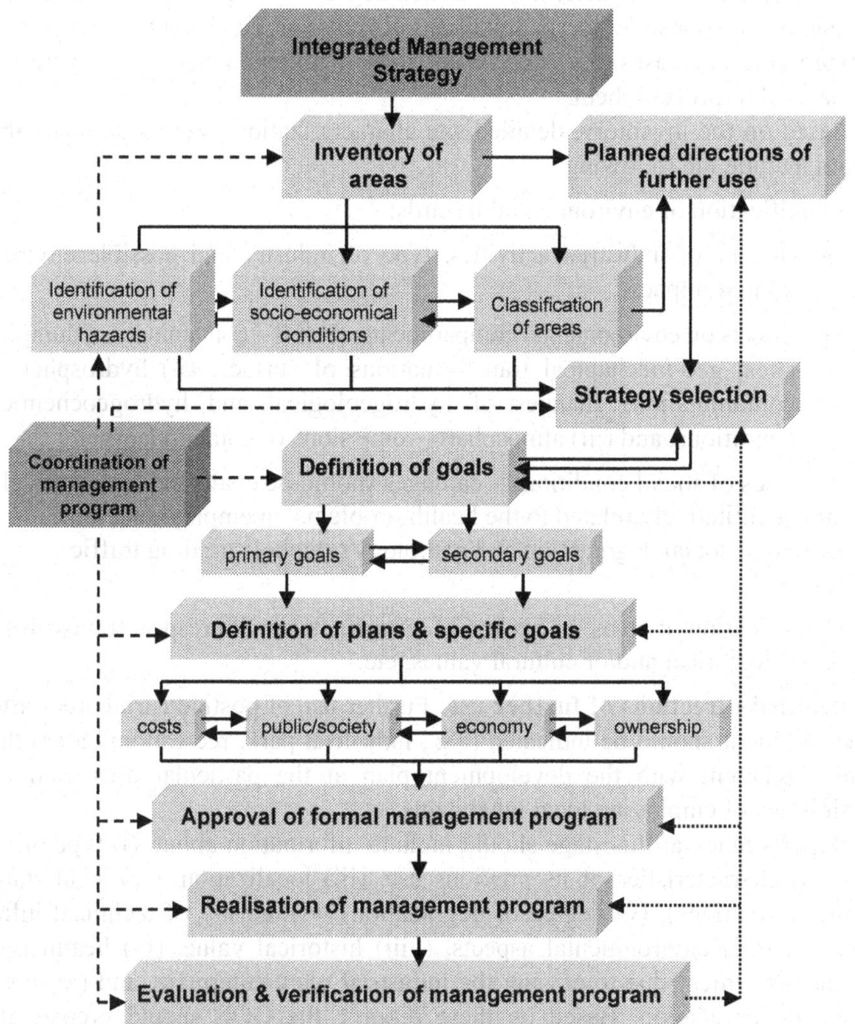

Figure 5. Model of the integrated management strategy for postindustrial areas.

Coordination: The management strategy should be coordinated by the group of stakeholders (GOS), assisted by the advisory group of (multidisciplinary) experts (GOE). GOS should comprise all bodies that have, or may potentially have, any interest at a given postindustrial area, and is responsible for any decision undertaken.

Inventory of postindustrial areas: To achieve the management goals, it is essential to make an inventory of postindustrial areas. This should be initiated at a commune level, and be based on maximal use of all existing information and data. It is critical to determine reasons, degrees of degradation, and sources and factors responsible for degradation. Results of the inventory should be transferred to databases (e.g., GIS) compatible with mathematical software that can be used to process them.

Based on the inventory, detailed site characterizations should be made that should include the following:

1. Identification of environmental hazards:
    - History of industrial activities, type of industry and possible environmental impacts
    - Effects on environmental compartments: (i) soil – contamination, dumping sites, geo-mechanical transformations of surface, (ii) hydrosphere – contamination, changes of hydrogeological and hydrogeochemical conditions, and (iii) atmosphere – emissions of contaminants
2. Analyses of social conditions – damages should be evaluated quantitatively and quantitatively related to the health problems, unemployment and increased social degradation and pathology (alcoholism, drug traffic, robberies)
3. Classification of areas – a review of infrastructure with respect to possible reuse, historical and/or cultural values, etc.

Planned directions of further use: Further use of postindustrial areas after redevelopment should be indicated (e.g., industrial park, recreational area) that is in agreement with the development plan at the particular area, and the possibilities of employing local inhabitants.

Reports made at this stage should include information about: (i) type of the area, (ii) characteristics of its previous use, (iii) localization, (iv) legal status (property relations), (v) degree of degradation, (vi) density of technical infrastructure, (vii) environmental aspects, (viii) historical value, (ix) health risk, (x) unemployment due to closing the industrial plant/enterprise, and (xi) areas of social degradation. Based on these reports, the GOE should prepare the management strategy for the given postindustrial area to be accepted by the stakeholders.

Strategy selection: The selection of a revitalization strategy should be based on the need to reduce risk for humans and the environment, and to find solutions that are economically and ecologically effective.

Primary and secondary goals definition: After diagnosis of the situation, the program should be elaborated at a regional scale, or as a target program for communes belonging to different provinces. Prioritization is given to starting revitalization projects at the areas that pose significant hazard to the environment. Conditions and mechanisms should be created for redevelopment of postindustrial areas (brownfields) according to the sustainable development principle, i.e., not to reduce nondegraded areas (greenfields). The primary goal should always focus on reducing health and environmental risks. The secondary goals should be directed to achieve economical profits due to reuse of redeveloped areas (e.g., remediation and revitalization, creation of new jobs, initiatives for reuse of the areas).

Definition of plans and specific goals: Any given postindustrial area has a set of specific features that, by definition, indicates its specific function, or limits the possibilities to have this function. Thus, the management of a postindustrial area is usually a result of local requirements and abilities, and possible ways of further use. Often, due to specific characteristics of the area and postindustrial infrastructure, reuse for industrial purposes is limited, as a result of high environmental demands in the case of newly established enterprises and industrial plants.

Costs: Financing is crucial for an integrated management strategy. Funds can come from different sources, including: (i) plants and enterprises according to the principle "polluter pays", (ii) national budget, if the industry was owned by the state, (iii) local governments, (iv) individual investors, who considering invested money can get areas with good location and infrastructure, (v) NFEP and regional funds for environmental protection, and (vi) EU special funds and programs.

Public/society: It is necessary to control and take care of inhabitants living at or in the vicinity of a postindustrial area by means of continuous monitoring of their health, and applying obtained results for defining specific and detailed goals of the management strategy, as well as organizing preventive–therapeutic trips for children. Other required actions should focus on: (i) new jobs related to the remediation and revitalization works, and in restructurized enterprises, (ii) qualification courses as a result of new employment and job demands, (iii) friendly conditions and stimulus for initiating small firms for local societies, (iv) cultural, recreational, and sport centers that reduce social degradation, (v) additional funds for ex-workers, who lost their working abilities due to occupational diseases (pneumoconiosis, lead poisoning, accidents, etc.), and (vi) increased ecological education.

Economy: Existing economy should be restructured (particularly mining and steel industry), and new modern industrial sectors should be initiated that will meet EU requirements and obtain integrated licenses. It is important to establish cooperation between governmental administrations and newly built enterprises that will allow for keeping balance between economic, social, and ecological issues.

Ownership and legal status: Postindustrial areas have different owners that are often not defined and unclear legal status. Therefore, diagnosis of the ownership problems is necessary prior to starting the liquidation of contamination sources and revitalization. Local societies should be continuously informed about current actions and plans. Creating mechanisms of cooperation among the present owners and the local authorities should allow for attracting external investors.

Formal approval of the management program: After identification of all aspects of a postindustrial area and defining ways of financing, methods and approaches, and responsible people, an action plan should be established and accepted by stakeholders.

Program implementation: The management program implementation should proceed according to the assumptions previously made. It can be a subject of changes and corrections as a result of nonpredicted problems, or new possibilities previously not available. It should take into account the BAT principle, including techniques that may improve the revitalization process, and should apply the research results.

Program evaluation and verification: An important element of any effective management system is permanent control of each step of activities. It allows the advisory group to analyze the coherence among actions. Control should be systematic, documented (through reports) and verified to provide an objective picture of activities. It should evaluate if the integrated management strategy applied for a given postindustrial area is in accordance with the approved program, and national and international legal standards.

4. Selected Case Studies

4.1. INTEGRATED MANAGEMENT STRATEGY AT THE FORMER CHEMICAL PLANT IN TARNOWSKIE GÓRY

At Tarnowskie Góry, the soils and the quaternary groundwater are heavily contaminated with heavy metals and boron, originating from uncontrolled waste disposal in the vicinity of the chemical plant closed down in 1995. It is hazardous to the Triassic reservoirs, the Major Groundwater Basins (MGWB) no. 330 and 327, the main sources of potable water, and it may potentially affect the population of more than 600,000 (Malina, 2004). The general management goal is to protect the drinking water supply function of the MGWB 330. Due to complexity related to the site conditions, contaminant characteristics, organization, regulatory aspects, and/or considerable costs, an integrated approach is recommended to manage risk for the defined receptors (Malina et al., 2006).

The conceptual model includes the primary (hazardous waste deposits) and secondary (contaminated soils and quaternary sediments) sources of contamination. Contaminants are leached out from the waste heaps and infiltrate through the unsaturated zone and aquifers, accumulate there, and/or migrate further and deeper through the hydrogeological windows (i.e., discontinuities of overlying nonpermeable clay layers). Based on: (i) frequency of occurrence, (ii) fate, mobility, persistence, and abiotic natural attenuation (NA) potentials, (iii) toxicity and impacts to the receptors, and (iv) data availability, the priority contaminants (As, Ba, B, Cd, Sr, and Zn) were selected. Three major risk clusters (zones in which the risks can be managed in a defined way) were identified that are oriented on receptors, pathways, boundary conditions, and stakeholders' interests. Within each cluster, the primary and secondary contamination sources were recognized. Based on the existing and/or predictable groundwater contamination, the Risk Management Zone (RMZ) was delineated (an area where the requirements of the WFD may not be fulfilled for a defined period of time). The RMZ provides space and time to manage risks for the receptors in a cost-effective way. Four planes of compliance were derived, i.e., the horizontal or vertical borders, at which a certain concentration or a mass flux have to be achieved, to comply with the site-specific threshold values.

The Risk Management Scenarios (RMS) were developed according to the following possible measures for risk clusters: (i) source-oriented, i.e., capping or removal of dumping sites (primary sources), contaminant immobilization in soils and quaternary deposits (secondary sources), (ii) pathway-oriented, i.e., the construction of internal/external hydraulic barriers, and (iii) receptor-oriented, i.e., monitoring and development of a specific water extraction regime at the receptors (wells). The effectiveness of these RMS in fulfilling the management goals was analyzed based on: (1) containment of the boron plume within the RMZ, (2) stabilization of the contamination plume within the current boundaries, and (3) maximum possible reduction of the contamination plume within a moderate time frame.

Three potential RMS were built up: S1-controlled NA (source removal + monitoring/control of the hydrological regime), S2-active groundwater remediation (source removal + groundwater cleanup + monitoring), S3-engineered NA (source removal + increased groundwater extraction within the RMZ + monitoring).

The final RMS was not selected due to uncertainties in boundary conditions. Currently, the scenario S1 is implemented; however, two other scenarios are not excluded. The management plan was accepted by the GOS that includes: (i) construction of a controlled landfill (area: 16 ha, capacity: 1.6×10^6 m^3) where waste from dumping sites, debris, and contaminated soils are deposited in a way that complies with the legal Polish and EU requirements, (ii) revitalization of the river valley, (iii) reclamation of previous dumps (~50 ha) with allocation of ~10 ha for further use, and (iv) continuous long-term monitoring of environmental compartments (soil, water, and air). The results of simulation studies show that waste removal and deposition in the controlled landfill should have a moderate effect on improving groundwater quality in short (30 years) and medium (60–80 years) time frames. It guarantees that some of the risks at the surface are moderated or eliminated, whereas the risk, due to the secondary sources of contamination, is attributed mainly to groundwater quality within the RMZ. The RMZ is formally included in the watershed management plan realized jointly by the local and regional administration, and the chemical plant liquidator. Monitoring focuses on the Triassic aquifers to trace migration of contaminants (pathways) from the dumping sites (primary sources) and quaternary sediments (secondary sources) to the water wells (receptors).

4.2. MANAGEMENT OF THE POSTINDUSTRIAL AREA IN BOGATYNIA

Strip mining of lignite in the area of Bogatynia resulted in formation of excavations and external dumping grounds. The 5000 ha area exploited by the brow coal mines (KWB) "Turów" is one of the largest in Europe. In 1962, the electric power plant was opened and became the largest source of pollution emitted to the atmosphere in the area. Moreover, the sources of pollution located at the area between Saxony (Germany), Czech Republic, and Poland caused the ecological disaster, and the area was named a "Black Triangle".

Investigations made in 1975–1976 indicated that above 75% of pollution at the Polish part of this area was originating from old electric power plants in the former East Germany and Czechoslovakia. This problem drew public attention only in the mid-1980s when forests died in the Karkonosze and Izerskie mountains. In 1990, the KWB and the power plant were put on the "List of 80". In 1991, an international program was created and signed by the ministries of environment of Poland, Germany, Czechoslovakia, and the EU. A working group was established to deal with the environmental problems. The program was sponsored by the US Pew Charitable Trust Foundation. The main goal was to improve the environmental quality to reach the status of this area comparable to EU standards, mainly with respect to air quality. Air contamination levels exceeded all possible standards and resulted in soil acidification. In order to stop further soil degradation, environmental monitoring was initiated. Analyses of soil, air, and water quality monitoring data allowed for assessing the associated risks.

Reclamation of dumping sites of the KWB "Turów" has been carried out since the 1960s (Krzaklewski and Wójcik, 1996; www.Turów.com.pl). It took into account acidification of overlying deposits, mainly carbon-containing condensed tertiary clays (bad aeration and deficit of water available for plants). On the dump slopes the following treatment was realized: neutralization of acidified soils, planting of papilionaceous plants and grass using traditional methods of seeding, as well as the avio-hydroseeding method, followed by forestation. Fast forestation of dump slopes after avio-hydroseeding allowed for good protection against erosion. These activities resulted to date in 186 ha of forests. Annually, ~80 ha of degraded areas are the reclaimed.

Besides reclamation of degraded areas, actions have been undertaken to prevent further environmental degradation since the 1990s. In KWB "Turów" monitoring has been working since 1997 to control all sources of waste generation. Waste management is based on preventing waste generation, recycling, and reuse. An integrated system of quality management and occupational hygiene according to ISO was introduced at the Turów power plant. Continuous reclamation of the areas in the vicinity of the mine and power plant resulted in their removal from the "List of 80" in 2001. For realizing the complex program of ecological reclamation, both plants were awarded the title of the Leader of Polish Ecology in 2002.

The electric power plant is currently run according to the sustainability principle, and is recognized as "environment friendly". In 2003 it obtained the integrated license for use of the environment. The mine and the power plant, together with the authority of Bogatynia, carry out intensive works to reclaim natural conditions in the Turoszów and Bogatynia regions. The integrated management strategy in the case of these postindustrial areas includes:

- Diagnosis of reasons and status of degradation (mining activity, emission of contaminants from power plants in Poland, Czech Republic, and Germany)
- Remediation of soil/land and groundwater, surface water cleanup, reduction of gases, and dust emissions to the atmosphere
- Continuous modernization of the power plant and the mine (new power units, transport of ashes, a wastewater treatment plant, reduction of noise)
- Continuous cooperation between Poland, Germany, and Czech Republic aimed at air protection within the former "Black Triangle" area (cofinancing, joint projects, etc.)
- Implemention of an integrated environmental monitoring, systems of quality control, and environmental and occupational hygiene management according to ISO

4.3. MANAGEMENT AT THE FORMER ZINC SMELTER IN PIEKARY ŚLĄSKIE

The zinc smelter "Waryński" in Piekary Śląskie that operated since 1927 was closed down in 1990. Currently, this postindustrial area is owned by the mining–metallurgy complex (KGH) "Orzeł Biały" in Bytom. Under the supervision of the US EPA, the revitalization program was established that included:

- Retrospective review of activities carried out at the area
- Detailed site characterization based on collection of existing data
- Development of revitalization scenarios
- Feasibility study

At the area of more than 60 ha, ~3 million tons of diverse industrial waste products were deposited, including additional waste from the copper smelter in Legnica. Results of the studies indicate soils/land contamination with heavy metals (As, Zn, Cd, Pb) and need for remediation. To date, all buildings and technical infrastructure have been demolished and the dumps reclamation was started. Except the funds of KGH, the Regional Fund for Environmental Protection in Katowice provides financing.

The present owner has declared to transfer the reclaimed areas to the commune of Piekary Śląskie. According to the local development plan, the area will be reused for: service and craft, recreation, trade centre purposes, and for a controlled dumping site.

5. Conclusions

The most important issue in managing postindustrial areas is to define reasons of degradation and to evaluate possibilities of their revitalization. It requires risk assessment, and diagnosis of contamination sources and levels. Identification of problems is a prerequisite for developing management mechanisms.

The integrated management strategy for postindustrial areas is based on conscious and intentional aggregation of all actions related to remediation, reclamation, and revitalization of industrially degraded areas, as well as restructuring of the economy and human resources. It is a long-term and complex process that comprises many fields and requires many tasks and activities to be carried out:

- Appointing the institution responsible for the management, and creating the group of principal stakeholders
- Elaborating the register of post industrial areas
- Preparing maps of spatial distribution of contaminants and degradation

- Developing databases (preferably GIS databases) including all information related to postindustrial areas
- Cooperating with countries having experience in dealing with postindustrial areas
- Applying R&D results in terms of innovative technologies in remediation, reclamation, and revitalization
- Organizing courses and workshops on management of postindustrial areas
- Participation of representatives of national and local governments, research institutes, foundations, NGOs, etc.
- Adjusting the legal and financial status of postindustrial areas, and defining ways of financing
- Cooperation between the owners of postindustrial areas and local governments for planning and taking decisions (i.e., stakeholder cooperation)
- Working out and applying systems of reduced tariffs and stimulus for potential investors
- Promoting postindustrial areas, informing local societies about the planned direction of further use, improving ecological education and awareness

The National Program for Postindustrial Areas gives the basis, and formal and legal framework for developing the integrated management strategy to reclaim and revitalize contaminated/degraded abandoned areas.

Acknowledgments

The Tarnowskie Góry case study was the subject of research within the frame of the 5 EU Project: Water, Environmental, Landscape Management at Contaminated Megasite – WELCOME (EVK1-CT-2001-00103). The author was one of the partners in the project consortium.

Some elements of the concept of system management are based on ideas worked out within the consortium team (led by Dr. M. Korcz, Dr. R. Strzelecki, and the author) in the frame of the Project: "Defining directions for public administration and economic sector in the scope of reclamation and revitalization of postindustrial areas", commissioned by the Ministry of Environment, and financed by the National Fund for Environmental Protection and Water Management (NFOŚiGW – project no. 307/DO/03).

Some considerations related to the presented model of the management strategy are the result of discussions with Mr. K. Bolingier during preparation of his MSc thesis supervised by the author.

References

Berkel, A., 1996, Ochrona środowiska na pograniczu Niemiec i Polski, *Aura* 4: 22–30.

Bolingier, K., 2004, Zintegrowany system zarządzania zdegradowanymi terenami post-industrialnymi, MSc thesis, Faculty of Environmental Protectection Engineering, Częstochowa University of Technology, Częstochowa (unpublished).

Ciesielski, R., Motak, E., 2001, Degradacja geotechniczna podłoża pod długotrwale eksploatowanymi obiektami przemysłowymi, *Inżynieria i Budownictwo* 12: 87–91.

COM, 2002, (179 final), Komunikat Komisji Europejskiej do Rady Europejskiej, Parlamentu Europejskiego, Komitetu Ekonomiczno-Społecznego oraz Komitetu Regionów – W Kierunku Tematycznej Strategii Ochrony Gleby, Komisja Wspólot Europejskich, Brussels.

GIOŚ, 2003, Raport – Stan Środowiska w Polsce w Latach 1996–2001, Biblioteka Monitoringu Środowiska, Warszawa.

GUS, 2003, Raport – Stan Środowiska w Polsce w Latach 1996–2001, Central Statistical Office (GUS), Warszawa.

Hamrol, A., Mantura, 1998, Zarządzania Jakością. Teoria i Praktyka, PWN Warszawa-Poznań.

Kozłowska, B., 2004, Dyrektywa IPPC i pozwolenia zintegrowane – założenia ogólne, Przegląd Komunalny 3(150): 44–45.

Kozłowski, S., 2002, Ekorozwój. Wyzwanie XXI wieku, PWN Warszawa.

Krzaklewski, W., Wójcik, J., 1996, Rekultywacja w przygranicznych kopalniach węgla brunatnego, *Aura* 4: 37–43.

Malina, G., 2004, Eco-toxicological and environmental problems associated with the former chemical plant in Tarnowskie Góry, *Toxicology* 205(3): 157–172.

Malina, G., 2005, Risk evaluation and reduction of groundwater contamination from petrol stations, in: *Soil and Sediment Remediation: Mechanisms, Technologies, Applications*, Lens, P. et al., (eds.), Integrated Environmental Series, Part IV, Chapter 23, IWA, London, pp. 437–457.

Malina, G., 2006, Integrated system for groundwater management at postindustrial areas, in: *Reclamation and Revitalization of Demoted Areas*, Praca zbiorowa, Futura, PZiTS Poznań, pp. 45–57.

Malina, G., Kwiatkowska, J., 2006, Towards sustainable soil protection strategy in Poland: national vs. EU legislations and recommendations", in: *International Workshop on: Soil Protection Strategy – Needs and Approaches for Policy Support*. Book of Abstracts, IUNG Pulawy, Poland, pp. 12–14.

Malina, G., Krupanek, J., Sievers, J., Grossman, J. ter Meer, J., Rijnaarts, H.H.M., 2006, Integrated management strategy for complex groundwater contamination at a megasite scale, in: *Viable Methods of Soil and Water Pollution Monitoring, Prevention and Remediation*. NATO Science Series. I. Twardowska et al. (ed.), Springer, Dordrecht, pp. 567–578.

Matuszak-Flejszman, A. (ed.), 2002, Od zarządzania środowiskowego po zintegrowane, Zakład Poligraficzny Mos-Luczak, PZiTS, Poznań.

Podskrobko, B., 1998, Zarządzanie Środowiskiem, PWE Warszawa.

Program Rządowy dla Terenów Poprzemysłowych, 2004, Ministry of Environment, Warszawa.

Siuta, J., Kucharska, A., 1997, Wieloczynnikowa Degradacja Ziemi w Polsce, Institute of Environmental Protection, Warszawa.

Wójcik, J., 1997, Zanieczyszczenie wód powierzchniowych Wałbrzycha, *Aura* 1: 24–30; www.Turów.com.pl

Założenia programu rządowego dla terenów poprzemysłowych, 2003, CP Kancelarii Prezesa RM, Warszawa.

8. INDUSTRIAL ECOLOGY IN TRANSITION COUNTRIES: HISTORICAL PRECEDENT AND FUTURE PROSPECTS

INGA GRDZELISHVILI*, ROGER SATHRE
Cooperation for a Green Future, Tbilisi, Georgia

Abstract: Industrial ecology uses the structure and processes of natural ecosystems as a model for organizing industrial activities, seeking to integrate wastes and by-products into the production process thus reducing the need for material extraction and waste disposal. In this paper we discuss the theory and practice of industrial ecology as developed in the former centrally planned economies, and make observations on the suitability of industrial ecology in current transition countries. We describe the Soviet concepts of "combined production" and "waste-free technology", and show that Soviet scientists were familiar with many fundamental elements of modern industrial ecology. Although the potential environmental benefits of industrial ecology were recognized by central planners, industrial ecology was pursued primarily as a means to increase production. We then discuss issues specific to the (re) implementation of industrial ecology in transition countries, such as appropriate policy instruments, the relation between economic growth and environmental impacts, the question of central planning versus spontaneous organization of industrial interactions, the valuation of resources and environmental externalities, and the underlying goals of industry and society in the context of changing material aspirations among the populations of transition countries.

Keywords: industrial ecology; transition countries; by-products; central planning; market economy; Soviet Union; environmental protection; production; consumption

*To whom correspondence should be addressed. Inga Grdzelishvili, Cooperation for a Green Future, Nutsubidze No.72 Apt.25, Tbilisi 0186, Republic of Georgia; e-mail: cgf.georgia@yahoo.com

R. N. Hull et al. (eds.), Strategies to Enhance Environmental Security in Transition Countries, 95–115.

1. The Modern Concept of Industrial Ecology

Environmental security literature in recent years has recognized dual roles for industrial production in modern society, both as a source of material welfare as well as a cause of ecological stress. The concept of industrial ecology is increasingly noted for its potential to reconcile manufactured abundance with minimized environmental impact (Frosch and Gallopoulos, 1989; Graedel and Allenby, 2003).

1.1. WHAT IS INDUSTRIAL ECOLOGY?

Industrial ecology is an approach to organizing industrial systems that uses natural biological ecosystems as a model. The structure and processes of natural ecosystems, including the complex interactions of organisms with each other, are used as a basis for creating industrial networks. The traditional pattern of industrial production/consumption is a linear flow, where raw materials are extracted from nature, processed in factories and used by consumers, and then disposed as waste back to the environment (Figure 1a). Industrial ecology seeks to integrate wastes and by-products back into the production process, thus reducing the demand on nature for material extraction and waste dumping (Figure 1b). This approach closes material cycles and reduces the need for raw material input and pollution output.

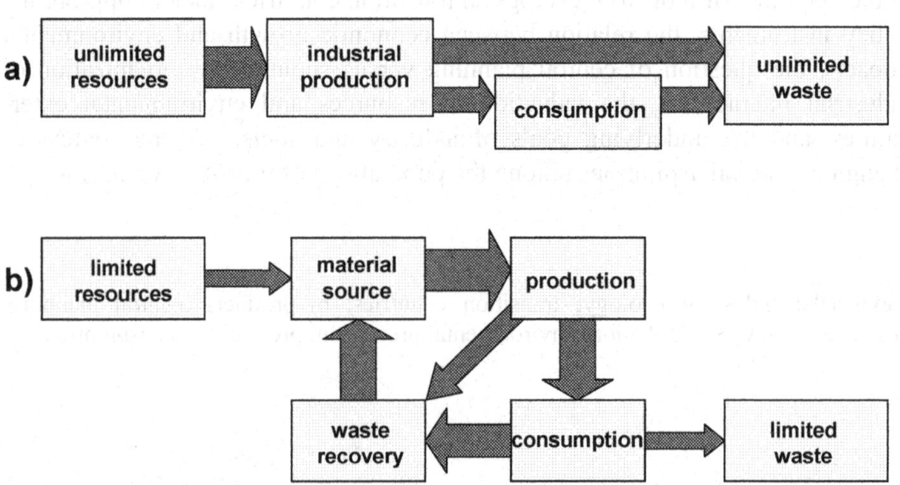

Figure 1. Material flows in (a) traditional linear industrial system and (b) industrial ecology system.

A key component of industrial ecology is the exchange of by-products or waste material among different firms for mutual economic and environmental benefit. In such a case, one enterprise gains by having ready access to a needed raw material, and the other benefits by either reducing waste disposal costs or increasing revenues. Ceteris paribus, the environmental impact is lower if the first enterprise extracted virgin material and the other released its waste into the environment. This integration and optimization of material and energy flows of industrial activities avoids the creation of pollutants rather than polluting first and then cleaning up.

Industrial ecology is often expressed as a symbiosis where groups of Industries work collaboratively through exchanges to reduce total natural resource consumption while simultaneously increasing individual profitability. In practice, it is best implemented in "eco-industrial parks", which build community networks of industries that interact by sharing utility inputs and making use of each other's by-products. There are numerous successful examples of industrial ecology in Europe, North America, and elsewhere.

1.2. EXAMPLE OF MODERN INDUSTRIAL ECOLOGY: KALUNDBORG

The best known example of industrial ecology practice is in Kalundborg, Denmark, an industrial port city of 20,000 people. A cooperative network has developed between several industrial companies and the Municipality of Kalundborg. One company's by-product becomes an important resource for other companies (Kalundborg Centre for Industrial Symbiosis, 2006). Resources exchanged by the firms include steam, heat, fly ash, gypsum, sulfur, and organic sludge that otherwise would have been unwanted waste (Figure 2).

Residual heat from the coal-fired Asnæs power station is recovered and used as district heating for the city of Kalundborg and as process steam for industrial production. The combination of heat and power production results in a 30% improvement of fuel utilization compared to separate production of heat and power. Approximately 4500 households in Kalundborg receive district heat from the power station. The Statoil oil refinery receives process steam and water from the power station. The steam covers about 15% of the refinery's total consumption of steam. The refinery uses the steam for heating oil tanks, pipelines, etc. The Novozymes enzyme factory and the Novo Nordisk pharmaceutical factory use steam from the power station for the heating and sterilization of the processing plants. Some of the cooling water from the power station is used by a fish farm producing 200 t of trout and salmon per year. The fish have better growth conditions in the heated water.

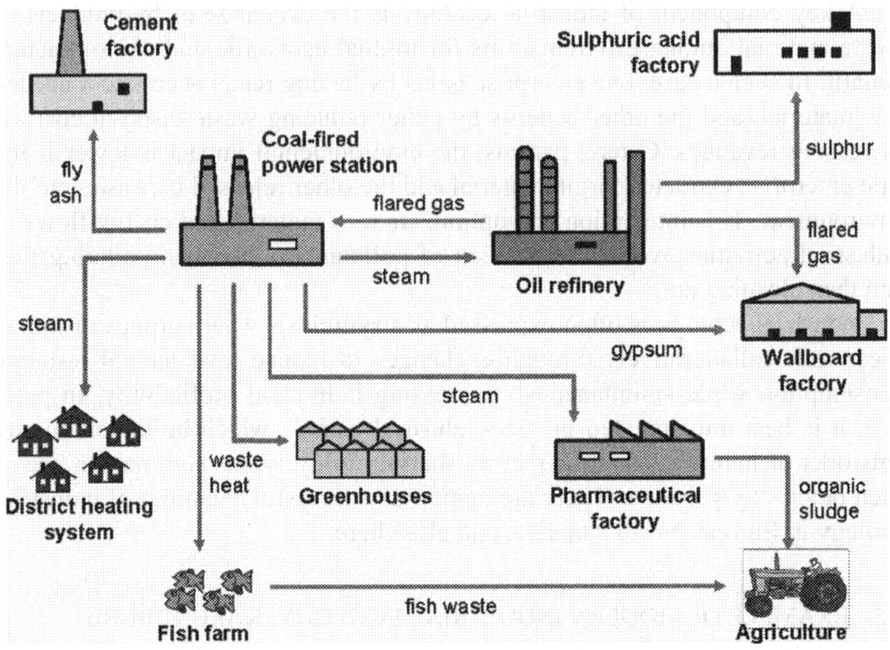

Figure 2. Material and energy exchanges in the eco-industrial park in Kalundborg, Denmark.

The desulfurization plant of the power station, which removes sulfur dioxide from the flue gas, produces about 200,000 t of gypsum per year. Desulfurization is a chemical process in which sulfur dioxide is removed while forming gypsum as a by-product. The gypsum is sold to BPB Gyproc, a company that manufactures plasterboard products for the construction industry. The gypsum from the power station significantly reduces the import of natural gypsum. Being more uniform and purer than natural gypsum, power station gypsum is therefore well suited for the plasterboard production. Gypsum recovered by the municipal recycling station of Kalundborg is also delivered to BPB Gyproc, thereby further reducing imports of natural gypsum and the amounts of solid waste for landfilling.

The power station removes fly ash from the smoke to reduce emissions to the atmosphere, thus producing about 30,000 t of fly ash per year. The ash is used in the cement industry as a cement-blending agent. The largest ash customer is Aalborg Portland. Ash deriving from the firing of orimulsion, a bitumen-based fuel, is recycled in a plant in Great Britain. Nickel and vanadium are reclaimed from this ash.

Enzyme production at the Novozymes enzyme factory is based on fermentation of raw materials such as potato flour and cornstarch. The fermentation process generates as by-products about 150,000 m^3 of solid biomass and about 90,000 m^3 of liquid biomass per year, containing nitrogen, phosphorus, and lime. After inactivation and hygienization, local farmers use the material as fertilizer in the fields, thereby reducing their need for commercial fertilizers.

The insulin production at the Novo Nordisk pharmaceutical factory produces yeast as a residual product, which is converted into yeast slurry that is used as feed for pigs. Sugar water and lactic acid bacteria are added to the yeast, making the product more attractive to pigs. The yeast slurry replaces ~20% of the soy proteins in traditional feed mixes. Over 800,000 pigs are fed per year on this product.

The desulfurization plant at the oil refinery reduces the sulfur contents of the refinery gas leading to significantly reduced SO_2 emissions. The by-product is used in the production of ~20,000 t of liquid fertilizer roughly corresponding to the annual Danish consumption. In addition, sludge from the municipal water treatment plant in Kalundborg is utilized at Bioteknisk Jordrens Soilrem as a nutrient in bioremediation processes.

The exchange of resources between industrial companies in Kalundborg provides a number of advantages. Through recycling, the by-product of one company becomes an important resource for another company. This leads to improved economy and reduced consumption of resources, e.g., water, coal, oil, gypsum, fertilizer, etc. The overall environmental strain is reduced, including discharges of wastewater, CO_2, and SO_2. The utilization of energy resources is improved through the use of residual heat and waste gases.

2. Theory of Industrial Ecology in Centrally Planned Countries

While generally considered to be a modern, state-of-the-art concept originally popularized in a 1989 article (Frosch and Gallopoulos, 1989), the fundamental approach of industrial ecology has existed for many decades in countries with both market economies (Erkman, 1997; Desrochers, 2005) and centrally planned economies (Sathre and Grdzelishvili, 2006). In this section we discuss the theoretical development of industrial ecology in the former centrally planned states, drawing information from original documents in Russian as well as international scientific literature.

2.1. REVIEW OF PREVIOUS LITERATURE

Several authors have noted the parallels between particular aspects of industrial ecology and the structure of industry in centrally planned economies. Erkman (2002), in an overview of the recent history of industrial ecology, acknowledged that examples of industrial ecology-like thinking existed in the former

Soviet Union, particularly in the areas of resource and waste optimization, but did not describe or document this in detail. Hewes (2005), in a study of the social relations and initiatives required for the successful development of eco-industrial parks, described examples of industrial symbiosis that existed in Soviet Ukraine, as well as efforts underway to reestablish those links within the current market economy. Gille (2000) described the establishment, implementation, and partial abandonment of elaborate mechanisms to reuse waste materials in Hungary, beginning in the 1950s and continuing to the market economy transition in the 1990s. This study shed light on the motivations, policies, and "pathologies" of industrial ecology in a centrally planned economy. Desrochers and Ikeda (2003) contrasted this Hungarian experience to historical efforts towards waste reuse in market economies, arguing that the most accurate means to determine the "least wasteful" uses for materials is through market prices that provide an index of scarcity for alternative materials and products. Sathre and Grdzelishvili (2006) provided a comprehensive overview of the theory and practice of industrial symbiosis in the former Soviet Union, upon which much of the historical information in the present paper is based.

2.2. COMBINED PRODUCTION

Soviet leaders asserted that large-scale industry would best create the material basis of a communist society (Blackwell, 1994). To increase industrial productivity, they promoted a form of industrial ecology termed "combined production" (*kombinirovanaia produksia*). A year before the Bolshevik revolution, Lenin outlined the economic benefits of combined production, a practice that he described as:

> the grouping in a single enterprise of different sectors of industry, which represents either consecutive steps in the processing of raw materials (for example, the smelting of iron ore into pig iron, the conversion of pig iron into steel, and the further manufacture of different products from steel), or cooperation between industrial sectors (for example, the utilization of waste materials or byproducts, the production of packing materials, etc.). (Lenin, 1916, p. 312)

Influenced by the directives of Lenin, Efimov and Zhukova (1969) categorized three forms of combined production in Soviet industry. They spoke first of combination through coordination of different processes of output, in which all or many of the industrial actors needed to produce a certain product are geographically proximate. An example of this is the Magnitogorsk Iron and Steel Works at which all stages of metallurgical production are located, from extraction of ores to production of rolled metal, as well as production of coke and other required inputs. Secondly, they noted combination through the complex utilization of raw materials, in which multiple finished or intermediate products are obtained from a given raw material. For example, coal, oil, or complex metal ores are subjected to various thermal and chemical processes to make usable all the elements of the input resource. Lastly, they recognized combination through the utilization of waste, which is similar to the complex utilization mentioned above, but focused on by-products of industrial processing of other principle materials. An example is the further use of wood processing residues that result from the manufacture of sawn lumber. Efimov (1968) described the advantages of combined production thus:

> Combination of production ensures the fullest possible use of raw materials and the waste left over after their primary processing, reduces investments, and raises production efficiency indices. It reduces expenses on shipments of raw materials and half-finished products, accelerates production processes, ensures utilisation of wastes, gives rise to absolutely new products and materials (synthetics, plastics, etc.) into existence, makes mining operations cheaper because it allows utilisation of poorer types of raw materials, permits fuller use of the raw materials already brought to the surface, and reduces the range of employment of organic raw materials and, consequently, the amounts spent to obtain them. (Efimov, 1968, p. 201)

Note that the benefits listed are those that increase production or decrease costs, and environmental benefits were not explicitly considered.

2.3. WASTE-FREE TECHNOLOGY

As environmental awareness began to grow in the 1970s, the Georgian scientist Davitaya (1977) perceived the analogy relating industrial systems to natural systems as a model for a desirable transition to cleaner production:

> Nature operates without any waste products. What is rejected by some organisms provides food for others. The organisation of industry on this principle—with the waste products of some branches of industry providing raw material for others—means in effect using natural processes as a model, for in them the resolution of all arising contradictions is the motive force of progress. (Davitaya, 1977, p. 102)

In Soviet theory, environmental problems had a social basis and were engendered by specific social conditions (Granov, 1980). The causes and solutions of environmental degradation depend on the social forces that prevail in society and on the interests that those forces pursue. Problems result because technology in the Western world is owned by a specific social class who use it to extract profits and enrich itself rather than benefiting society as a whole. Communism alone, according to this theory, is free of private ownership and other selfish interests and is therefore capable of finding the optimal solution to meeting the aspirations of all social strata. It was recognized, however, that:

> [u]nder socialism the urgent problems of environmental protection do arise in the course of scientific and technological progress. This happens, particularly due to the fact that the socialist countries have not yet developed the new productive forces to the desired extent, which would make the environmental pollution minimal, above all, by creating a closed-cycle, no-waste production process. (Granov, 1980, p. 93–94)

Thus closing material loops came to be seen as an ultimate solution to environmental problems, to be achieved through continued economic and technological development. This affirmation of the validity of the environmental Kuznets curve[1] allowed the continued pursuit of increased industrial production while accommodating growing concern for environmental issues. The specific approach used by Soviet scientists and engineers was termed "waste-free technology". This concept, introduced in the 1980s, shared much in common with the earlier ideas of combined production, though it placed emphasis not only on raising the indices of resource utilization but also on lowering indices of waste emission:

> Soviet policy encourages the wide use of waste-free technological processes and the introduction of effective methods for complex use of raw materials and waste. Specialists emphasise the modern scientific technological potential to transition from extensive methods of utilization of natural resources to intensive, resource-saving technologies. This transition is important because reduced waste ad waste-free technologies successfully solve problems of environmental degradation and pollution. Thus, criteria for the selection of new technologies will be not only the economic effect, but will take into account all socio-economic consequences of the incomplete use of resources. (Turkebaev and Sadikov, 1988, p. 18)

Towards this goal of industrial production with reduced environmental impact, the development of complex industrial ecosystems and the importance of a diversity of actors were acknowledged by Soviet scientists who stated that:

[1]The environmental Kuznets curve, an inverted U-shaped relationship between income and pollution, suggests that environmental degradation increases during initial stages of economic growth but then declines with further economic development.

> [t]he principle of waste-free technology cannot always be introduced in an individual enterprise, but is definitely applicable in a big industrial economic system, e.g., a territorial industrial complex with multi-sectoral structural management. Here, based on deep inter-sectoral cooperation, cyclical use of raw materials and wastes in different industries could develop. This kind of resource use offers tight interweaving of the flows of initial resources and waste materials between regional sectors of industrial and agro-industrial complexes. The ideal symbiosis of industry in the near future should be a cycle where there is no beginning and no end. (Turkebaev and Sadikov, 1988, p. 18–19)

Such progressive thought and eloquent expression would not be out of place in a journal article or conference proceedings written today by modern practitioners of industrial ecology.

3. Practice of Industrial Ecology in Centrally Planned Countries

In this section we briefly illustrate the implementation of industrial ecology techniques in command economies by presenting examples of combined production and waste-free technology in several sectors: a metallurgical facility in Ukraine, the forest products sector in Latvia, and heat cascading in many areas of the former Soviet Union. We then discuss the motivations and limitations of industrial ecology as practiced in centrally planned economies.

3.1. EXAMPLES OF CENTRALLY PLANNED INDUSTRIAL ECOLOGY

3.1.1. Metallurgy

The Nikopol manganese ferroalloy facility in Ukraine was a large electrometallurgical production complex that included smelting, sintering, and electrode manufacturing operations, plus electricity production, transport operations, and ancillary workshops (Zubanov and Velichko, 1988). The complex produced about 1.5 million tons of slag annually, which had been simply dumped into slag piles covering previously fertile agricultural soil. This was representative of other ferrous metallurgy facilities throughout the USSR, where in total more than 75 million tons of slag was produced annually during iron and steel manufacture. A project entitled “Waste” (*Otkhody*) was implemented at the complex during the 1980s, with the stated goals of efficiently using waste materials from production, and reducing air pollution problems.

Usage of four types of waste materials – gas, dust, sludge, and slag – was pursued (see Figure 3). Blast furnace gas generated during the ferroalloy production was collected. After appropriate cleaning, the gas was used as industrial process fuel and for the heating of domestic buildings. Other by-products suitable for reprocessing were dust and sludge from the dry and wet cleaning of blast furnace gas. Sludge from the sintering plant was pelletized and returned as raw material for ferroalloy production. Some sludge was used in concrete building materials, and future plans were developed for using the sludge as an additive in the production of manganese-rich fertilizer. Scientists found that the crystallized slag had qualities of high wear resistance, heat resistance, mechanical strength, thermal stability, and dielectric character, and so was suitable for various uses. They determined four main directions for reprocessing the slag. Some was crushed and used as gravel for road construction and building foundations, ballast for railways, and aggregate for concrete. Some slag was crushed and separated, yielding a metal concentrate that included about 60% manganese. It was used in the production of ferroalloy, and also was sold as raw material containing manganese. Molten slag granulated by contact with water was used as an additive for reinforced and mass concrete constructions, for cinder blocks used as thermal insulation, and for backfilling. Lastly, "slagstone" was produced by the casting or moulding of molten slag. The key to this method was the regulation of the crystallization of slag. Slagstone products included fire blocks, refractory linings, pipes, and panels.

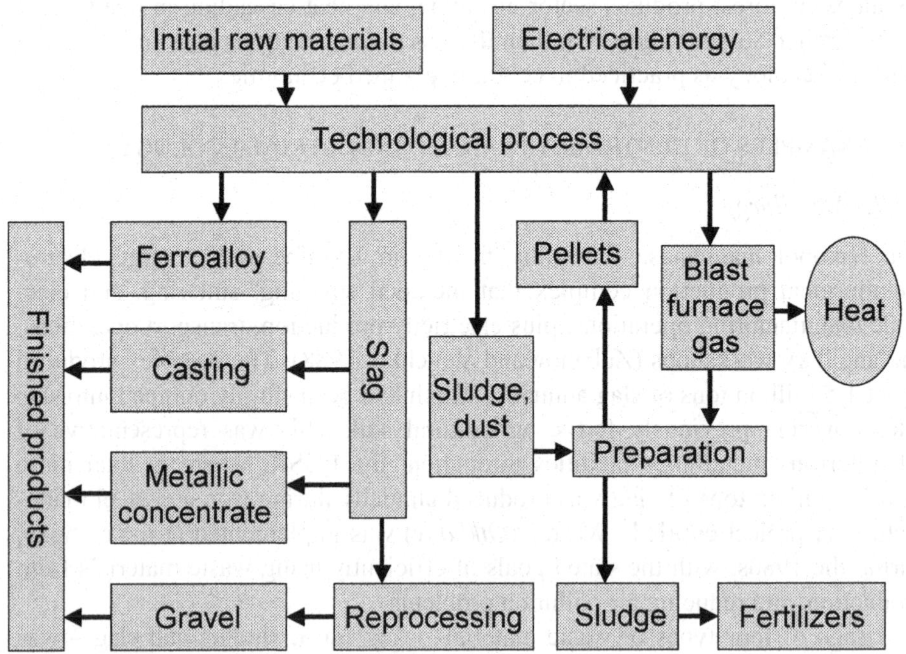

Figure 3. Schematic diagram of utilization of by-products from the Nikopol metallurgical complex in Ukraine. (Translated and redrawn from Zubanov and Velichko (1988)).

3.1.2. *Forest Products*

Vikulina (1983) described efforts in Soviet Latvia to increase the overall efficiency of forestry and wood processing industries. Part of this initiative involved the recovery and utilization of by-products of forestry activities. In particular, logging residues including branches, needles, and treetops, which had previously been left in the forest after the harvesting of commercial stemwood, were recovered. The potential resource available was on the order of 20 million tons annually. Uses for the logging residues included cellulose fibre for paper, vitamin flour (a feed supplement for cattle made from needle-bearing conifer twigs), and pharmaceuticals including chlorophyll carotene paste and petroleum ether.

One problem encountered was that the various components of the logging residues were mixed, reducing its suitability for specialized applications. Manual separation was slow and inefficient. Based on technology used in seed-cleaning machines in the agricultural sector, Soviet technicians developed a cyclone separator to divide the various fractions of logging residues for different uses. The lightest fractions such as needles, leaves, and twigs were used for vitamin flour and pharmaceuticals. The heavier fractions could be used for cellulose pulp production or as biofuel. Research continued on development of machinery to separate this fraction into high value pulping chips and lower value biofuel. A reported advantage of the increased efficiency of forest product use was the potential reduction in the total forest harvest needed.

3.1.3. *Heat cascading*

Cascading of heat resources was widely implemented in the Soviet Union in the forms of combined heat and electricity generation, and the recovery of heat from industrial processes. As early as the 1920s, power plants in Ukraine not only produced electricity but also provided hot water to heat greenhouses, fish farms, and other adjacent enterprises (Hewes, 2005). By 1975, about 29% of all electricity generated in the Soviet Union came from combined heat and power plants, operating at overall conversion efficiency of up to 70%, compared with 40% efficiency of steam plants supplying only electricity (Dienes and Shabad, 1979). The comparatively high usage rate has been partially attributed to differences in property rights between the Soviet Union and some Western countries, leading to differences in transaction costs and treatment of externalities that favor utilization of district heating, and combined heat and power (Mcintyre and Thornton, 1978).

The recovery of industrial process heat, particularly in the metallurgical sector, was widespread. The production of 1 t of ferrous metal, for example, required the heating of 100 t of water that could then be used for secondary purposes. Additional waste heat was found in slag and blast furnace gas. It is reported that 55% of recoverable heat in the metallurgical sector was utilized as a secondary resource (Efimov, 1968).

3.2. MOTIVATION FOR IMPLEMENTING INDUSTRIAL ECOLOGY

Notwithstanding the progressive theories developed by scientists in the Soviet Union and other socialist states, and despite the growing awareness of the severity of environmental problems, in practice the primary raison d'être of industrial ecology in centrally planned economies continued to increase production. Faced with growing societal demand for increased material welfare but with limited means to satisfy all demands, industrial symbiosis was part of an attempt by central planners to get by "on the cheap" (CIA, 1989, p. iii). In the Five Year Plan of economic and social development for the period of 1986–1990, Soviet Communist Party officials declared that:

> [r]esource conservation should be transformed into the main means of satisfying the growing demand in the national economy. Satisfaction of 75–80% of the growing demand for heating, energy and materials should be covered by using the resources more economically…[through]…complex use of natural and material resources, maximal reduction of losses and non-rational consumption, and wider integration of bio-resources and secondary resources into the economic circulation. (KPCC, 1986, p. 274)

It thus appears that industrial ecology was viewed by Soviet leaders as a panacea to satisfy the growing material demands of society, becoming the last in a series of proposed technological panaceas that included electrification, mechanization, chemicalization, automation, and cybernetics (Hutchings, 1976). In Soviet thinking,

> [t]echnological progress offers a possibility of making a more effective use of economic resources and of obtaining a bigger economic effect with relatively low expenditure. This creates an opportunity for substantially increasing the rates of growth of the output of consumer goods… (Efimov, 1968, p. 15)

Thus Soviet industrial symbiosis was Jevons' "rebound effect" by design – increasing the efficiency of production with the specific intent of raising total production and consumption. This illustrates the contradiction of the approach to industrial ecology as implemented in the centrally planned economies, reflecting the fundamental goals of industry and of the society it served. As the core objective was to increase material consumption (i.e., physical growth) the marginally increasing the technical efficiency of industrial processes could not eliminate environmental problems, but at best postpone them.

4. Implementation of Industrial Ecology in Transition Economies

Can the transition countries take advantage of the legacy of centrally planned industrial ecology, as they shift from a state-planned organizational structure to an industrial system planned and implemented by private firms within a policy context defined by government? In this section we seek to understand the conditions necessary for the success of industrial ecology activities, the potential obstacles to the realization of industrial production with reduced environmental impact, and how industrial ecology might be fomented in transition countries to contribute to the development of sustainable industry. The transition countries are diverse both in terms of their varied industrial and environmental experiences during the centrally planned era, as well as their more recent developmental pathways towards a market economy. Here we do not focus on this diversity, but instead highlight general themes that may be common to many of the transition countries, acknowledging that not all observations will apply to every country.

4.1. POLICY AND ECONOMIC FRAMEWORKS

4.1.1. Policy instruments to encourage industrial ecology

As discussed above, the actual environmental performance of centrally planned industry was less than optimal, in spite of progressive theoretical considerations. Furthermore, the complex intersectoral linkages mandated by central planners were dissolved by the collapse of the command economies, resulting in industrial disruption (Oldfield, 2000). Today, significant improvements in the environmental performance in transition economies can result from improved resource management and technological modernization linked to new organizational arrangements. An appropriate environmental policy framework, if enforced rigorously, can encourage investment and operational decisions to incorporate environmental consideration (OECD, 2000). Much attention is currently focused on the management and policy aspects of industrial ecology, seeking to develop appropriate interorganizational and intraorganizational structures (Korhonen et al., 2004). In transition countries, such structures fulfill the roles that were previously played by central planners and state enterprises.

A policy framework to promote industrial ecology can include various types of instruments including regulatory, economic, and information sharing (OECD, 1999). Regulatory or legal instruments including environmental standards should encourage continual improvement rather than formalistic compliance. Current environmental legislation in most transition countries provides incentives for end-of-pipe techniques, rather than process changes and intersectoral integration that characterize industrial ecology. One example of legislation that favors industrial symbiosis is umbrella licensing, which implies licensing a group of firms instead of a single firm (Eilering and Vermeulen, 2004). The umbrella license establishes a ceiling for the various forms of environmental burdens such as emissions to air or water, within a geographic area covered by the license. Within the area, the licensed firms agree amongst themselves, the most efficient way to collectively obey the limits, which may involve exchanging or selling emission rights between the firms covered by the umbrella license.

Economic instruments applied by governments can also play important roles as incentives or as deterrents (OECD, 1999). The phasing out of subsidies that encourage the inefficient use of energy, water, and raw materials can lead to both environmental and economic benefits. Measures to create incentives for enterprises to adopt industrial ecology, e.g., research and development schemes and soft loans, could be used.

An open flow of information on the causes and potential solutions to environmental degradation can facilitate the implementation of industrial ecology (OECD, 1999). Efforts to improve communication within the industrial sector can increase the local knowledge of opportunities for effectively closing material cycles. Raising public awareness of environmental issues in general, and impacts caused by industry in particular, can stimulate the implementation of cleaner production techniques. This is especially relevant in the context of consumer decisions informed by a government-administered environmental labeling system for consumer products (Banerjee and Solomon, 2003).

4.1.2. Economic requirements in transition countries

The successful implementation of industrial ecology is more likely under conditions of a stable economy, reliable cooperation between companies, predictable supply of industrial inputs, and uniform and transparent environmental regulations (OECD, 1999). Industry in transition countries is undergoing a shift from centrally planned interaction to self-organization of industrial actors. More generally, the role of government in representing environmental interests' vis-à-vis economic agents and interests is shifting. This is a question of public versus private responsibility for pollution emission and resource depletion, as well as involving more mechanical issues of availability of knowledge for effective decision making (Sathre and Grdzelishvili, 2006).

In principle, market economies are coordinated by the flow of price information, and decisions on industrial activities are made individually by market agents. In contrast, economic decisions in command economies were made centrally, with industrial activity being coordinated by central authorities. Waste products of one enterprise, which in a market economy might be valued and used by an adjacent enterprise, may simply have been discarded by a factory manager obeying material flow decisions made by distant planners who were unaware of the local conditions (Åhlander, 1994). Thus the specific knowledge of by-product availability and potential usage may be locally available, but the authority and decision making in centrally planned economies was distant, a point also raised by Desrochers and Ikeda (2003).

Different levels of decentralization in the transition countries affect the outcome of economic measures and their impact on industrial restructuring. For example, it is often considered that the Czech Republic, Hungary, Poland, and the Baltic countries have been the more active reformers in many key aspects of intergovernmental fiscal relations (Dabla-Norris, 2006). Many other countries including Russia and some Central Asian states still need substantial reforms of the incentive structures that govern intergovernmental fiscal relations in order to obtain an efficient and well-functioning multitier system of government. The current movement towards democratic forms of governance is a means of increasing democratic participation in the decision-making process, thereby enhancing accountability and transparency of government actions.

Nevertheless, market conditions in transition countries are not always favorable for the implementation of industrial ecology activities (OECD, 1999). In the transition from a centrally planned economy to a market economy, many enterprises struggle for survival, facing unpredictable markets for their products. The level of production is difficult to predict, making it difficult to assess the needs for inputs and the availability of by-products.

4.2. PRIORITIZING INDUSTRIAL ECOLOGY

4.2.1. Valuation of resources and environmental externalities

Environmental damage that took place in the past in transition countries is now widely acknowledged, as is the continuing environmental disruption in many of these countries. However, populations generally consider environmental damage a secondary problem compared to the other hardships of the transition period such as high inflation, persistent unemployment, lessened social cohesion, financial arrears, adverse terms of international trade, and political instability (OECD, 2000). Due to such conflicting priorities, the measures implemented by other policy sectors, for example, economic policy and agriculture policy, often work contrary to environmental policy (Glasbergen, 1996).

One way this is manifested is the pricing or valuation of natural resources and external costs related to industrial activities. The use of economic principles to guide the efficient allocation of scarce resources depends on the existence of appropriate measures of value. In the command economies, central planners developed nonmarket methodologies to assign values to natural resource deposits and products (Thornton, 1978). This method has been criticized for inefficient allocation contributing to environmental degradation (e.g., Goldman, 1972). Near the collapse of the Soviet Union, Soviet scientists admitted that:

> [b]ecause of economic growth, there are also increasing problems with the utilization of waste materials. But in solving this utilization problem, there appears a negative side of the existing economic stimulation system of industry. Usually, industries seek immediate reduction of expenses, are not interested in complex use of raw materials, and do not take environmental damage into account. The negative influence of people on the environment is growing, and in some regions it has passed a critical point. (Turkebaev and Sadikov, 1988, p. 8)

The *Metodika* guidelines were adopted in 1987 by the former USSR State Committee for Environmental Protection, in an effort to calculate the true costs of environmental damage (OECD, 2000). The approach of the *Metodika* was to aggregate different pollutants into standard units and to assign economic value to a unit of aggregate pollution. The national guidelines used specified conversion coefficients to convert metric tons of emissions into "standard" tons of emission, based on epidemiological data on the relative harm caused by pollutants, emission source location, height of pipe, speed of emissions, and climatic conditions.

The significance of the choice of valuation method is illustrated in an example comparing the estimates of environmental damage costs as calculated by the *Metodika* and the OECD-recommended methodology based on dose-response function. As part of the analytical work to define priority environmental actions in the Regional Environmental Action Plan, a team of Russian and international specialists assessed the cost of environmental damages from pollution. The results for three major air pollutants are shown in Table 1. The Russian methodology consistently undervalued the cost of environmental damage.

Nevertheless, market-based industrial systems have also been criticized for lacking appropriate valuation of natural resources and environmental externalities. In the practice of modern resource economics, the "optimal" use of environmental resources and the most "efficient" level of pollution can be consistent with environmental unsustainability (Ekins, 1996). Environmental externalities, which were belatedly recognized by Soviet authorities as detrimental to social

TABLE 1. Estimates of rates of health damage from air pollutant emission in Sverdlovsk Oblast, using the *Metodika* methodology and the OECD-recommended methodology. (From OECD, 2000)

Damage (thousands of dollars/t)	OECD method	*Metodika*
Particulates	5	0.2
Lead	123	13
SO_2	0.2	0.1

well being (Goskompriroda, 1989), are potentially significant in modern industrial systems as well. Other valuation issues include intergenerational equity questions linked to discount rates choices, intragenerational equity based on the ability to participate in the market economy, and differentiation of nonrenewable and renewable resources. For these reasons it has been argued that, while the centrally planned system lacked appropriate short-term to medium-term signals to properly allocate natural resources, the market-based industrial system may lack the necessary signals to properly allocate resources in the medium to long term (Sathre and Grdzelishvili, 2006).

4.2.2. Levels of production and consumption

Possibilities of optimizing industrial processes and systems are limited if overall levels of consumption are not addressed, with consideration of the underlying goals of industry and of the society it serves. Consumption level is a critical issue, particularly in the context of evolving material aspirations of the populations of transition countries. Shevchenko (2002) observed the changing consumption patterns in Russia during the transition from state socialism to a market economy. Parallel to the dramatic increase in availability of consumer goods, numerous socioeconomic factors encouraged a rise in consumption. These include the use of goods as a form of investment and savings in the face of unstable currency valuation, a form of insurance in case of future supply disruption, and the purchase of security equipment as a response to uncertain police protection. The introduction of a market economy has led to increased car transportation causing urban air pollution, and rapidly increasing waste production by households (Pavlínek and Pickles, 2004).

A significant increase in overall consumption levels has the potential to trivialize marginal increases of production efficiency gained through industrial ecology. This can be represented by the well-known equation

$$I = P \times A \times T$$

where I is environmental impact, P is population, A is affluence (or per capita consumption), and T is the effect of technology. Industrial ecology activities improve the efficiency of technology, T, allowing reduced environmental impact for a given production level. If, however, per capita consumption is simultaneously increasing, the total environmental impact may rise or remain unchanged. It is acknowledged that population change will also affect the outcome of the equation, a significant issue in some transition countries.

Thus the contribution of industrial ecology to societal sustainability will depend not only on the technological increases in resource use efficiency gained within the production sphere, but also on the extent of overall production and consumption within the context of societal metabolism.

5. Conclusions

Here we have outlined the theories of industrial ecology as developed under centrally planned economies, and presented brief examples of combined production and waste-free technology as practiced in the former Soviet Union. We find that elements of industrial ecology were appreciated and practiced from the early years of the Soviet Union and other command economies. By the time of the collapse, industrial ecology was viewed by central planners as an important approach to increase industrial efficiency. Soviet theories of Industrial symbiosis were well developed, and included such elements of modern industrial ecology as the analogy between natural ecosystems and industrial ecosystems, and the benefits of having a diverse range of actors within an industrial ecosystem. However, while the potential environmental benefits of industrial ecology were eventually recognized and valued, central planners pursued industrial ecology primarily as a means to increase production.

In the transition countries today, the potential exists to integrate the modern environmentally focused technologies of Western countries together with the extensive practical industrial experience of the centrally planned era. Large-scale eco-industrial parks could be created based on existing infrastructure, by extending cooperative links between individual facilities (Coyle, 1996). Perhaps the implementation of industrial ecology seems unrealistic in the current climate of economic fragility, persistent unemployment, and political instability experienced in many transition countries. However, as the industrial systems in the transition countries are (re)developed in the coming years, it appears beneficial to incorporate industrial ecology practices for both economic and environmental benefits (Grdzelishvili and Sathre, 2005).

Rehabilitation of industrial infrastructure in a way that encourages efficient utilization of material and energy resources is a timely and appropriate task for transition countries needing to build flexible, vibrant economies while protecting natural heritage. Industrial ecology is a progressive framework that guides the structuring of industrial activities to provide needed materials and services while minimizing impact on the environment. While industrial ecology is not a technological panacea that allows unlimited, eco-friendly production, it could be one element in a larger strategy of transforming society's production and consumption patterns toward more sustainable forms.

References

Åhlander, A.-M.S., 1994, *Environmental Problems in the Shortage Economy: The Legacy of Soviet Environmental Policy*, Edward Elgar, Hants.

Banerjee, A., Solomon, B.D., 2003, Eco-labeling for energy efficiency and sustainability: a meta-evaluation of US programs, *Energy Policy* 31(2): 109–123.

Blackwell, W.L., 1994, *The Industrialization of Russia: A Historical Perspective*, 3rd edn., Harlan Davidson Inc., Arlington Heights, IL.

CIA (Central Intelligence Agency), 1989, *Industrial Pollution in the USSR: Growing Concerns But Limited Resources*, Intelligence Assessment, CIA Directorate of Intelligence, Langley, VA.

Coyle, R., 1996, Managing environmental issues at large-scale industrial estates: problems and initiatives in central and eastern Europe and the former Soviet Union, *UNEP Ind. Environ.* 19(4): 45–47.

Dabla-Norris, E., 2006, The challenge of fiscal decentralisation in transition countries, *Comp. Econ. Stud.* 48: 100–131.

Davitaya, F., 1977, Changes in the atmosphere and some problems of its protection, in: *Society and the Environment: A Soviet View*, Progress Publishers, Moscow, pp. 99–110.

Desrochers, P., 2005, Learning from history of from nature of both? Recycling networks and their metaphors in early industrialisation, *Prog. Int. Ecol.* 2(1): 19–34.

Desrochers, P., Ikeda, S., 2003, On the failure of socialist economies to close the loop on industrial by-products: insights from the Austrian critique of planning, *Environ. Polit.* 12(3): 102–122.

Dienes, L., Shabad, T., 1979, *The Soviet Energy System: Resource Use and Policies*, V.H. Winston, Washington, DC.

Efimov, A.N., 1968, *Soviet Industry*, Progress Publishers, Moscow.

Efimov, A.N., Zhukova, L.M., 1969, *Kontsentratsiya, Spetsializatsiya, Kooperirovaniye i Kombinirovaniye v Mashinostroenii (Concentration, Specialization, Cooperation and Combination in Machine Industry)*, Visshaia Shkola Publisher, Moscow (in Russian).

Eilering, J.A.M., Vermeulen, W.J.V., 2004, Eco-industrial parks: toward industrial symbiosis and utility sharing in practice, *Prog. Ind. Ecol.* 1(1/2/3): 245–270.

Ekins, P., 1996, Towards an economics for environmental sustainability, in: *Getting Down to Earth: Practical Applications of Ecological Economics*, Costanza, R., Segura, O., Martinez-Alier, J., (eds.), Island Press, Washington, DC.

Erkman, S., 1997, Industrial ecology: an historic view, *J. Clean. Prod.* 5(1–2): 1–10.

Erkman, S., 2002, The recent history of industrial ecology, in: *A Handbook of Industrial Ecology*, Ayres, R.U., Ayres, L.W., (eds.), Edward Elgar Publishing, Cheltenham, UK.

Frosch, R.A., Gallopoulos, N.E., 1989, Strategies for manufacturing, *Sci. Am.* 261(3): 144–152.

Gille, Z., 2000, Legacy of waste or wasted legacy? The end of industrial ecology in post-socialist Hungary, *Environ. Polit.* 9(1): 203–231.

Glasbergen, P., 1996, Learning to manage the environment, in: *Democracy and the Environment: Problems and Prospects*, Lafferty, W.M., Meadowcroft, J., (eds.), Edward Elgar Publishing, Cheltenham, UK, pp. 175–193.

Goldman, M.I., 1972, *The Spoils of Progress: Environmental Pollution in the Soviet Union*, MIT Press, Cambridge.

Goskompriroda (USSR State Committee for the Protection of Nature), 1989, *Sostoyanie prirodnoi sredy v SSSR v 1988 g (Environmental condition in the USSR in 1988)*, VINITI, Moscow (in Russian).

Graedel, T.E., Allenby, B.R., 2003, *Industrial Ecology*, 2nd edn., Prentice Hall, Upper Saddle River, NJ.

Granov, V., 1980, The ideological struggle and ecological problems, *Int. Aff.* (USSR Foreign Ministry), December 12: 87–95.

Grdzelishvili, I., Sathre, R., 2005, Industrial ecology: a strategy for efficient resource use in the Caucasus, *Caucasus Environ.* 2(11): 10–13.

Hewes, A.K., 2005, *The Role of Champions in Establishing Eco Industrial Parks*, PhD dissertation, Department of Environmental Studies, Antioch New England Graduate School, New Hampshire.

Hutchings, R., 1976, *Soviet Science, Technology, Design: Interaction and Convergence*, Oxford University Press, London.

Kalundborg Centre for Industrial Symbiosis, 30 May 2006; http://www.symbiosis.dk/

Korhonen, J., von Malmborg, F., Strachan, P.A., Ehrenfeld, J.R., 2004, Management and policy aspects of industrial ecology: an emerging research agenda, *Business Strategy and the Environ.* 13: 289–305.

KPCC (Komunisticheskaia Partia Sovetskogo Soiasa: Communist Party of the Soviet Union), 1986, *Record of the XXVII Meeting of the Communist Party of the Soviet Union*, Moscow (in Russian).

Lenin, V.I., 1916, *Polnoe Sobranie Cochinenii (Lenin's Collected Works)*, Vol. 27, printed 1980, Political Literature Publisher, Moscow (in Russian).

Mcintyre, R.J., Thornton, J.R., 1978, Urban design and energy utilization: a comparative analysis of Soviet practice, *J. Comp. Econ.* 2(4): 334–354.

OECD, 1999, *Cleaner Production Centres in Central and Eastern Europe and the New Independent States*, Report CCNM/ENV/EAP(99)25, Task Force for the Implementation of the Environmental Action Programmes for Central and Eastern Europe.

OECD, 2000, *Background Paper on Valuing Environmental Benefits and Damages in the NIS: Opportunities to Integrate Environmental Concerns into Policy and Investment Decisions*, Report CCNM/ENV/EAP/MIN(2000)3, Task Force for the Implementation of the Environmental Action Programmes for Central and Eastern Europe.

Oldfield, J.D., 2000, Structural economic change and the natural environment in the Russian Federation, *Post Commun. Econ.* 12(1): 77–90.

Pavlínek, P., Pickles, J., 2004, Environmental pasts/environmental futures in post-socialist Europe, *Environ. Polit.* 13(1): 237–265.

Sathre, R., Grdzelishvili, I., 2006, Industrial symbiosis in the former Soviet Union, *Progress in Industrial Ecology*, 3(4): 379–392.

Shevchenko, O., 2002, "Between the holes": Emerging identities and hybrid patterns of consumption in post-socialist Russia, *Eur. Asia Stud.* 54(6): 841–866.

Thornton, J.A., 1978, Soviet methodology for the valuation of natural resources, *J. Comp. Econ.* 2(4): 321–333.

Turkebaev, E.A., Sadikov, G.X., 1988, *Kompleksnoye Ispol'zovaniye Syr'ia i Otkhodov Prom'ishlennosti (Complex Use of Raw Materials and Waste Materials in Industry)*, Kazakstan Publishers, Alma-Ata (in Russian).

Vikulina, E., 1983, Specific wood quality in industrial plantations, *Teknika* 11: 43–53 (in Russian).

Zubanov, V.T., Velichko, B.F., 1988, *Effekt Bezotkhodnoyi Tekhnologii (Effect of Waste-free Technology)*, Promin Publisher, Dnepropetrovsk (in Russian).

9. POTENTIAL OF STRATEGIC ENVIRONMENTAL ASSESSMENT AS AN INSTRUMENT FOR SUSTAINABLE DEVELOPMENT OF THE COUNTRIES IN TRANSITION

OLENA BORYSOVA*
Kharkiv National Academy of Municipal Economy, 12 Revolutzii St., Kharkiv, Ukraine

Abstract: Strategic Environmental Assessment (SEA) is the formalized, systematic, and comprehensive process of evaluating the environmental impacts of a policy, plan or programme and its alternatives, including the preparation of a written report on the findings of that evaluation, and using the findings in publicly accountable decision making. Currently, SEA is applied primarily as a means of minimizing the adverse environmental effects of the implementation of proposed strategic actions. However, its potential shall also be explored as a sustainability development tool. Given the flexible, transparent, participatory nature of the SEA process and procedures, SEA implementation could play a significant capacity building role in a sustainability-driven development process. For the post-Soviet and post-socialist countries surviving the economic pressure of the transitional process and an environmental legacy of unsustainable use of natural resources, SEA, as a preventive planning tool fit for the open market-oriented economy, may become of significant relevance. SEA implementation, especially if developed in a synergism with other tools such as environmental planning, risk assessment and management, and public participation in environmental decision making, may decrease and/or mitigate environmental risks of economic development and enable transitional societies to better meet environmental security and sustainability challenges.

Keywords: Strategic Environmental Assessment, sustainable development, countries in transition, planning context, capacity building

*To whom correspondence should be addressed. Olena Borysova. Department of Environmental Engineering and Management Kharkiv National Academy of Municipal Economy, Revolutzii str., 12, Kharkiv 61002, Ukraine; e-mail: oborysova@velton.kharkov.ua

R. N. Hull et al. (eds.), Strategies to Enhance Environmental Security in Transition Countries, 117–126

1. Background

Strategic Environmental Assessment (SEA) is the formalized, systematic, and comprehensive process of evaluating the environmental impacts of a policy, plan or program and its alternatives, including the preparation of a written report on the findings of that evaluation, and using the findings in publicly accountable decision making (Therivel et al, 1992). SEA originally emerged as an extension of project-level environmental impact assessment (EIA) and aimed at evaluating the environmental impacts of the policies, plans, and programs (so-called PPP concept) that had been already proposed. Most recently, SEA is being seen as shifting away from the traditional "object" of assessment (draft PPPs) towards a more encompassing view of the policy process and its political dimensions, with special attention to decision making (Bina, 2003). This shift significantly strengthens the potential of SEA as an instrument of sustainable development. Gibson et al. (2005) call environmental assessment and the pursuit of sustainability "the marriageable pair...of two of the major concepts introduced over the past few decades to improve the odds of continued human survival on this planet".

Unlike the project level EIA that has a certain prototype in almost every national environmental protection system, SEA is seen as a new development, especially for the transitional countries (for the purpose of this paper, we will use the term "transitional countries" to describe the diverse group of the ex-socialist states of Europe and Central Asia that are undergoing the process of economic and political reform). The level of SEA introduction into national legal and planning systems and its implementation varies significantly within the group. On the one extreme, some countries have already gained considerable practical experience in compliance with the European Union (EU) SEA Directive (2001/42/EC) (e.g., the Czech Republic, Slovakia, and Poland). On the other extreme, countries of Central Asia (such as Kyrgyzstan, Tajikistan, and Uzbekistan) keep the environmental assessment procedures almost unreformed and, according to Cherp (2001), there is little or no application at the strategic level. A number of ex-Soviet states (Ukraine, Armenia, Georgia, and Moldova) signed the United Nations Economic Commission for Europe (UNECE) SEA Protocol (hereafter, SEA Protocol) to the Espoo Convention (UNECE Convention on the Environmental Impact Assessment in a Transboundary Context) at Kiev in 2003 and plan to ratify it within the course of the next 4 years. Some other states in the region (e.g., Belarus) are now considering possible accession to the SEA Protocol. These countries are steadily, though with various scope and speed, moving towards practical implementation of SEA by bringing the elements of international good practice into national environmental assessment systems.

2. Planning Context

The former Soviet Union countries in Eastern Europe, Caucasus, and Central Asia are going through a transition from centrally planned to open market-based economies. During this process they mainly abolished the highly sectoralized and standardized system of economic planning that had existed, but the alternative has not been sufficiently developed yet. The example of Ukraine that is presented below might be, to a certain extent, considered as one of the most extreme cases of planning "negativism" and is, in our view, quite illustrative.

The analysis reported below is based on the findings of the UNDP/REC Strategic Environmental Assessment – Promotion and Capacity Building (Ukraine) project (2005–2006) that resulted in development of the National Manual on SEA Protocol Implementation in Ukraine.

The Ukrainian system of the state prognosis of economic and social development was analyzed as the key framework in which SEA should operate and where major innovations should be sought. The following features of the planning and prognosis system in Ukraine were identified:

- The legislative basis, strategic and tactical tasks of the system of prognosis and indicative planning of socioeconomic development of the country are identified and regulated by the Constitution of Ukraine, acts of legislative and executive power, and an annual message of the President of Ukraine to Verkhovna Rada of Ukraine. The long-term prognosis of economic development is carried out by research institutions with participation and assistance of the executive power bodies and is submitted to the Cabinet of Ministers of Ukraine for agreement and confirmation.
- The national planning system as a whole is under serious transformation. Planning, as it is understood in the higher income countries, does not exist in Ukraine. There is an annual programming of the state and governmental policy in the field of socioeconomic development. This programming document is adopted together with the forecast of the state budget use and is a budget-related document.
- The programme of socioeconomic development includes the components of environmental strategic planning and prognosis, where SEA implementation shall, in principle, take place. However, the process of program development and adoption is not open and transparent, and decision making is not flexible and is heavily dominated by economic incentives.
- Framework legislative, regulative, and organizational provisions and management conditions for SEA system introduction to the general structure of national programming and prognosis of economic development are

contained in the Laws of Ukraine "On state prognosis and development of the programs of economic and social development of Ukraine" (2000), and "On the state targeted programs" (2004). These laws contain the requirements and conditions of development of the short-term and medium-term national governmental programs but are not supported by the regulative framework, and sufficient methodological basis.

- The practice of environmental planning and planning for sustainability is nearly nonexistent.
- Sectoral agencies, such as the Ministry of Health Protection, Ministry in the Affairs of Protection of the Population Against Natural Disasters and Consequences of the Chernobyl Catastrophe, and Ministry of Transportation, all have their own planning and prognosis systems that may use SEA.

3. Planning Theory

In order to proceed with further analysis and recommendations on the prospect of SEA implementation in transitional countries, it is important to take a broader look at the origin and possible development of the present situation. As has been shown by Cherp (2001) and Cherp and Lee (1997), a significant part of the environmental assessment system operating in post-Soviet countries is a legacy of the centrally planned economy. This to a lesser extent applies to the countries of Eastern and Central Europe, although part of the EA professional community are still those who were trained and gained professional experience in the conditions of a planned economy. At the same time, the very term "planning" has been heavily discredited by the inefficient performance caused namely by the centralization phenomena. This causes serious psychological difficulties and, in some cases, professional resistance towards strengthening of planning elements and introduction of planning instruments into national assessment frameworks. Meanwhile, as seen from the decision cycle shown in Figure 1, the notion of planning does not have any "ideological" loading and is not necessarily related to the centralized and heavily regulated economic system. Instead, planning is a necessary component of the strategic development process and is generally understood as a means to (Partidario, 2006):

- Make development objectives and strategies explicit into strategic actions
- Allocate resources to activities and determine the use of different areas

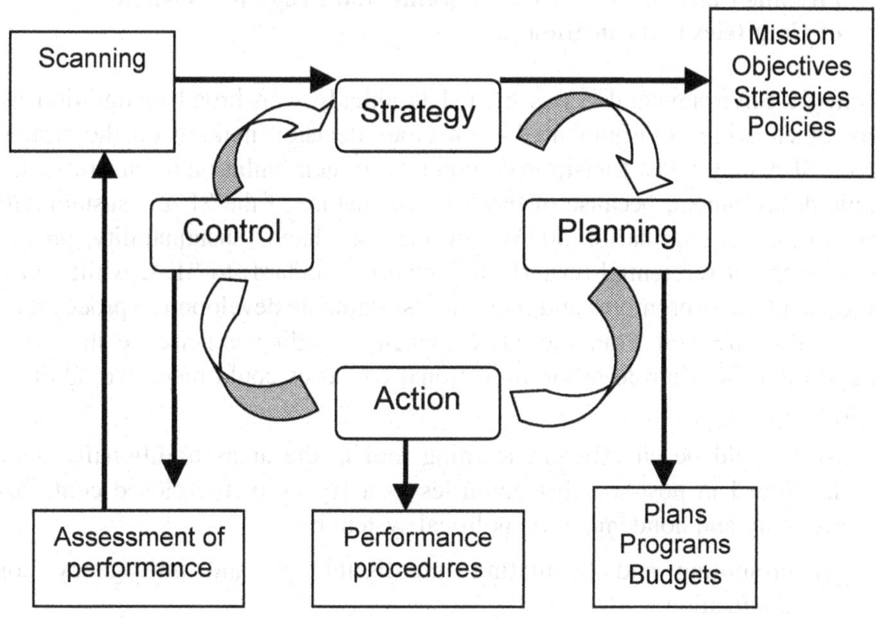

Figure 1. Decision cycle of Strategic Environmental Assessment. (From Partidario, 2006).

- Establish rules for development and zoning regulations
- Set up an action program

In more theoretical terms, the process of planning constitutes three fundamental steps: value formulation, means identification, and effectuation (Davidoff and Reiner, 1991). It is exactly on the stage of decision of transforming values into policy commitment where SEA shall take its point of entry in order to ensure its mission as understood by the recent literature, namely to put upstream and mainstream environmental consideration in the decision-making process (OECD DAC, 2006). It is important to underline here that planning *per se* is seen as a vehicle, which collects, analyzes, and publicizes information (such as forecasts and assessment of third-party costs and benefits) required to make reasoned decisions (Davidoff and Reiner, 1991), and could be designed to serve both as market aid and market replacement. This notion reinforces the idea of SEA being an effective modern planning tool that could be applied in the conditions of various levels of regulation of the economic system.

4. Possible Place of SEA in the Planning and Prognosis System of Countries in Transition

Strategic Environmental Assessment has already won broad recognition as a tool of ensuring environmentally conscious decision making on the strategic level. SEA may assist transitional countries in their ambition to ensure sustainable development, because of the two-way nature of the SEA – sustainability relationship (Partaridario, 2006): on the one hand, sustainability provides an assessment referential (criteria framework, standard) to SEA, while, on the other hand, SEA promotes and facilitates sustainable development processes.

Lessons derived from the SEA capacity building experience in Ukraine suggest that SEA's role in the transitional countries could have two additional aspects:

1. SEA could be an efficient learning tool in the areas traditionally underdeveloped in post-socialist countries as a legacy of the closed centralized economy and nondemocratic political system by:
   (a) Promoting and facilitating interdisciplinary and interagency communication
   (b) Requiring methodologies to be developed to take into account long-term, cumulative, indirect environmental consequences of strategic actions
   (c) Supporting and facilitating public participation in strategic decision making
   (d) Requiring population health issues to be looked at together with the environmental consequences of strategic decisions
2. SEA could be a means for "rehabilitation" of the notion of planning in transitional countries; in other words, SEA may be used to demonstrate the benefits of strategic planning and may thereby play the role of reintroduction (reinforcement) of the planning process, bringing along such up-to-date concepts as environmental planning and planning for sustainability.

In these respects, it could be speculated that to a certain extent the introduction of SEA could play a more significant role in governance development and improvement in transitional countries than in the developed ones; however, this is to be left to further research.

5. Capacity Building Recommendations

When introducing and developing national SEA systems, transitional coun tries will need to pay thorough attention to the existing systems for environmental assessment of plans and programs. This is necessary to both bridge the difference between national assessment systems and the SEA requirements, and, at the same time, to root the emerging SEA systems in the existing institutions to ensure their smooth functioning (UNDP, REC, and UNECE, 2006).

Capacity development processes shall ideally rely as much as possible on the existing in-country practice, knowledge, expertise, and skills. As initial capacity needs analysis[1] has shown, sound prerequisites exist in post-Soviet countries for SEA implementation. In order to develop these prerequisites into active agents of the SEA implementation process, the system, institutional, and human capacities should be developed.

It is important to avoid a simplified understanding of the capacity building process (capacity building is often wrongly perceived as a number of training) and to apply a systematic approach that has proven efficient when used in developing countries (Kirchhoff, 2006). It has the potential to identify the most pressing needs and prioritize key elements that would produce the greatest results given a tight budget (often the case in transitional countries).

Recommendations that follow are mainly based on the research conducted for Ukraine; since environmental assessment provisions remain generally similar across the post-Soviet region, a common approach to the capacity building development is justified. Relevance of the capacity building recommendations to the countries of Eastern and Central Europe is limited and depends on the level of their SEA system development. While recommendations on the system capacity building may not be applicable, institutional and human capacity development processes may still benefit from the suggested measures in the above countries.

5.1. SYSTEM CAPACITY

System capacity is created by frameworks that enable institutions and individuals to operate and interact with each other. Development of system capacity aims to enhance effectiveness of these interactions on different levels of institutional and political hierarchies (UNECE, REC, 2006).

The SEA development objective for system capacity in post-Soviet countries could be suggested as:

[1]"Capacity Building Needs Assessment for the Implementation of the UN/ECE Strategic Environmental Assessment (SEA) Protocol" that has been elaborated by national experts within the project "SEA Promotion and Capacity Development" jointly implemented by the UNDP, REC (implementing agency), and UNECE, and funded by the UNDP and the Environment and Security Initiative.

> [T]o bring legislative and regulatory bases in line with the UNECE SEA Protocol requirements; to develop the mechanisms of the SEA-related legislative and regulatory basis implementation and enforcement; and to designate and make effective a body for review of SEA practice effectiveness, practical guidance development, and best practice dissemination.

One of the biggest problems to be addressed within the system capacity building is development of an interactive practice between the SEA actors in "horizontal" (sectoral and planning) and "vertical" (local–regional–countrywide) dimensions. System capacity building could be first of all focused on facilitating consultations, dialogue, and cooperation between various representatives of authorities and society in the process of strategic decision making. If implemented successfully, SEA by nature may become a powerful tool of finding consensus between the interests of society, state, business, and the environment. It will set a precedent of cooperation of various interest groups that does not have an analogue in the history of the independent and market-oriented economy of the country.

5.2. INSTITUTIONAL CAPACITY

Institutional capacity is the ability of an organization to effectively operate within the given system. Development of institutional capacity aims to enhance overall organizational performance. This requires ability of an organization to adapt to changes (UNECE, REC, 2006).

The SEA institutional capacity objectives for transitional countries could be suggested as:

> [T]o develop a national practice of SEA based on existing environmental assessment and planning and programming processes and by implementtation and adjustment of SEA specific tools and approaches.

SEA institutional capacity development activities would be focused on streamlining and strengthening SEA processes, by supporting (methodologically, logistically, and by information and knowledge dissemination) and motivating country professional bodies responsible for on-the-ground SEA implementation. An important task here will be to create the network of the "SEA centers of excellence" that will disseminate best international and national practice, be the training centers, serve for cross-auditing for performance, and help setting "rules of the game" in the national context. These centers shall also ensure that all available institutional and human capacity (not just in environmental, but also in managerial, technological, etc. fields) is used for SEA implemen tation purposes, and, vice versa, that all SEA-generated benefits are clearly demonstrated to and made use of by potential customers. The centers shall also be the resource centers for international donors, providing information on institutional and human capacity, legal framework, and running activities supported by international assistance, in order to avoid duplication of efforts and provide for synergism between actions.

5.3. HUMAN CAPACITY

Human resources play a central role in SEA. It comprises the skills and expertise of persons. Capacity development of human resources aims to change attitudes and behaviors, develop skills, and support long-term motivation and commitment (UNECE, REC, 2006).

The SEA human capacity development objective for transitional countries could be formulated as:

> to prepare a critical mass of professionals (representing authorities, research institutes, consultancies, etc.) and public members able and willing to participate fully and in constructive ways in the implementation of SEA practice and able to maintain newly adopted capacity features.

SEA human capacity development activities might have a binary focus: first, on transferring global good practices of SEA to national assessment systems; second, on applying already existing skills, knowledge, expertise, and infrastructure to the new tasks. SEA-related human capacity development activities shall ideally be related to the labor market situation (increasing competitiveness of the practitioners), fit into the educational conception of the country (in order to achieve synergism), and have a well-specified target audience. Human capacity in SEA shall develop not only in terms of professional skills, but also in terms of the ability for team work, delegating responsibilities, and accepting trade-offs.

6. Conclusions

If in-country expertise in planning is transformed and updated by SEA capacity-building processes in order to meet the requirements of the SEA Protocol, not only will SEA begin to be implemented, but also the concepts of "environmental planning" and "planning for sustainability" will be introduced into planning and prognosis systems of transitional countries. Important features of the SEA process include in-depth consideration of public health issues, strict requirements for public participation, and provisions for intersectoral and interinstitutional dialogue. Lessons learned in Ukraine may be applicable to other transitional countries implementing an SEA Protocol or SEA Directive requirements.

Given the flexible, transparent, participatory nature of the SEA process and procedures, SEA implementation could as such play a significant capacity-building role in a sustainability-driven development process. For the post-Soviet and post-socialist countries surviving the economic pressure of the transitional process and an environmental legacy of unsustainable use of natural resources, SEA, as a preventive planning tool fit for the open market-oriented economy, may become of significant relevance. SEA implementation, especially if developed in a synergism with other tools such as environmental planning, risk assessment and management, and public participation in environmental decision making, may decrease and/or mitigate environmental risks of economic development and enable transitional societies to better meet environmental security and sustainability challenges.

References

Bina, O., 2003, Re-conceptualising Strategic Environmental Assessment: Theoretical Overview and Case Study from Chile. Ph.D. thesis, University of Cambridge, Cambridge.

Cherp, A., 2001, SEA in newly independent states, in: *Proceedings of International Workshop on Public Participation and Health Aspects in Strategic Environmental Assessment*, Dusik J. (ed.), Regional Environmental Center for Central and Eastern Europe, Szentendre, Hungary.

Cherp, A., Lee, N., 1997, Evolution of SER and OVOS in the Soviet Union and Russia (1985–1996). *EIA Rev,* 17: 177–204.

Davidoff, P., Reiner, T.A., 1991, A choice theory of planning, in: *A Reader in Planning Theory*, Faludi, A., comp, Pergamon Press, London.

Gibson, R.B., Hassan, S., Holtz, S., Tansey, J., Whitelaw, G., 2005, *Sustainability Assessment: Criteria and Processes*, Earthscan, London.

Good Practice Guidance on Applying Strategic Environmental Assessment in Development Cooperation, 2006, OECD DAC Task Team on SEA; http//:www.seataskteam.net

Kirchhoff, D., 2006, Capacity building for EIA in Brazil: preliminary considerations and problems to be overcome. *J. Environ. Assess. Policy Manag.* 8(1): 1–18.

Partidario, M.R., 2006, Strategic Environmental Assessment (SEA). Experience and good practices in strategic approaches to assist decision-making. *Training Materials, IAIA'06 Pre-Conference Training*, Stavanger, 2006.

Resource Manual to Support Application of the UNECE Protocol on Strategic Environmental Assessment, Draft for consultations, UNECE, REC, 2006; http://www.unece.org/env/sea/

SEA Protocol Initial Capacity Development in Selected Countries of the Former Soviet Union. UNDP-REC-UNECE SEA Bulletin #2, 2006.

Therievel, R., Wilson, E., Thompson, S., Heaney, D., Pritchard, D., 1992, *Strategic Environmental Assessment*, Earthscan, London.

10. THE INTEGRATED INDICATOR OF SUSTAINABLE DEVELOPMENT AND POLICY TOOLS

ALEXANDER GOROBETS*
Sevastopol National Technical University, Management Department, Streletskaya Bay, Sevastopol 99053, Ukraine

Abstract: Sustainable development is based on both the intergenerational and intragenerational socioeconomic and ecological justice of society and focuses on a permanent livability (ecocentric rationale) and human well-being: the necessary material minimum, freedom and choice, health and bodily well-being, clean diverse environment, good social relations, personal security, and the conditions for physical, social, psychological, and spiritual fulfillment. In this paper, human health is proposed as the integrated indicator of sustainable development. The specific policy tools focusing on socioecological economics are suggested as the necessary conditions to achieve sustainability while the internal human sustainability based on the ecocentric rationale is considered as its sufficient condition that can be achieved through educational policy. Although these tools can significantly contribute to environmental security in all countries, including transition countries, they should be necessarily complemented by adoption of precautionary and responsibility principles.

Keywords: sustainable development; health; socioecological economics; education

1. Introduction

Present ecological, socioeconomical, and political situations in many countries, including developed and transitional ones, confirm that they have failed to change significantly the extensive type of economic development and to ensure successful transition to sustainable development (UNEP, 2006). While fresh water deficit, deforestation, natural resources depletion, environmental pollution,

*To whom correspondence should be addressed. Alexander Gorobets, Sevastopol National Technical University, Management Department, Streletskaya Bay, Sevastopol 99053,Ukraine; e-mail: alex-gorobets@mail.ru

R. N. Hull et al. (eds.), Strategies to Enhance Environmental Security in Transition Countries, 127–134.

population growth, gap between rich and poor, and military conflicts are mostly relevant to developing countries, and high material consumption and energy use, waste generation, overweight, and blood circulation diseases are usual for developed and transitional ones, the problems of climate change (UNEP, 2006) and biodiversity loss (WWF, 2004) are global and have very dangerous systematically stable trends.

The main obstacles countries are facing on the way to sustainable development are:

(a) Poor understanding of the basic ideological approaches to sustainable development policy both by the government authorities and by the public due to a lack of qualified staffing potential
(b) Incompetent decision making (weak management) on all levels and psychological (sociocultural) and institutional barriers (inertia) of taking measures in a new, rapidly changing, and complex environment
(c) Absence of clear goals and specific well-developed national programs of sustainable development and contradictions between existing indicators

Therefore, the goal of this research is to develop the integrated consistent indicator of sustainable development and appropriate policy tools that will be consonant with the national strategies of sustainable development and millennium development goals: poverty reduction, quality lifelong education, environmental sustainability, improved health, and reduced HIV/AIDS and other diseases, gender equality, and global partnership.

This paper proceeds as follows. In the second section the general overview of some mainstream indicators is given and the new integrated characteristic of sustainable development is proposed. The third section presents the specific policy tools to address this indicator and finally the paper closes with conclusions.

2. The Indicators of Sustainable Development

It seems that construction of a simple indicator, which will reflect the evolution of a country's sustainability, is possible. Attempts to calculate integrated indicators based on ecological parameters are made often. The integrated "living planet index" (LPI), aimed to assess the state of the world's ecosystems, is calculated annually in the international World Wildlife Fund report (WWF, 2004). It shows very dangerous stable biodiversity loss in the last decades.

Another constructive indicator is “the Ecological Footprint” (EF), which computes the pressure of human population on the environment (WWF, 2004). According to this indicator, because of “material growth”, humans consume (deplete) the natural capital at the rate above the carrying capacity of earth simultaneously with high growth of municipal wastes and emissions of carbon dioxide (causing climate change with severe consequences – global warming, agricultural change, ice melting, flooding, droughts, etc.) and air pollutants affecting human health. Therefore, “material growth” caused by population and economic growth cannot be ecologically sustained because of the limited capacity of natural flows and cycles for providing the resources and absorbing or assimilating the waste (such as heavy metals or carbon dioxide). However, the ecological footprint is distributed unequally in the world regions, developed countries, i.e., North America and Europe are the major natural resources consumers (WWF, 2004) but the consequences are transmitted to all countries around the globe.

The Statistical Department of the United Nations Secretariat and United Nations Environment Program proposed a system for integrated environmental and economic accounting (SEEA), aimed at accounting of ecological factors in national statistics (UNSTAT, 2006). This system describes the correlation between the state of the environment and a country’s economy. “Green” accounts are based on the correction of traditional economic surveys (GDP per capita) when the costs of extracted natural recourses and socioecological damage from pollution are taken into consideration. The environmentally adjusted net domestic product (EDP) forms the core of ecological transformation of national accounts.

The Genuine Progress Indicator and the Index of Sustainable Economic Welfare are intended to improve the gross domestic product (GDP) by including such factors as income distribution and costs associated with pollution and resource depletion and other economically unsustaining costs (crime, family breakdown, etc.) (Daly and Cobb, 1994).

The Human Development Index (HDI) is often considered among the integral indexes of sustainable development. The index is elaborated within the framework of the United Nations Development Program (UNDP, 2005). The HDI consists of three interrelated components:

1. Health, measured by life expectancy at birth
2. Education – adult literacy rate (two-thirds weight) and the combined primary, secondary, tertiary gross enrollment ratio (one-third weight)
3. Standard of living – GDP per capita at purchasing power parity in US dollars

However the HDI has some weak points. It concentrates on the quantitative parameters of human life according to the principle "more is better" (especially for GDP per capita) rather than on its qualitative growth. Clearly human health might be considered as the biggest value but it should not be measured only by life expectancy (especially in developed countries with strong medical systems) but also by sickness rate and physical and intellectual capacity. Furthermore, education and standard of living have direct impact on health, especially in developing countries, so all three components are strongly correlated and can be replaced by one integrated characteristic of human health, which is a state of complete physical, mental, and social well-being.

Human health depends not only on human biology (genetic peculiarities), the public health system, but is also concerned with the state of the economy, the social, physical, spiritual, and moral life (lifestyle), and the state of the environment.

According to the World Health Organization (WHO), environmental risk factors play a role in more than 80% of the diseases (Prüss-Üstün and Corvalán, 2006). The major findings are as follows:

1. Globally, nearly one quarter of all deaths and diseases can be attributed to the environment (water pollution 7%, air pollution 5%), although for children it is even higher
2. The environmental disease burden is not distributed evenly across the world, and developing regions carry a disproportionately heavy burden for communicable diseases and injuries
3. Public and preventive health strategies that consider environmental health interventions are very important, cost-effective, and yield benefits that also contribute to the overall well-being of communities

Therefore, sustainable development should be based on both the intergenerational and intragenerational socioeconomic and ecological justice of society (Pearce, 1987), and focused on a permanent livability (adoption of ecocentric rationale instead of dominating anthropocentric one) and ecosystems and human well-being (Dodds, 1997): the necessary material minimum, freedom and choice, health and bodily well-being, clean diverse environment, good social relations, personal security, and the conditions for physical, social, psychological, and spiritual fulfillment.

As a result, in this paper, the integrated characteristic of human health (physical and intellectual capacity, life interval, sickness rate, psychological health) is proposed as the major indicator of sustainable development.

3. Policy tools

To achieve a high level of human health the following policy tools have been developed:

1. Integrated economic responses:
   - Adoption of up-to-date scientific socioecological economics theory (Costanza, 1991, 1996), based on the fundamental thermodynamic laws governing natural ecosystems, and focused on qualitative human development and ecosystem well-being instead of dominating environmental economics, which is a part of old-fashioned destructive neoclassical economics concentrating on material growth that can not be ecologically sustained (Bergh, 2000).
   - Changing the economic type of development from industrial to post-industrial, knowledge-based ecological economics by "green" taxation, i.e., heavy taxation of natural resource use and no taxation on intellectual products.
   - High investments in science, education, and cultural institutions oriented on human development.
   - Restructuring economy: give more priority to "green" sectors, i.e., producing bicycles, wind turbines, renewable energy industry (instead of oil, coal), recycling industry, organic agriculture and information, and ecoefficient technologies (Brown, 2001).
   - Transition to a service-sector oriented economy with low material–energy throughput (Daly and Cobb, 1994).
2. Integrated science and technologies and transfer:
   - Ecoefficiency (material and energy efficient clean technologies) (Brown, 2001)
   - Reusing and recycling materials (industrial ecology, including waste reduction)
   - Green efficient public transport system (fueled by biogas, hydrogen, ethanol)
   - "Solar" buildings using low energy alternatives
   - Green energy – solar, wind, hydrogen, biofuel, and waterpower use in all sectors – very consonant with ratified Kyoto Protocol to reduce CO_2 emissions
3. Integrated legal and institutional responses:
   - Adoption of SEEA, aimed at accounting of ecological factors in national economic statistics (UNSTAT, 2006)

- Obligatory incorporation of environmental auditing in all business sectors (especially industry and transport), adoption of international standards ISO 14000 and ISO 9000
- Changing consumption and production patterns through regulation (forbidding laws), setting right prices (including environmental costs), ecotaxation (adoption of polluter pays principle instead of consumer pays)
- Creating the appropriate ecocentric institutions to integrate human activity and natural cycles (e.g. ecovillages and ecocities)
- Institutional reforms of decision making to establish sustainable scientific and ethical control of ecocentric rationales and socio-ecological economics theory (Gorobets, 2006)
- Empowerment of communities, women, youth, and NGOs
- Creation of the institutes of civil control – community participation in decision making at all levels to achieve transparency and avoid corruption

However, these economic, scientific (technological), and institutional responses are only the necessary conditions of sustainable development while the internal human sustainability (mentality) based on the ecocentric rationale and qualitative human development (instead of material economic growth) is considered as its sufficient condition (Gorobets, 2006). This can be achieved through appropriate integrated educational policy:

1. Development of educational institutions for local communities to change people's values, mentality, and behaviour from anthropocentrism and consumerism, which are unsustainable socioecologically and psychologically (Maiteny, 2000) towards eco-centrism (socioecological well-being), and personal physical, intellectual, and spiritual development by:
   - Raising awareness about the critical socioecological state of the world and showing the multiple advantages of new, green lifestyles (particularly, stronger health and ecosystem well-being)
   - Using the mass media to influence individual and community mind-sets toward issues of the environment and sustainable behavioral patterns
2. Reorienting existing education programs. Rethinking and revising education from nursery school through university to include more principles, knowledge, skills, perspectives, and values in all dimensions of sustainability – social, environmental, and economic should be done in a holistic and interdisciplinary manner

3. The development of specialized training programs to ensure that all sectors of the workforce (from public to decision makers) have the knowledge and skills necessary to perform their work in a sustainable manner (the UN Decade of Education for Sustainable Development) (UNESCO, 2003)

Although these tools can significantly contribute to environmental security in all countries, including transition countries, they should be necessarily complemented by adoption of precautionary and responsibility principles (the best way to solve a problem is to prevent the problem).

4. Conclusions

In this paper, the integrated characteristic of human health, i.e., physical and intellectual capacity, life interval, sickness rate, and psychological health is proposed as the major indicator of sustainable development.

The specific economic, technological, and institutional policy tools focusing on ecosystems and human well-being oriented socioecological economics are developed as the necessary conditions to achieve sustainability while the internal human sustainability (mentality) based on the ecocentric rationale and human development (physical, intellectual, and spiritual) is considered as its sufficient condition that can be achieved through well developed educational programs for all populations.

The advantage of developed policy tools is that they are universal for application in every specific country, especially in transition and developing countries, being only subject to a local environment, cultural traditions, and socioeconomic situation.

References

Bergh, J.C., van den, J.M., 2000, Ecological economics: themes, approaches, and differences with environmental Economics, *Tinbergen Institute Discussion Paper* TI 2000-080/3; http://www.tinbergen.nl

Brown, L.R., 2001, *Eco-Economy: Building an Economy for the Earth*, W.W. Norton, New York, 352 p; http://www.earth-policy.org/Books/Eco_contents.htm

Costanza, R., (ed.), 1991, *Ecological Economics: The Science and Management of Sustainability*, Columbia University Press, New York.

Costanza, R., 1996, Ecological economics: reintegrating the study of humans and nature, *Ecol. Appl.* 6(4): 978–990.

Daly, H., Cobb, J., 1994, *For the Common Good. Redirecting the Economy Toward Community, the Environment and a Sustainable Future*, 2nd edn., Beacon, Boston.

Dodds, S., 1997, Towards a 'science of sustainability': improving the way ecological economics understands human well-being, *Ecol. Econ.* 23(2): 95–111.

Gorobets, A., 2006, An eco-centric approach to sustainable community development, *Community Dev. J.* 41(1): 104–108.

Maiteny, P., 2000, The psychodynamics of meaning and action for a sustainable future, *Futures* 32: 339–360.

Pearce, D., 1987, Foundations of an ecological economics, *Ecol. Modell.* 38: 9–18.

Prüss-Üstün, A., Corvalán, C., 2006, *Preventing Disease Through Healthy Environments: Towards An Estimate Of The Environmental Burden Of Disease: Executive Summary*, World Health Organization, WA 30.5., Geneva; http://www.who.int/en/

UNDP, 2005, Human Development Report 2005, Human Development Index, UNDP, New York; http://hdr.undp.org

UNEP, 2006, The GEO Data Portal. United Nations Environment Programme; http://geodata.grid.unep.ch

UNESCO, 2003, *United Nations Decade of Education for Sustainable Development (2005–2014)*, Framework for the international implementation scheme, UNESCO 32 C/INF.9. Paris; http://unesdoc.unesco.org/images/0013/001311/131163e.pdf

UNSTAT, 2006, *Handbook of National Accounting: Integrated Environmental and Economic Accounting – An Operational Manual*, UNSTAT F, No.78., New York; http://unstats.un.org

WWF, 2004, Living Planet Report 2004, WWF International, Gland; http://www.panda.org

11. SOCIOECONOMIC AND COMMUNICATION TOOLS TO ENHANCE ENVIRONMENTAL SECURITY: A CASE STUDY IN EGYPT

NERMINE ZOHDI*
International Development and Environment Associates (IDEA), 37 Syria St., 4th Floor, Apt.#8, Mohandessin, Giza, Egypt

Abstract: An initiative recently completed in Egypt as part of a climate change project illustrates several socioeconomic and communications tools that can be used to enhance environmental security. The primary purpose of the climate change initiative (CCI) project was to reduce the levels of greenhouse gases (GHGs) emitted during production at brick factories in Egypt. This was accomplished through the conversion of the present system, which employs heavy oil (mazot), to natural gas. While the reduction of GHG emissions in Egypt is an important national and global goal, it is relatively meaningless to those most directly affected by the project (i.e., the brick factory owners, workers, and local residents of the surrounding communities). Therefore, project activities not only covered technical matters (related to the conversion of the fuel systems) but extended to environmental and social matters as well. Significant effort was extended to identify ways by which the envisaged technical improvements could facilitate or enhance the social, health, environmental, and economic conditions of those most impacted by exposure to atmospheric emissions, namely brick factory workers and the populations of villages situated in close proximity to them. A socioeconomic assessment was designed with the understanding that technical interventions are more successful when they take into account the social and economic context in which the project is executed. The socioeconomic assessment provided the key link from the technology transfer side of the project to social development priorities (including basic education, basic health and nutrition, and child protection), in order to benefit vulnerable groups (such as children) that are directly impacted by the environmental problems the project sought to resolve. The socioeconomic assessment enabled a holistic understanding of the context within

*To whom correspondence should be addressed. Nermine Zohdi; e-mail: znerm@hotmail.com

R. N. Hull et al. (eds.), Strategies to Enhance Environmental Security in Transition Countries, 135–151.

which the CCI project was to be executed. It identified issues related to brick factory production and possible actions/actors to address them, produced proposed programming responses, and brought forward a number of potential partners for the implementation of the proposed interventions. The study demonstrated the importance of linking technology solutions to the socioeconomic conditions of people and that the two are inextricably linked. It is a proposed methodology for all environmental projects, not just ones related to climate change. The outcomes were sustained through a strong and effective capacity development program and an effective communications strategy, which will enhance the capacity of targeted organizations to participate in GHG reducetion initiatives; enhance awareness of environmental, health and socioeconomic costs and benefits of brick factory retrofits; and result in the implementation of sustainable and replicable emissions reduction projects.

Keywords: socioeconomic; communication; Egypt; greenhouse gas emissions

1. Introduction

An initiative recently completed in Egypt as part of a climate change project illustrates several socioeconomic and communications tools that can be used to enhance environmental security. The Climate Change Initiative (CCI) was an emissions reduction project executed through the Canadian International Development Agency (CIDA). The primary purpose of this project was to reduce the levels of greenhouse gases (GHGs) emitted during production at brick factories in Egypt. This was accomplished through the conversion of the present system, which employs heavy oil (mazot), to natural gas. A group of 50 out of a cluster of 135 brick factories in the Arab Abu Saed area within Giza Governorate was selected for this pilot project. While the reduction of GHG emissions in Egypt is an important national and global goal, it is relatively meaningless to those most directly affected by the project (i.e., the brick factory owners, workers, and local residents of the surrounding communities).

The rural area of Arab Abu Saed is situated 30 km south of central Cairo and roughly 5 km south of Helwan, a region considered as one of the most industrialized areas in greater Cairo and home to a variety of industrial activities. The Arab Abu Saed area falls between two different districts, Helwan, to the north and El Saf, to the south. The area has a cluster of 135 brick factories out of approximately 2000 all over Egypt. The Arab Abu Saed area is subjected to relatively higher levels of emissions than other villages in the impact zone, because of the concentration of brick factories surrounding the area. It is located directly in the path of year-round northeasterly winds that carry emissions from factories in Helwan and other areas in Cairo.

Project activities not only covered technical matters (related to the conversion of the fuel systems) but extended to environmental and social matters as well. In the context of this demonstration project, emphasis was accorded to the ways by which the envisaged technical improvements could facilitate or enhance the social, health, environmental, and economic conditions of those most impacted by exposure to atmospheric emissions, namely brick factory workers and the populations of villages situated in close proximity to them (Roche Ltd. Consulting Group, December 2002). A socioeconomic assessment was designed with the understanding that technical interventions are more successful when they take into account the social and economic context in which the project is executed. The socioeconomic assessment provided the key link from the technology transfer side of the project to the social development priorities of CIDA (including basic education, basic health and nutrition, and child protection), which brought out the need to identify ways by which the project could benefit vulnerable groups (such as children) that are directly impacted by the environmental problems the project sought to resolve.

The overall objective of the extensive socioeconomic analysis of the brick factory workforce and surrounding communities was to obtain a holistic understanding of the socioeconomic context, especially for vulnerable groups such as children, in which the project is executed that facilitates technological work, assesses potential social and economic benefits of the intervention, and identifies socioeconomic issues related to brick factory production and possible actions/actors to address them. The aim of the study was to collect socioeconomic and health data about workers in the brick factories as well as communities nearest to the brick factories for input into the health risk assessment, and to determine the need for and opportunity to enhance the quality of life for people in association with a technology-transfer project.

2. Methodology used in Carrying out the Socioeconomic Study

The study attempted to understand the key elements affecting employment and labor issues, living and working conditions of workers, health, and educational status, and occupational safety. It focused on understanding the gender and age-specific division of labor and the degree of exposure and occupational risks that this labor entails.

The methodology used in completing the baseline report included field work, focus groups, and research from secondary sources. A sample of 10 factories in the initial cluster of 50 were surveyed. The selection criteria included factories, which employ women and children; were mechanized in transport of bricks; demonstrated superior working and living conditions to the norm; or demonstrated inferior working and living conditions.

Information on the workers in the brick factories of Arab Abu Saed was gathered by a field survey administered in the ten factories. The total population of workers in all ten factories is estimated at 1661 individuals. In each factory about 111 workers were met, interviewed, and/or observed throughout the duration of the study. Thus, the total number of workers met, interviewed, and/or observed in all ten factories was 1114. This number represents 67% of workers for all ten factories in the survey. This one sample of workers represents 20%, which coincides with the percent of factories surveyed (10 out of 50 in the cluster).

Field visits were conducted over a period of 4 months. Semistructured interviews were conducted with local officials, community members, and key informed persons. Two of the meetings were attended with some of the brick factory owners, where socioeconomic and health data about the factory workers were obtained and verified.

Additionally, the methodology included visits to six villages because of their proximity (2–4 km) to the brick factory area and the dispersion of emissions from brick factory smoke stacks as well as proximity to each other. Three of the villages located in Giza Governorate are affected by the same northeasterly winds, which carry brick factory emissions throughout the year across the boundary into Cairo Governorate. The other three villages are in the Cairo Governorate and are home to other major polluting industries, such as cement, coke, etc. This has implications for the provision of water and wastewater services: complicated legal, financial, and administrative hurdles need to be overcome if a service being extended to neighborhoods in the impact zone in Cairo is to be extended to other neighborhoods in the impact zone, but in Giza Governorate.

Information was organized according to the following categories:

- Age cluster: workers are separated into three age groups: children are defined as persons 15 years and younger; youths are defined as persons ranging between the ages of 16 and 25 years; and adults are defined as persons over the age of 25
- Gender
- Occupational cluster: according to the stage of the brick-making process, or occupational group. This was done because the workers in each group seemed to share similar characteristics, such as geographic origin, income, and working conditions
- Survey instrument: organizing the observations and comments according to the different possible areas of intervention

The research team faced many challenges in conducting these surveys:

- Since workers are paid on a per brick basis, production was their highest priority. Therefore the field research team had to operate in ways that did not delay or pose a hindrance to the manufacturing process. Interviews with workers on the kiln (oven) had to be done during a rest period when the workers were smoking *sheeshah* (pipe) or drinking water.
- There were a few inconsistencies in the information collected indicating the inaccuracy of replies given by workers interviewed and their mixed perceptions. This has been addressed by cross-referencing with other responses from the survey sample, and with other sources of information from Egypt at large.
- Younger workers tended to be more cooperative whereas older workers were sometimes suspicious of the team's intentions. Much information provided by key informants was based on personal observation, thus accuracy could not always be verified.
- Since many of the workers are around loud machinery (brick molding conveyor belt, trucks, blowers, etc.) and are constantly moving around, it was difficult to find an appropriate time or method for the interviews. This was overcome by meeting with them during tea breaks, lunch breaks, and in the evening after working hours.

Certain research studies provided did not demonstrate the rigor of research methods deemed adequate for the team to rely on their findings.

3. Tools for Conducting the Socioeconomic Study

The process for conducting the socioeconomic assessment included the following seven tools:

- Comprehensive baseline evaluation of socioeconomic conditions of target groups
- Focus groups to analyze results and identify social issues
- Assessment of real and potential social and economic benefits of emissions reductions
- Identification of possible measures to address socioeconomic issues
- Roundtable discussions for dissemination of results and recommendations
- Reporting of roundtable discussions

- Summary of all aspects of the project.

Each of these tools is described in greater detail below.

3.1. COMPREHENSIVE BASELINE EVALUATION ON SOCIOECONOMIC CONDITIONS OF TARGET GROUPS

The baseline report provided a detailed socioeconomic portrait of the people associated with the brick factories and in the surrounding impact zone. The operations of the brick factories and the roles of workers, including women and children, were described. Living conditions, wages, health, and safety risks were identified. In addition, the report described the socioeconomic conditions of the surrounding areas, identified basic services available to residents and workers, and identified the key decision makers within those contexts (Roche Ltd. Consulting Group, April 2005).

The analysis was intended to identify social and economic issues that, though related to brick production, may not be solved solely through the reduction of atmospheric emissions. Preliminary assessment suggested that the issues of importance were human health risks related to exposure to mazot emissions, occupational safety risks associated with work in the factories, risks associated with living conditions in brick factories, and issues related to the work of children.

3.2. CONDUCTING FOCUS GROUPS TO ANALYZE RESULTS AND IDENTIFY SOCIAL ISSUES

The means used to assess the potential social and economic impacts included a combination of specialized studies and analytical focus groups. The aim of the analytical focus groups was to identify social and economic issues linked to brick factory production of relevance to the diverse stakeholders (particularly vulnerable groups) that would be unresolved by the technology transfer undertaken by the CCI project.

Three focus groups (Table 1) were conducted with: the brick owners and workers; women and children; and decision makers. The approach of these focus groups varied with the target audience. However, the main objectives were to present the results of the baseline study, identify unresolved socioeconomic issues associated with brick production that were of concern to the stakeholders, arrange these concerns in order of priority, develop consensus on these concerns, and to identify ways of dealing with the issues in a collaborative manner.

TABLE 1. Three focus groups to analyze results and identify social issues

Focus group	Title	Participants
1	Focus group with brick factory owners	Brick factory owners
2	Focus group with brick factory workers	Sample of brick factory population (labor contractors, women, children)
3	Presentation and discussion of issues with government officials and decision makers	Community leaders, brick factory owners, and decision makers

The first two focus group meetings took place over a period of 2 days while the presentation to government officials occurred in a separate meeting and included:

- A presentation of the synthesis of results of the socioeconomic study
- A presentation of the analytical framework
- An analysis of the results and recommendations through small focus group work
- Consensus on the results of group analysis and recommendations

The focus group discussions were conducted in small groups to ensure a closed small setting, where everyone felt comfortable, and where all participants had an opportunity to speak without fear of being reprimanded for their opinions. The focus groups for owners and workers were on separate days to avoid issues such as influence and intimidation. The focus group for workers included men, women, and children. Decision makers were contacted and visited individually prior to the third focus group to orient them on the topic and prepare them to address community issues.

3.3. ASSESSMENT OF REAL AND POTENTIAL SOCIAL AND ECONOMIC BENEFITS OF EMISSIONS REDUCTIONS

The assessment summarized the overall costs and benefits of emission reductions from brick factories and incorporated findings from the Environmental Health Risk Assessment (EHRA) and the cost of the new technology. It was concluded that the conversion to natural gas would result in improved air quality and reduction of risk to health of people by 99% for arsenic and benzo (a) pyrene, 95% for chromium, and 93% for cadmium, 83% for dioxins, 97% for coarse particulate matter, 99% for fine particulate matter, and 78% for sulfur dioxide; reduction of risks to terrestrial plants and invertebrates by 60–99%; reduction of risks to wildlife by 50–99%, and reduction of risks to livestock by 80–90%. This could lead to a significant reduction in cardiorespiratory illnesses and cancers, and could improve the productivity of workers as the incidence of exposure to disease causing vectors would be greatly reduced (Roche Ltd. Consulting Group, April 2005).

3.4. IDENTIFICATION OF POSSIBLE MEASURES TO ADDRESS SOCIOECONOMIC ISSUES WITHIN CCI AND BY OTHER ORGANIZATIONS

Possible measures were identified to address socioeconomic issues on two levels: within the brick factories, and in the surrounding communities. Programming responses (potential measures to improve conditions of the labor force in brick factories) that correspond to the issues identified by the different groups were identified. In general, issues related to infrastructure were identified as a governmental responsibility, including:

- Provision of clean water
- Provision of adequate emergency and regular health services
- Upgrading of wastewater systems
- Paving and lighting of main roads

Measures to address issues identified by the brick factory owners focused on strategies to bring together brick factory owners, organized in their two community-based organizations, to improve and upgrade living and working conditions for workers in the brick factories of Arab Abu Saed.

The following measures could be implemented at the brick factory community level:

- Water: Extending water supply needed for personal hygiene near bedrooms, by equipping the factories with more tanks and maintaining their hygiene by involving workers in a food for work program.
- Sanitation: Constructing cleaner toilet facilities and showers closer to bedroom quarters, using local appropriate technology and involving workers in constructing their own facilities to increase ownership and maintenance of facilities.
- Living conditions/accommodations: Building makeshift beds, spreading local floor covering, providing basic furniture as well as reducing the number of workers living in each room, either by better distribution and/or by adding more rooms in overcrowded situations. The workers would provide input to the design and provide the required labor.
- Health and occupational safety: Establishing an emergency first aid centre in the brick factory area to deal with emergencies as it can take time to

reach the nearby health unit; providing first aid emergency services to workers on the premises; launching an advocacy campaign for workers to obtain health, social, and life insurance especially for workers over 40 years who are released from work because of poor health conditions. The project would network with other nongovernmental organizations (NGOs) that have effectively begun lobbying on that issue. As well, awareness campaigns could be designed to heighten workers' preventive measures regarding diseases transmitted to them by not maintaining sanitary conditions in toilets, and the importance of keeping animals and animal pens clean as a potential source of diseases transmitted to humans.

- Nonformal learning program: This would include literacy, primary health care, recreation, culture and arts, science curriculum revolving around brick making for children working at the brick factories. It would build on existing replicable models in Egypt and along the lines of UNESCO's education for all goals as well as the millennium development goals.
- Canteen component: Piloting an intervention where women would be managers of the canteen, raising hygiene standards, quality of food, and extending credit to women to engage in this business.
- Animal treatment and safety program: Care of animals and cooperation with Brooke Animal Hospital for the provision of vaccinations and veterinary services.

3.5. ROUNDTABLE DISCUSSIONS FOR DISSEMINATION OF RESULTS OF SOCIOECONOMIC ANALYSIS AND RECOMMENDATIONS CONCERNING PROGRAMMING ACTIONS

The purpose of the event was to disseminate and share the findings of the socioeconomic analysis conducted in the brick factories and surrounding communities with the various stakeholders potentially interested in the CCI.

The objectives included the following:

- Information sharing and capacity development to promote replication by others interested in undertaking similar environmental activities.
- Promoting dialogue among different donors on issues within environmental project design, such as child labor protection, occupational health and safety, gender issues, and social protection for workers.
- Presenting the proposed programming responses shared with stakeholders, as well as with governmental and NGOs and donors with experience and interest in addressing the issues identified.

The activity involved presentations to donor groups, government officials, and other parties that may be able and interested in executing the programming responses.

3.6. REPORT ON ROUNDTABLE DISCUSSIONS

The roundtable report provided a summary of the roundtable event on "Brick-Making in Arab Abu Saed: Socio-Economic Profiles of the Workforce and the Surrounding Communities". The report highlighted the presentations made, discussions which ensued, and the recommendations offered by participants relevant to the issues presented (Roche Ltd. Consulting Group, April 2005).

3.7. SUMMARY REPORT

The report brought together and presented all components of the socioeconomic assessment conducted as part of the CCI project. It included a summary of all activities undertaken starting with the baseline socioeconomic analysis, focus groups, proposed programming responses to address the socioeconomic issues identified in the study, the roundtable event to present proposals to stakeholders, and the subsequent findings and conclusions. This report took a further step in proposing and presenting specific measures in the form of modules and potential implementation partners for such an initiative with the hope of bringing such modules into the realm of implementation (Roche Ltd. Consulting Group, April 2005).

4. Socioeconomic Assessment Findings

The specific objectives of the socioeconomic analysis had the following findings:

- Potential social issues to be addressed by the CCI project
- Potential target groups for social development actions
- Possible actions and potential partners for implementation of social development activities

Each of these findings is discussed briefly below.

4.1. THE POTENTIAL SOCIAL ISSUES TO BE ADDRESSED BY THE CLIMATE CHANGE INITIATIVE

Findings indicated that the most critical issues to be considered in reducing human health risks, occupational safety, risks associated with living conditions in brick factories, issues related to work of children, and promoting social development of those involved in, or impacted by the CCI project are:

- Improvement in the situation of children employed in the targeted brick factories through provision of basic services rendered through non-formal education activities related to their work
- Improvement of occupational health and safety measures, and accompanying education and information within the factories targeted
- Capacity development and public awareness to promote the results of the initiative and the need for replication of this work at other brick factories

4.2. POTENTIAL TARGET GROUPS FOR SOCIAL DEVELOPMENT ACTIONS

Findings collected from field visits conducted over a period of 4 months, semistructured interviews with local officials, community members, and key informed persons and focus group meetings, indicate that the focus of social development efforts should be centered on a population that may be divided into two groups: those involved directly in the on-site workplace production and those inhabiting the surrounding area in closest proximity to the brick production.

4.2.1. Factory Workers

The socioeconomic study ascertained that the labor force in brick factories is composed predominantly of men, who perform the most strenuous tasks, of children (mostly boys) involved in activities such as transportation and drying of bricks, and in some cases of women, whose tasks centre mainly on stacking bricks to dry. Further analysis focused on understanding more about gender and age-specific division of labor in the production process, the degree of exposure to chemical emissions, and occupational risks that this labor entails and the identification of gender and age-appropriate measures designed to reduce health and safety risks, and improve quality of life.

4.2.2. Inhabitants of Surrounding Area

Often those most impacted by harmful effects of pollution receive the least amount of information on the risks they are incurring and what may be done to reduce those risks. This is most probably the case for the populations living closest to the brick factory production, who in many cases are workers also employed in the factories and thus susceptible to risks associated with exposure both at work and at home. The populations that are most impacted by emissions are therefore the main target for the project measures proposed.

4.3. POSSIBLE ACTIONS AND POTENTIAL PARTNERS FOR IMPLEMENTATION OF SOCIAL DEVELOPMENT ACTIVITIES

The socioeconomic analysis investigated ways by which actions may be implemented to impact positively or significantly reduce identified risks for the target groups. The social development results obtained have become an integral part of the CCI model put forth for replication. The types of actions proposed are derived from the results of the analysis and the discussion and validation by a wider audience in the development community.

The analysis of the data and information collected through specialized studies and focus groups has identified social and economic benefits of the CCI project, as well as issues that remain unsolved but that are directly or indirectly caused or exacerbated by brick production. The socioeconomic issues relating to brick production that have been identified, will not be addressed by the technology conversion but will need to be addressed within the context of human development of the workforce of these factories. Actions suggested were those that may be undertaken to both resolve the issues and mitigate the harmful condition.

5. Capacity Development

The CCI project focused on the development of local capacities to understand and sustain the improvements that result in emissions reductions, making replication possible and desirable on a larger scale. One such element was to build the technical capacity of stakeholders who may be expected to be part of any such replication. The CCI approach to capacity development is based on an understanding that whereas there is a specific individual client for any given technical initiative, there are as well various other stakeholders whose inputs are necessary for successful project implementation. Furthermore, since donor funded programming always strives to achieve sustainability beyond the continued operation of the specific technical initiative – that is, through stand-alone replication of the tools, techniques, and process mechanisms of project development, if not the project itself – it is important to understand that the specific client is unlikely to replicate the project development process beyond his/her operations. The vehicle for this strategy was the formation of an Implementation Project Team (IPT) consisting of representatives from geographic areas with significant clusters of brick factories where potential replication could reasonably be expected. This included brick factory owners, gas companies, Governorate Environmental Management Units, regional branch offices of the Egyptian Environmental Affairs Agency (EEAA), local health units, and community NGOs (Roche Ltd. Consulting Group, December 2002).

Accordingly, two types of capacity-development were carried out to ensure sustainability: (1) client-focused capacity development associated with the skill sets required to operate the technical initiative; and (2) stakeholder (including brick factory owners, gas companies, Governorate staff) focused capacity development associated with the process of development and implementation, which can subsequently be used to assist other clients in implementing other projects. The client-focused capacity development activities involved hands-on training and support related to operation and maintenance of the new gas-fired systems. The stakeholder-focused capacity development activities included all of the significant steps and milestones involved in developing and implementing such a project as well as activities aimed at increasing the awareness of the need for such a replication, and the motivation to do so. This was carried out through focused workshop presentations and field demonstrations, with the IPT. CCI held seven IPT sessions over the course of the project, generally at milestones, to share with them the progress/results achieved to date, and to advise them of planned activities over the coming period. One of these was a field session at Arab Abu Saed so that they could observe firsthand the four technologies being tested. One of the sessions was an interactive workshop where the IPT provided advice to the project regarding potential locations for replication workshops, potential attendees, and how to contact them, as well as content and format of the presentation material. Feedback from the IPT was strongly positive, indicating that they believed they had acquired valuable capacity to engage in subsequent replication workshops (Roche Ltd. Consulting Group, September 2004).

6. Communications Strategy

The CCI communications strategy aimed at including ongoing stakeholder consultation and networking throughout the project implementation by generating discussions around identified issues and proposed responses. The strategy involved a variety of stakeholders, technical implementation partners, as well as potential collaborators in the field of social and economic development. Regular meetings and workshops helped facilitate the:

- Sharing of findings and recommendations
- Validation of the findings and feasibility of implementing the recommendations from the point of view of different stakeholders
- Reflection on how social development actions may be implemented making linkages to the CCI activities and the need for and availability of additional resources
- Identification of potential implementation partners in the community
- Identification of responsibilities and commitment, resources needed, and potential sources of funding

The project carried out a number of analytical focus groups with the brick factory owners, workers, women, and children in order to assess the potential social and economic impacts of the CCI project. The aim was to identify issues of priority on site and in the surrounding community that were linked to brick factory production. These issues included those of relevance to the diverse stakeholders that would not be resolved by technology transfer and proposed socioeconomic responses.

Focus group discussions were conducted to supplement field work, interviews, and research from secondary resources. Separate discussions were scheduled for each group: owners, workers, women, and children, in order to allow each group to voice their opinion without feeling intimidated. The focus groups identified the most significant issues/needs of the representative groups in order to improve their socioeconomic conditions, and extended to community concerns affecting the community at large and not just the brick factories in Arab Abu Saed. The process subsequently moved towards presenting findings to local government officials.

A number of stakeholders were found to be impacted by the same issues such as water, being an issue for the owners, workers, and the community. Efforts were invested in mobilizing these stakeholders around social development issue. This was done by identifying stakeholders impacting/impacted by socioeconomic issues; identifying common interests; mobilizing a variety of resources to address the issues (i.e., resource mapping,), and adopting a holistic process to leverage resources needed to improve social conditions of the disadvantaged groups. Implementing partners would complement these efforts by providing missing financial resources and technical assistance. This could be accomplished through a nonprofit, community-based organization.

The CCI communication strategy was based on using the results from the technology transfer and the associated studies of the environmental health risk assessment and the socioeconomic assessment, to provide relevant information and tools to stakeholders that would promote the replication of the project in other regions of Egypt, where brick making is a significant industry (Roche Ltd. Consulting Group, September 2004). Technology transfer refers to natural gas combustion technology transferred to the cluster of brick factories to achieve reductions in GHG emissions. The general criteria for technology transfer considered in the CCI project related to broad developmental and environmental criteria that were part of the overall project design and are as follows:

- Technologies must be environmentally sound
- They must have GHG reduction potential
- Cost effective
- Support sustainable development
- Transfer must include capacity development
- Transfer must involve a participatory approach (Roche Ltd. Consulting Company, December 2002)

Environmental, human health, and socioeconomic aspects (gender disaggregated) of brick factory emissions were communicated to targeted organizations with advocacy potential for replication. The need for and the cost-benefits of converting brick factories from mazot to natural gas were communicated to stakeholders in other regions with potential for replication.

Risk assessments were conducted to assist the people in understanding the importance of conversions on human health and to motivate them to replicate the project. Studies completed formed the basis of the communication material, and replication was promoted through focused workshops with the different stakeholders. Communication and promotion focused on targeting specific audiences with specific messages as the communication approach was fashioned according to the target group. For example, the health risk impact data targeted local health units and environmental NGOs while the community socioeconomic impact data targeted community services organizations and brick factory owners.

Different information was presented in focused workshops and meetings with the different stakeholders and beneficiaries (children, women, and youth of the 50 brick factory workers in Arab Abu Saed) of the project in order to familiarize them with the steps involved in the execution of such a conversion program. Specific elements pertinent to the self-interest of specific IPT representatives and their constituencies were presented, such as environmental and regulatory implications of emissions on human health and the costs to the health-care system associated with the emissions and implications of brick factory operations on the workforce and surrounding communities. It is important to note that the workshops were intended to engage the IPTs to assist in the development of targeted communications plans for their communities and constituencies as well.

The data and information collected were synthesized and presented in such a way as to facilitate their use by other actors, interested parties, and stakeholders in the development community. The response received from the community was a further indication of the willingness of civil society organizations to participate in holistic projects affecting the lives of people and their willingness to put forward their resources to work collaboratively on something which is not necessarily on its core program. Some of the actions suggested were implemented within the context of CCI while others may interest organizations (governmental and NGOs, donors involved in types of issues identified) with the mandate, experience and resources necessary to plan and execute the actions. Brick factory owners have been shown the importance of participating in actions and decisions related to implementation, as they are the primary beneficiaries from improved conditions of their labor force. As has been experienced in similar projects, once the public becomes aware of such positive impacts, they become a strong advocacy voice promoting replication of these success stories.

7. Conclusions

On one hand the socioeconomic assessment conducted at the forefront provided input into the risk assessment, while on the other hand, the risk assessment provided an added value to the socioeconomic assessment with respect to the programming responses. The socioeconomic assessment enabled a holistic understanding of the context within which the CCI project was to be executed. It identified issues related to brick factory production and possible actions/ actors to address them, produced proposed programming responses and brought forward a number of potential partners for the implementation of the proposed interventions. The assessment led to the identification of socioeconomic issues related to brick production, which would not be addressed by the technology conversion, but which needed to be addressed within the context of human development of the work force of these factories. The study demonstrated the importance of linking technology solutions to the socioeconomic conditions of people and that the two are inextricably linked. It is a proposed methodology for all environmental projects, not just ones related to climate change.

On a project level, sustainability usually refers to sustaining project outcomes. The CCI outcomes are sustained through a strong and effective capacity development program as previously described and an effective communications strategy, which will enhance the capacity of targeted organizations to participate in GHG reduction initiatives; enhance awareness of environmental, health, and socioeconomic costs and benefits of brick factory retrofits; and result in the implementation of sustainable and replicable emissions reduction projects.

Acknowledgments

The case study presented above has been the outcome of my work with the Climate Change Initiative (CCI), a Canadian Project funded by the Canadian Climate Change Development Fund (CCCDF) and executed through the Canadian International Development Agency (CIDA).

Special thanks and acknowledgment goes out to Mr. Richard Szudy, President of IDEA-Egypt to whom I am indebted for his continued support and guidance throughout all the years of working together. His energetic enthusiasm for my work with socioeconomic issues never lagged and is greatly appreciated. I am grateful to Ms. Ruth Hull, Senior Scientist at Cantox Environmental for inviting me to attend the workshop, and providing me with the opportunity to participate in the theme of the workshop by presenting this case study on Egypt. I sincerely thank both Mr. Richard Szudy and Ms. Ruth Hull for their comments, suggestions, and editorials that further enhanced this paper. Very special thanks to Dr. Laila Iskander, Managing Director of Community and Institutional Development (CID) and her team for carrying out the socioeconomic analysis for the CCI project and for continuing to find ways to implement some of the programming responses identified.

References

Roche Ltd., 2002, Consulting Group, Climate Change Initiative, Project Implementation Plan, Vol. 1, December 2002.

Roche Ltd., 2005, Consulting Group, Climate Change Initiative, Project Progress/Performance Report, October 2004–March 2005, April 2005.

Roche Ltd., 2004, Consulting Group, Climate Change Initiative, Project Workplan, August 2004–end of project, September 2004.

12. SENSORS AND COMMUNICATIONS IN ENVIRONMENT MONITORING SYSTEMS

WALDEMAR NAWROCKI* AND TADEUSZ NAWALANIEC
Poznan University of Technology, Institute of Electronics and Telecommunications,ul. Piotrowo 3A,PL-60965 Poznan, Poland

Abstract: The chapter describes a study of a distributed system for monitoring important physical and chemical quantities of the environment, including meteorological information. The monitoring system consists of: measurement stations with sensors, communication interface, and system controller with data acquisition, data processing, storage, and presentation. From a technical point of view, it is necessary to know the dynamics of the whole monitoring system and its reliability. The sensors measure air and water quality and other parameters like water levels. Monitoring of infrastructure (road, power and communications networks, sewage systems) is as important as natural environment monitoring for population security. We present a traffic monitoring system as an example of infrastructure monitoring.

Keywords: monitoring; environment; distributed measurement system; sensors

1. Introduction

Environmental monitoring is an important tool for maintaining environmental security and sustainability. Both the natural environment – comprising air, potable water reserves, soil, the natural electromagnetic field, etc. – and the man-made environment – referred to as infrastructure and including road, power and communications networks, sewage systems, generated electromagnetic fields, etc. – are important for humans. A growing number of parameters of both these

* To whom correspondence should be addressed. Waldemar Nawrocki, Poznan University of Technology, Institute of Electronics and Telecommunications,ul. Piotrowo 3A,PL-60965 Poznan, Poland; e-mail: nawrocki@et.put.poznan.pl

R. N. Hull et al. (eds.), Strategies to Enhance Environmental Security in Transition Countries, 153–165.

environments are monitored for environmental security. Natural environment parameters include physical or chemical quantities such as air temperature, river water level, metal content in water (including mercury, lead, or other heavy metals), radiation, and many others. Important infrastructure parameters include, among others, road traffic intensity (especially the formation of bottlenecks), performance of power and water supply systems, and potable water quality parameters related to the water supply system. Note that in metropolitan areas an efficient infrastructure is as important as a clean natural environment. Cities such as Warsaw, New York, or Bucharest could not function properly without power, water supply, sewage, and communications systems. Therefore, infrastructure monitoring is as important as natural environment monitoring for population security.

An environmental monitoring system uses a number of sensors for physical or chemical quantity measurements, a communications network, and a controller (Figure 1).

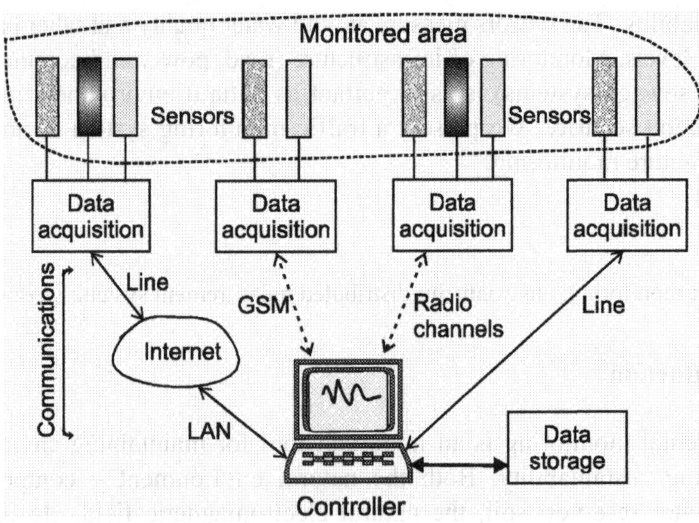

Figure 1. Environment monitoring system.

Sensors are used for data acquisition. Automated monitoring systems use sensors with electric output signals: examples include electric sensors for temperature, pressure, wind speed, and wind direction; laser range finders; lightning detectors; vehicle detectors; and many others. Also, monitoring system sensors include image converters and infrared detectors built in visual and infrared cameras, respectively. Visual cameras are used for monitoring traffic intensity and detecting traffic jams or conditions of increased risk. They can also be useful in flood risk evaluation as well as in crime detection in public places. Infrared, or thermal, cameras are used for detecting high-temperature areas. Information acquired by means of such cameras is used in fire risk evaluation or for detecting a risk of machine failure in systems of social importance.

The acquired data are processed in the controller. The results are visualized, and the data stored for long-term analyses and statistics. Each segment of the system has an effect on the quality and usefulness of the information acquired.

Environmental monitoring systems are seldom fully automated. Some of the sensors used for environment parameter measurements are not electric and thus require visual reading, necessary, for example, in the case of nonelectric rain or staff water gauges, or results of chemical analysis of water quality.

The dynamics of processes occurring in the natural environment tend to be low enough for available means of communications to provide online reading of all the data from a single measurement point. However, a monitoring system typically uses data acquired at a number of measurement points (e.g., collected from 1000 stations). The time necessary for communication with each station within a data collection cycle must be taken into account in the design of such multisensor monitoring systems. Enhanced communications are necessary in storm monitoring systems, designed to acquire information on the formation and movement of storms with atmospheric discharges, as well as in systems for monitoring technical infrastructure, especially power grid operation and city traffic in rush hours.

2. Communications in Monitoring Systems

Monitoring large surface areas, covering thousands of square kilometers, involves the use of distributed monitoring systems. Such systems can use communications networks based on various wire or wireless technologies. Data transfer in distributed measurement systems used for monitoring is performed on a bit-by-bit basis. The required rate of data transfer in a monitoring system depends on the dynamics of the monitored processes as well as on the amount of data acquired in and transferred from a single measurement point within a data collection cycle. The total amount of data depends on the number of measurement points and on the size (expressed in bytes) of a single data message.

The following means of communications (Nawrocki, 2005) are used in natural environment and infrastructure monitoring systems:

- Public switched telephone network (PSTN)
- Radiocommunication (based on radiomodems)

- Computer network and Internet-based communication (currently tested, not yet in common use)
- GSM mobile phone network.

PSTN-based communication is suitable for systems monitoring areas covered by the telephone network (Figure 2). A measurement station and the controller can be connected by a leased line or by a switched one; in the latter case communication requires a dial-up similar to that preceding a phone call. Though using a switched line is cheaper, some systems use leased lines for security reasons. A leased line also has the advantage of providing enhanced reliability of operation of the measurement/monitoring system involved, by eliminating the risk of connection failure due to a busy signal, securing stable connection quality, and saving the time otherwise necessary for switching.

The rate of digital data transfer in a telephone network depends on the network quality: analogue networks provide data rates up to 56 kilobits per second (kbps), while digital networks (ISDN) allow data transfer of hundreds of kilobits per second. The main standards for digital data transfer (through a modem) in telephone networks are RS-232C and RS-530, together with RS-485.

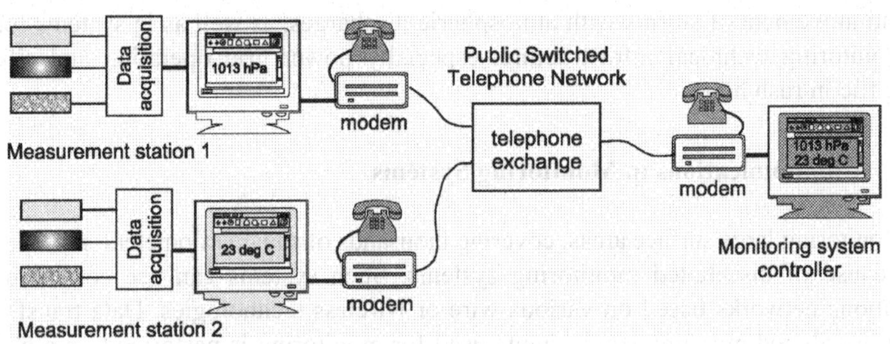

Figure 2. Monitoring system using a PSTN for communication.

Radio communication is a good solution when measurement stations are far from telephone lines or when access to them is difficult. A radio communication-based monitoring system is shown in Figure 3 (Nawrocki, 2005). A radio communication-based system consists of radiomodems, or transceivers equipped with modulators and demodulators for processing output and input signals, respectively. Monitoring systems of national importance use special bands allocated to civil government institutions for radio communication. Other systems use licensed bands allocated to industrial radio communication, with frequencies from the 430–860 MHz band being the most commonly used.

Radio communication-based systems allow transfer rates up to 20 kbps, but actual rates tend to be lower. The communication range, typically up to 50 km, is determined by the radio transmitter power output and the antenna height.

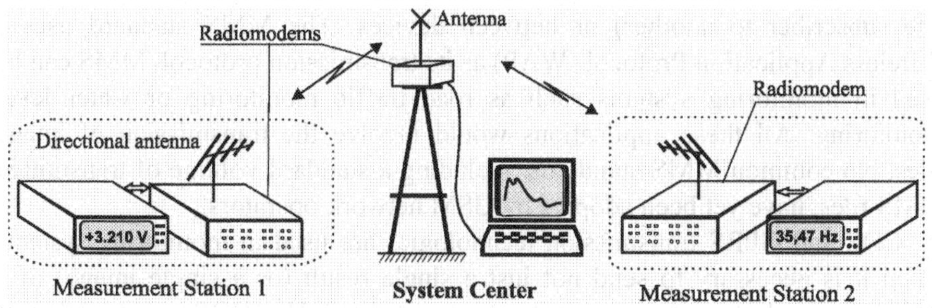

Figure 3. Radiocommunication-based monitoring system.

A cellular telecommunication system (GSM) has many advantages when used as the interface in systems for monitoring of the environment (Nawalaniec, 2003, Mackowski, 2006). GSM systems offer the following digital data transmission services:

- Short message service (SMS), which is the transmission of alphanumeric messages of up to 160 characters
- Multimedia messaging service (MMS), which is the store-to-forward transmission of text, graphic, and sound files
- Circuit switched data (CSD) transmission, which is the switched transmission of digital data with speeds up to 9.6 kbps via traffic channel; an option of the CSD mode is a high speed circuit switched data (HSCSD) with speeds up to 57.6 kbps
- General packet radio service (GPRS), a packet data transmission mode

Where there is a very light channel load, SMS is the cheapest mode of data transmission. A delivery report can also be sent from recipient to sender, to verify that the message has been delivered. In the case of a lack of communication with the recipient at the moment of transmission, the message is stored in an SMS center, and forwarded to the recipient after communication is reestablished. The typical message delivery time is a few seconds (5 s in our measurements – Nawrocki, 2005) from the moment of sending. However, delivery delays can be much longer. A message can be delivered after several hours or days, and in occasional cases can remain undelivered (which was the reason for creating the delivery report option). Our investigations show that on

New Year's Eve, several messages were not delivered (Nawrocki, 2005), which is a bad factor for the reliability of the monitoring system.

MMS is a device-to-device transmission method, like SMS, of multimedia files via the GSM network. Files can be transferred between users (i.e., from one subscriber to another) or between devices. The MMS standard uses a Wireless Application Protocol (WAP) as its transmission protocol. MMS can be used in monitoring systems such as road traffic monitoring or water level monitoring. All these applications would involve the transmission of image files. No common MMS standards, including a standard volume of transmitted MMS files, have yet been adopted by GSM network operators.

CDS and GPRS transmission technologies are used in monitoring systems when it is necessary to send not just a single result (or a single image) of a measurement but a stream of measurement data. The CDS transmission is online, without any delay, but its speed is rather slow – only 9.6 kbps. The GPRS is based on packet (file) switching instead. A GPRS session can be activated in the "always connected" mode, and data can be transferred during phone calls without interference. Each packet, or set of data transferred, is an integrated whole, and can be transmitted independently of the other packets. A great advantage of GPRS is the theoretical speed of data transmission – up to 115.2 kbps. As measurements show, GSM networks (Figure 4) can currently transmit data at a maximum speed of 26.4 kbps (GPRS) and 28.8 kbps (HSCSD) for data transmission and at a speed of 53.6 kbps (GPRS) and 57.6 kbps (HSCSD) for data receiving (Mackowski, 2006).

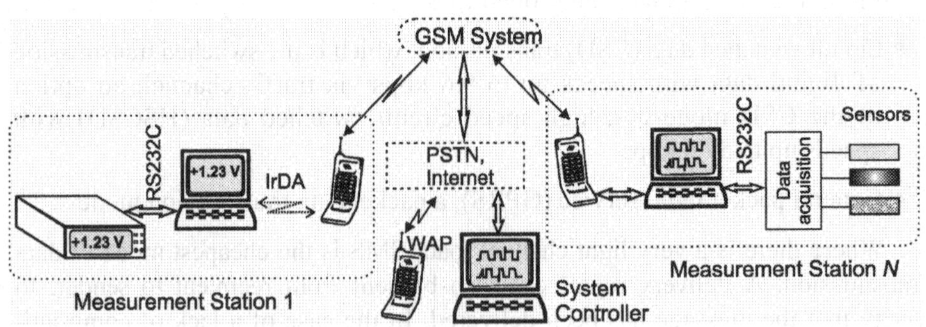

Figure 4. GSM mobile phone network-based monitoring system.

3. Weather and Water Level Monitoring in Poland

An example of a nationwide environment monitoring system is the system used for monitoring weather and water levels in Poland (Figure 5), operated by the state

Institute of Meteorology and Water Management, and covering the area of Poland, or 312,000 km^2.

The system comprises the following six subunits (IMWM, 2006):

- Weather monitoring system, with 211 weather stations
- System for monitoring the water level in rivers, lakes, and other water bodies, with 1000 gauging stations
- Radar system for monitoring clouds and waves of precipitation, with eight radars
- Lightning monitoring subsystem for detecting and registering atmospheric discharges, with nine measurement stations;
- System for monitoring air radioactivity, with nine measurement stations
- Satellite system for monitoring based on images using waves in or beyond the visible light range (photographic or infrared imaging, respectively)

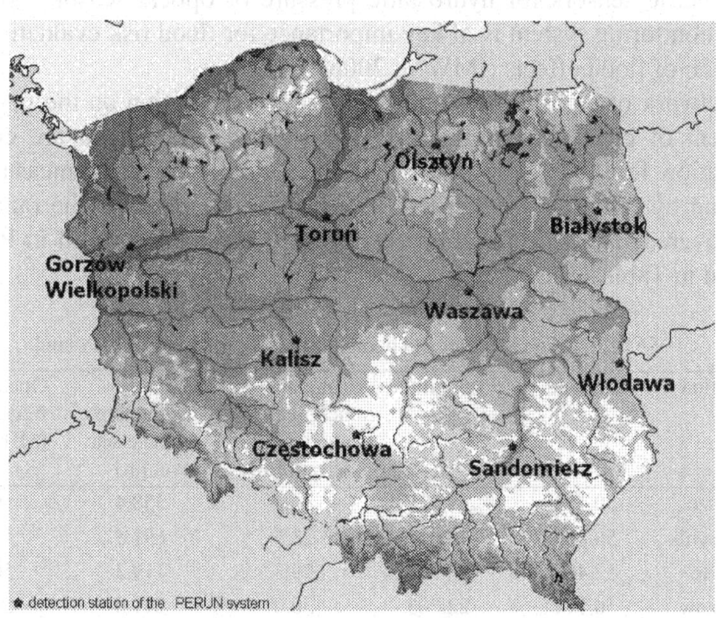

Figure 5. Geographic map of Poland.

The network of 211 weather stations allows measurement of tens of basic weather parameters, the most important of which include air temperature, atmospheric pressure, air humidity, wind speed and wind direction. Among the other parameters measured are cloud base height (laser-measured), amount of precipitation (rain or snow), visibility, and air temperature near the ground. Measurements are performed in enclosed areas within the weather stations.

The basic equipment of a weather station is comprised of a facility referred to as Meteorological Automatic Weather Station (MAWS). A MAWS is equipped with electric sensors: a Pt100 resistance sensor for temperature, a capacitance sensor for barometric pressure, a semiconductor sensor for air humidity, sensors for solar radiation and for visibility, a pulse sensor for wind speed, and a wind direction indicator with a Gray code disk (IMWM, 2006).

Besides a MAWS, a weather station uses a number of other devices: a laser ceilometer for determining the height of the cloud base; a rain gauge (electric or pulse), a radiation gauge, etc. Recognized manufacturers of meteorological instruments include Vaisala (Finland), Impulse Physik (Germany), and Young (USA). The weather stations in Poland use MAWS instruments manufactured by Vaisala (IMWM, 2006).

Hydrological monitoring is based on data collected from 1000 gauging stations in rivers, lakes, and water reservoirs. Information on the water levels is supplied in 10 min cycles. Water level gauges used by hydrological stations are either electric sensors for hydrostatic pressure or optical sensors. The hydrological monitoring system is of key importance for flood risk evaluation and for prevention of flood effects (IMWM, 2006).

A network of eight radar stations provides information on the position and movement of clouds, as well as on wind speed. These data are crucial for precipitation forecasts. The radars use the Doppler effect to measure clouds speed and by this way wind speed. Their observation range (the radius of the circle covered) is 200 km. Parameters of the radar stations used in Poland are specified in Table 1.

TABLE 1. The location and parameters of weather radars in Poland

Location	Latitude	Longitude	Antenna height above average terrain [m]	Antenna height above average sea level [m]	Operating frequency [MHz]
Orzesze	50.152	18.727	36.4	358.4	5660
Pastewnik	50.883	16.04	23.5	691.5	5660
Warsaw	52.40	20.931	29.3	119.3	5650
Rzeszow	50.114	22.003	30.0	241.0	5650
Poznan	52.41	16.832	35.0	123.0	5660
Swidwin	53.79	15.831	30.0	146.0	5660
Gdansk	54.384	18.456	20.0	158.0	5650
Brzuchania	50.394	20.08	35.0	453.0	5660

The thunderstorm monitoring system is very important, and is used for determination of the location and intensity of atmospheric discharges, or lightning strikes. Its nine stations distributed over the territory of Poland detect and register intracloud and cloud-to-ground discharges. The detection of intracloud discharges is of use for forecasting cloud-to-ground discharges. The site of the discharge is determined with resolution ranging from 1 km to 4 km, like that of radar system measurements.

Data on the intensity of lightning as well as on storm movement is of use for aviation, the power industry (for assessing the need of emergency power cut-off), as well as for high-altitude work management.

4. Data Flow Management in the National Monitoring System

The general scheme of the Information System of the National Environment Monitoring (ISNEM) is presented in Figure 6. National Environment Monitoring (NEM) collects, processes, and provides access to such data as measurements of characteristics of particular environment components, measurements of some emissions, and measurements of natural characteristics, e.g., measure indicators of air quality such as SO_2, NO_2, O_3, CO reading from the Meteorological Automatic Weather Station; meteorological indicators, e.g., noise or water surface

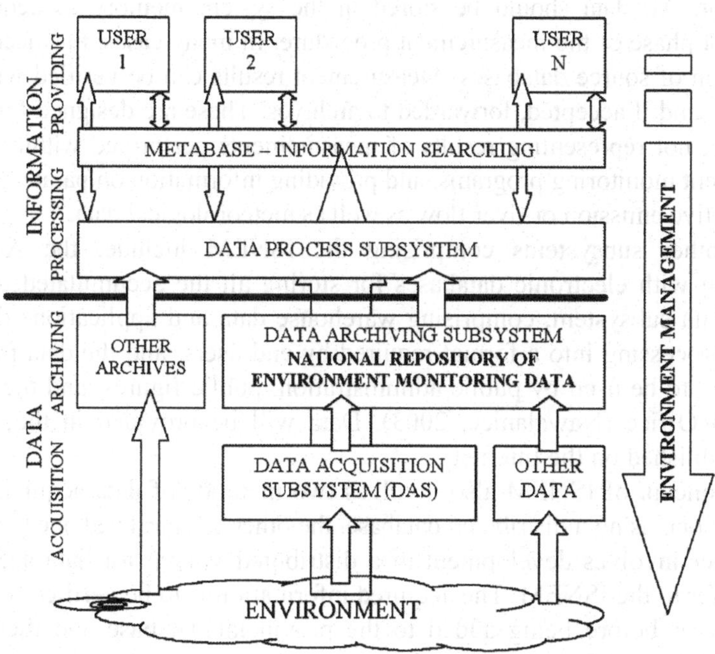

Figure 6. General functional scheme of ISNEM

indicators including water level in different river legs, which can be helpful in dealing with crisis situations. This information is collected by Provincial Environment Protection Inspectorates (PEPI), their sections and also research institutes and universities, which the Environment Protection Inspection (EPI) cooperates with. An ISNEM for data and information storage should be created in Poland to provide wide, easy and inexpensive access to data, as well as efficient updating (Kraszewski, 2001).

Below we present an ISNEM consisting of four interlinked subsystems, the correlation of which is schematically depicted in Figure 6. The four subsystems include:

- A Data Acquisition Subsystem (DAS)
- An Archiving Subsystem
- A Process Data Subsystem
- A Data Provision Subsystem

The DAD, comprises mechanisms of data acquisition from the environment by means of measurement devices, observation, or sampling and laboratory analyzes. A procedure of data input should be developed for every environmental component. All data should be stored in the system memory as acquired in the earliest phase of the measurement procedure. In many cases, this necessitates the creation of source databases. Measurement results can be verified within the subsystem and if accepted, forwarded to archives. These are designed for storing other data, not representing results of measurements performed within national environment monitoring programs, and providing information on parameters such as radioactive emission or river flow as well as meteorological data.

The other subsystems comprising the ISNEM include: the Archiving Subsystem with electronic databases for storing all the accumulated data; the process data subsystem, comprising warehouse data and applications designed for data processing into a format required by end users; and the data provision subsystem, to be used by public administration, public figures, and the Central Statistical Office (Nawalaniec, 2003). Data will be provided in the form of reports published on the Internet.

The concept of ISNEM also involves two methods of data acquisition and archiving: one using a distributed database, the other a centralized one (Figure 7). The former involves development of a distributed system for data acquisition and transfer to the ISNEM. The acquired information is to be verified by a PEPI administrator before being added to the provincial database and then to the central database. If environmental measurements and observations are only conducted on the national basis, a single centralized database is to be used.

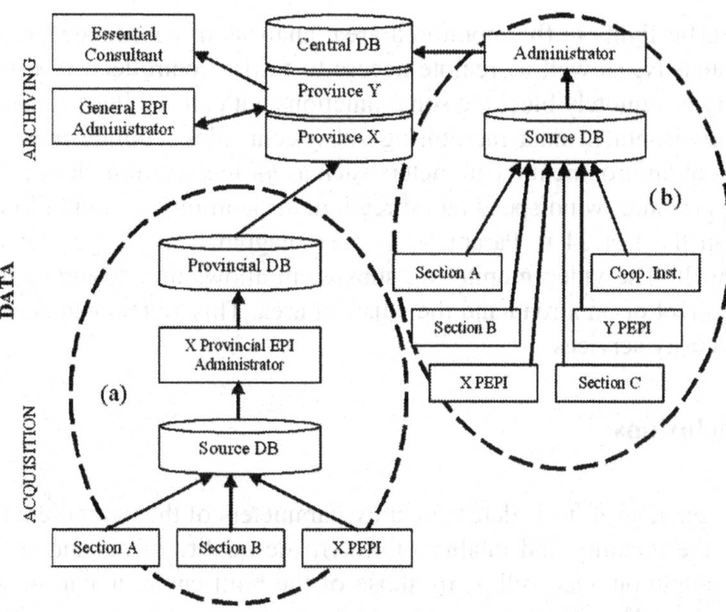

Figure 7. Organization of the data acquisition and archiving subsystems: (a) distributed and (b) centralized.

Regardless of the organization of the data acquisition and archiving subsystems, part of the information will come from automatic measurements of air and water pollution concentrations at a number of measurement stations, and from automatic water level gauges in individual river segments, which are of use for flood management. Because of considerable distances between measurement stations and data acquisition centers in this distributed system, the best solution for communication is a nationwide 2G cellular phone network, covering over 96% of the country's area.

5. City Traffic Monitoring

Three kinds of monitoring subsystem have been integrated into a single supervisory control and data acquisition (SCADA) system in the Ostrow Wielkopolski district of Poland. The system consists of the following units (Nawalaniec, 2006):

- A subsystem for supervising junction traffic lights in the city
- A subsystem for monitoring environment data
- A video monitoring subsystem

The district SCADA system is mainly designed for traffic management. The traffic light subsystem allows online monitoring of the operation and status of all the traffic lights in the monitored area, analysis of coordinated signals and traffic intensity, as well as remote access to traffic controllers. It also gives a possibility of remotely blocking some junctions with all-red lights, if necessary.

The environment data monitoring subsystem allows online monitoring of the status of environmental parameters such as air temperature, dew point, wind chill, air pressure, wind speed and direction, air humidity, or rainfall in the last hour or in the last 24 h. Parameter history diagrams can be generated at any time as well. The video monitoring subsystem allows monitoring of the situation in a street or on a road and the adjacent area. This system can be of use for public security services.

6. Conclusions

Environmental security is determined by parameters of the natural environment (such as the quantity and quality of water, the quality of air and soil, or the weather conditions) as well as by those of the built environment, or technical infrastructure. Therefore, parameters of both types of environment should be involved in environmental monitoring aimed at maintaining environmental security and sustainability. Parameters such as road traffic intensity should be monitored along with air pollution, flood, or hurricane indicators.

Environment monitoring is based on the use of sensors and communication systems, and is vital for environmental security. Though the dynamics of environmental processes is assumed to be relatively slow in comparison with the average performance of high-speed communication systems, more thorough studies show that the operation of monitoring systems using a number of field measurement stations involves time limits that should be taken into account for the system to operate efficiently. Besides the number of field measurement stations, the dynamics of a monitoring system is also affected by the number of sensors used by each station, and, obviously, by the rate of data transfer in the communication system employed.

Natural environment and infrastructure monitoring is very important for environmental security. Note that only the infrastructure – the built component of the environment – besides being monitored can also be controlled and modified by humans. The natural environment can only be monitored, and its status, e.g., the weather conditions, taken into consideration in action planning.

References

Institute of Meteorology and Water Management (IMWM), 2006, Warsaw, www.imgw.pl

Kraszewski A., Lobocki L., 2001, *The Idea of Information System of the National Environment Monitoring*, Final Report, Vol. 2.0 with supplement, Technical University of Warsaw 2001, Warsaw (in Polish) (unpublished).

Mackowski M., 2006, Application of the measuring systems with data transmission in GSM cellular network and Internet, *Kwartalnik Elektroniki i Telekomunikacji* 52: 245–266 (in Polish).

Nawalaniec T., Urbaniak A., 2003, *Distributed System for Data Acquisition and Transfer to Information System of the Environment Monitoring*, Proceedings of ICC Conference, Tatranská Lomnica, pp. 210–214.

Nawalaniec T., 2005, unpublished report.

Nawrocki W., 2005, *Measurement Systems and Sensors*, Artech House, Boston/London.

13. TEACHING AND RESEARCH IN ENVIRONMENTAL MICROBIOLOGY/BIOTECHNOLOGY TO ENHANCE ENVIRONMENTAL SECURITY IN TRANSITION AND OTHER COUNTRIES

ZDENEK FILIP, KATERINA DEMNEROVA*
Department of Biochemistry and Microbiology, Institute of Chemical Technology Prague; Technicka 5, 166 28 Prague 6, Czech Republic

Life is first and foremost a microbial phenomenon.
F.M. Harold (2001)

Abstract: In European countries that suffered under totalitarian regimes for several decades but also in some other regions of the world, deteriorated environmental quality endanger the existence of several species, and the well-being of human population, too. In this chapter we point out the key role of microorganisms in the ecological processes, which are of basic importance for terrestrial life, and emphasized both the necessity and possibilities of the utilization of environmental biotechnology to improve our environment. Through an extensive education and training at school and university levels, not only a microbial literacy should be enhanced but especially the public awareness of the fact that the sustainability of human life depends widely on better understanding and systematic governing of microbial activities in natural and anthropogenic environments. Priority scopes and individual topics in teaching environmental microbiology/biotechnology have been suggested.

Keywords: environmental microbiology/biotechnology; environmental security; transition countries

*To whom correspondence should be addressed. Katerina Demnerova, Department of Biochemistry and Microbiology, Institute of Chemical Technology Prague; Technicka 5, 166 28 Prague 6, Czech Republic; e-mail: katerina.demnerova@vscht.cz

R. N. Hull et al. (eds.), Strategies to Enhance Environmental Security in Transition Countries, 167–184.

1. Introduction

Since 1989/90, several Central European and Eastern European countries have largely adopted market-oriented economies, implying major changes in their production and trade systems, and sometimes also a variation within European boundaries. Due to these changes, environmental issues hidden for at least four decades became a matter of serious discussions. Climate change, air, water, and soil pollution, deforestation, loss of biodiversity, desertification, stratospheric ozone depletion – all these environmental issues mainly caused by anthropogenic activities showed negative consequences on land productivity, food and fodder supply, and freshwater quality. In some transition countries the quality of the environment culminated to a critical point because of a long-term politically caused environmental ignorance (Mejstrik, 1993; Moldan and Billharz, 1997). Many of the problems transcended national boundaries, and thus an international cooperation is ultimately required now in order to improve the unfavorable situation. Also, to avoid a risk of serious environmental and perhaps also social crisis, a sole anthropocentric view on the life security should be transformed in favor of an ecocentrism view. This is because the terrestrial ecosystems in their entity are undoubtedly capable of a sustainable existence even without being populated by humans. However, the existence of humans on this planet can flourish only upon a narrow limit of physicochemical and biological conditions. Thus, no sustainable existence of human population appears possible without a health terrestrial ecosystem (Filip, 2004).

2. Environmental Situation in a Former "Black Triangle" Region and some Other Countries

In Central Europe, the western part of the former Czechoslovakia, now the Czech Republic, the eastern part of the former German Democratic Republic, now the "New Länder" – part of the Federal Republic of Germany, and the southern part of Poland, all of them heavily industrialized regions, were commonly designated as a "Black Triangle". In the Czech Republic, e.g., in this and a few other heavily industrialized regions, about 55% of the total population was concentrated, and 96% of fuels, 94% of metallurgical products, 71% of chemicals, and 86% of electrical power was produced (Mejstrik, 1993). Emitted by the industry, the regions received a permanent flux of pollutants such as SO_2, NO_x, O_3, and heavy metals for several decades. The main source of the emissions was a low quality brown coal, which was used as the main energy source. Still in 1990, brown coal covered more than 65% of the country's total energy demand. In Table 1 the amounts of two capital emissions in the "Black Triangle" and some other European countries are listed.

Data in Table 1 show that emissions were much lower in Western European countries in comparison to the "Black Triangle" countries.

TABLE 1. Emission of sulfur dioxide and nitrogen oxides (in 10^3 t) in some European countries in 1990. (From Mejstrik, 1993.)

Country	Sulfur dioxide	Nitrogen oxides
Czechoslovakia	1400	1250
German Democratic Republic	2620	1005
Fed. Rep. of Germany (old)	530	2860
Poland	2250	1620
Belgium	210	300
Denmark	130	254
France	667	1772
Norway	33	226
Switzerland	31	184

There is no doubt that the total atmospheric deposition of pollutants is a value of high importance from the point of view of environmental toxicology and human health. The adverse effects of air pollution became apparent in at least 60% of the Czech forests. Acidity of forest soils was increased, nutrients were washed out, toxic elements became released, and thus plant available and the whole forest ecosystem became significantly degraded. A considerable loss in a natural resistance against external harmful effects, such as insects and pathogens was observed (Mejstrik, 1990). Even more, up to 50% of higher plant and vertebrate species in total became endangered in the Czech Republic, the incidence of respiratory diseases in both children and adults was growing steadily, the human population in total demonstrated deficiencies in their immune system, and the mortality, especially in males of 45–49 years, increased by 39% in comparison to 1964 (Moldan, 1990; Mejstrik, 1993). It has been stated that without international assistance and cooperation, the "Black Triangle" region would be doomed to permanent instability and internal destruction both economically and socially due to a heavy destruction of natural environments (Mejstrik, 1993).

A similar environmental situation, not necessarily with the same acute risk for humans, appeared in other European transition countries and in part even worldwide. The North Atlantic Treaty Organization (NATO) Committee on the *Challenges of Modern Society* adopted a syndrome approach as developed by the German Government's Advisory Council on Global Change, and the Potsdam Institute for Climate Impact Research. This approach provides a number of patterns of environmental stress identified on an international scale (Lietzmann and Vest, 1999).

In Table 2 the syndrome-based concept is summarized, which is based on an assumption, a repeatedly confirmed one, that environmental stress is part of dynamic interactions between humans and nature. Different types of such interactions occur in various environmental and geopolitical regions of the world, and they are divided into three subgroups "resources utilization", "development", and "sinks".

TABLE 2. An overview of global change syndromes. (From Lietzmann and Vest, 1999.)

Syndromes	Causes and/or effects phenomena
Utilization syndromes	
Sahel Syndrome	Overcultivation of marginal land
Overexploitation Syndrome	Overexploitation of natural ecosystems
Rural Exodus Syndrome	Environmental degradation through abandonment of traditional agricultural practices
Dust Bowl Syndrome	Nonsustainable agro-industrial use of soils and water bodies
Katanga Syndrome	Environmental degradation through the extraction of nonrenewable resources
Mass Tourism Syndrome	Destruction of nature for recreational ends
Scorched Earth Syndrome	Environmental destruction through war and military actions
Development syndromes	
Aral Sea Syndrome	Environmental damage of natural landscapes as result of large-scale projects
Green Revolution Syndrome	Environmental degradation through the introduction of inappropriate farming methods
Asian Tigers Syndrome	Disregard of environmental standards in the course of rapid economic growth
Favela Syndrome	Environmental degradation through uncontrolled urban growth
Urban Sprawl Syndrome	Destruction of landscapes through planned expansion of urban infrastructures
Major Accident Syndrome	Singular anthropogenic environmental disasters with long-term impacts
Sink syndromes	
Smokestack Syndrome	Environmental degradation through large-scale diffusion of long-lived substances
Waste Dumping Syndrome	Environmental degradation through controlled and uncontrolled disposal of waste
Contaminated Land Syndrome	Contamination of environmental assets at industrial locations

The importance of the syndrome-based approach for scientists and policy makers is that it may serve as a starting point for the development of the respective indicators, and for early intervention in order to avoid escalation of the environmental problems, and possibly also of related conflicts.

3. Microorganisms as the Basis of Life

Microorganisms are the basis of the entire biosphere. They represent the well-spring from which all other life has arisen, and a necessary support system for a continued existence of life on earth. It has been estimated that there are about 5 $\times 10^{31}$, (in weight over 50 quadrillion metric tons) of microbial cells on this planet. It means that microorganisms constitute more than 60% of the biomass of the whole biosphere. Plants and animals definitely emerged within a microbial world and have retained an intimate connection with, as well as dependency upon microorganisms. Humans are definitely latecomers to this planet, and even more, over 90% of the cells in a human body are micro-organisms (AAM report, 2004). Bacteria and fungi heavily populate our intestine, skin, mouth, and other orifices and maintain benign or even beneficial relationships with individual body organs. In this way, they influence both our physical and mental health.

Recent investigations of the interrelatedness of microbes brought to the light the model of the evolutionary relatedness of all life on earth – the tree of life (Figure 1).

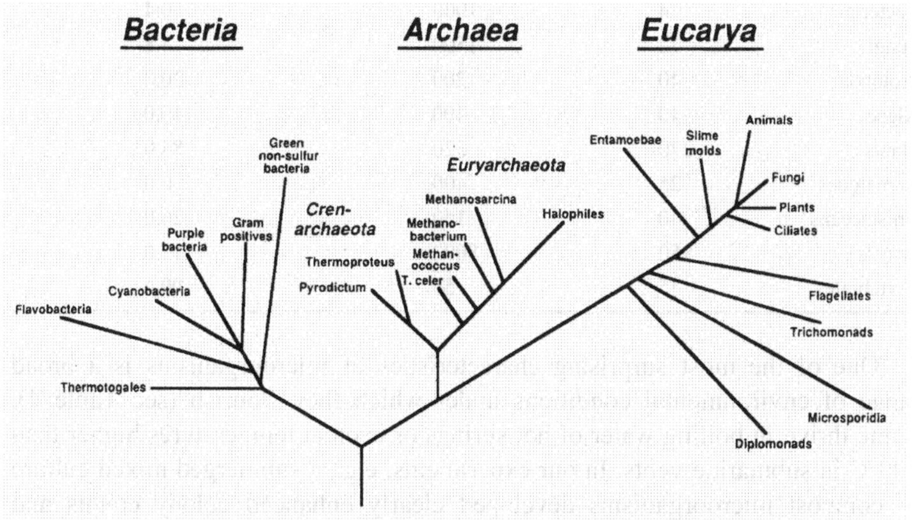

Figure 1. The tree of life – a model of the evolutionary relatedness of life on earth (From Leadbetter, 2002).

Figure 1 shows basically that the biological world is not divided into five kingdoms, as was held by most biologists until about 1970s, but is actually comprised of only three kingdoms or "domains". Two of the domains *Eucarya* (cells with nuclei, and including plants, animals, and fungi), and *Bacteria* had previously been described, but the third one, *Archea*, was entirely unheard-of. *Archea* have some characteristics of both *Bacteria* and *Eucarya*. For example, the process of DNA replication is similar between *Archea* and *Eucarya*, while both *Archea* and *Bacteria* lack membrane-bound structures within their cells (AAM report, 2004). Also, *Archea* have been found to be a diverse, versatile, ubiquitous group of microorganisms that can tolerate a wide range of habitats, including some of the most inhospitable environments on earth. The in-between adopted idea that, two of the three domains of life are microbial, have caused a radical paradigm shift for biology showing that most of the spectrum of genetic diversity undoubtedly belongs to microorganisms. Life forms visible to our naked eye, such as fungi, plants, and animals represent only a small fraction of biological diversity. Indeed, most of the microorganisms have not been grown in culture and characterized, yet. In contrast to plants and vertebrates of which about 90% have been described, it seems that less than 1% of the bacterial and 5% of fungal species are currently known (see Table 3).

TABLE 3. Approximate numbers ($\times 10^3$) of species in major groups of described and/or estimated organisms. (From Staley et al., 1997.)

Organisms known	No. of known species	Estimated no. of species in total	Percentage of species from total estimated no.
Bacteria	4	1000	0.4
Fungi	72	1500	4.8
Protozoa	40	200	20.0
Algae	40	400	10.0
Plants	270	320	84.0
Nematodes	25	400	6.0
Crustaceans	40	150	26.0
Insects	950	8000	12.0
Vertebrates	45	50	90.0

One of the most surprising characteristics of microorganisms is a broad range of environmental conditions under which they flourish (see Table 4). Some thrive in boiling water of hot springs or even at temperatures higher than 100°C in submarine vents. In our experiments, e.g., a submerged mixed culture of compost microorganisms developed clearly enhanced colony counts and metabolic activity (CO_2 release) at 70°C in comparison to controls kept at 30°C, if organic or especially mineral colloidal particles (humic matter and bentonite, respectively) were present in a growth medium (Filip, 1978). Different living microorganisms can be found in sea ice of Antarctica; some produce sulfuric and nitric acids, and many need no oxygen for their life. Others live in saturated salt brines, and some are resistant to high levels of radioactivity (Staley et al., 1997). Again, in some of our experiments, different soil bacteria not only survived but also exerted quite strong enzyme activities in soil samples heavily contaminated by man-made chlorinated solvents (Kanazawa and Filip, 1986, 1987).

Apparently, microorganisms can inhabit even hostile environments everywhere on our planet. They have quite a bizarre physiological capacity and very versatile metabolic activity, all of which attest to the richness of their genetic resources.

TABLE 4. Examples of extreme environmental conditions under which microorganisms live and metabolize. (From Staley et al., 1997)

Environmental parameter		Natural habitat	Microbial inhabitants
Extreme temperature	Less than 1°C	Polar sea ice	Psychrophilic Eubacteria
	More than 110°C	Hot springs; Marine hydrothermal vents	Hyperthermophilic Archea
Extreme pH values	Less than pH 1	Sulfur springs; Acid mine drainage	Acidophilic Eubacteria Archea
	More than pH 12	Soda lakes; Desert soils	Alkaliphilic Eubacteria Archea
Absence of oxygen		Sediments; animal intestine	Eubacteria, Archea, Anaerobic fungi, Protozoa
Saturated saline		Salt lakes; Brines	Extreme halophilic Archea
Low water activity		Desert soils; Saps, Brines	Xerophilic fungi
High hydrostatic pressure More than 1000 atm		Deep sea	Barophilic bacteria
High radioactivity		Radioactive soils	Radioduric bacteria (Deinococcus sp.) dark pigmented fungi

Microorganisms play an important role in geochemical processes that govern the turnover of elements and transformation of substances in all natural environments. For example, the global nitrogen cycle includes unique processes such as nitrogen fixation, i.e., bioconversion of atmospheric dinitrogen gas to organic cell nitrogen of living cells. Furthermore, microorganisms carry out oxidation of ammonia and nitrite to nitrate, and nitrate reduction with formation of di-nitrogen and nitrous oxide gases. Similar important and unique roles are played in the cycles of carbon, sulfur, phosphorus, and other bioelements. If it were not for microorganisms, substances such as cellulose and lignin would not be recycled but accumulate in the environment with disastrous consequences for entire life on earth. The microbial turnover of carbon amounts to some 43 Gt C/yr, which is approximately equal to a total net biomass production (Smith and Paul, 1990).

In difference to many higher organisms capable of existence and activity rather individually, the microbial life and activities in natural environments occur in communities, i.e., groups composed of single or multiple species that exist at the same place. Figures 2 and 3 shows small communities of bacteria and yeasts as established on a mineral surface (clay particle), and on oil droplet, both carriers apparently offering a semisolid base and nutrient source for the attached microorganisms.

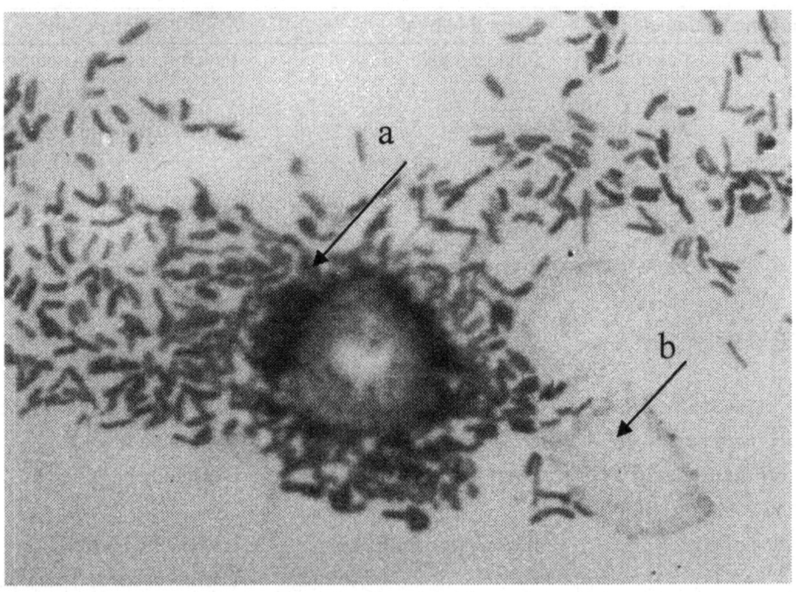

Figure 2. Clay particle saturated with nutrients and attracting bacteria: (a); quartz particle with only a few bacteria attracted (b) (Photo by Z. Filip).

The final structuring of a microbial community will be dictated by opportunities for metabolic and/or co-metabolic utilization of nutrients. Under natural conditions, e.g., in soil as the most heavily microbial inhabited environment, biologically based processes take place in a concerted action of different organism groups (micro-, meso-, and macro-edaphon), each of which is capable of delivering a specific contribution to the ecologically important processes (see Figure 4).

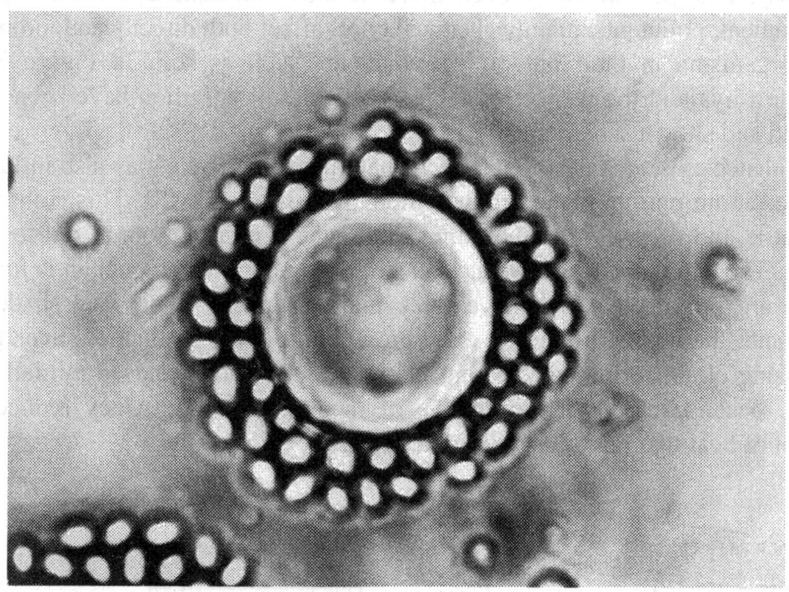

Figure 3. Oil droplet inhabited by yeast cells (Photo by Z. Filip/W. Pecher).

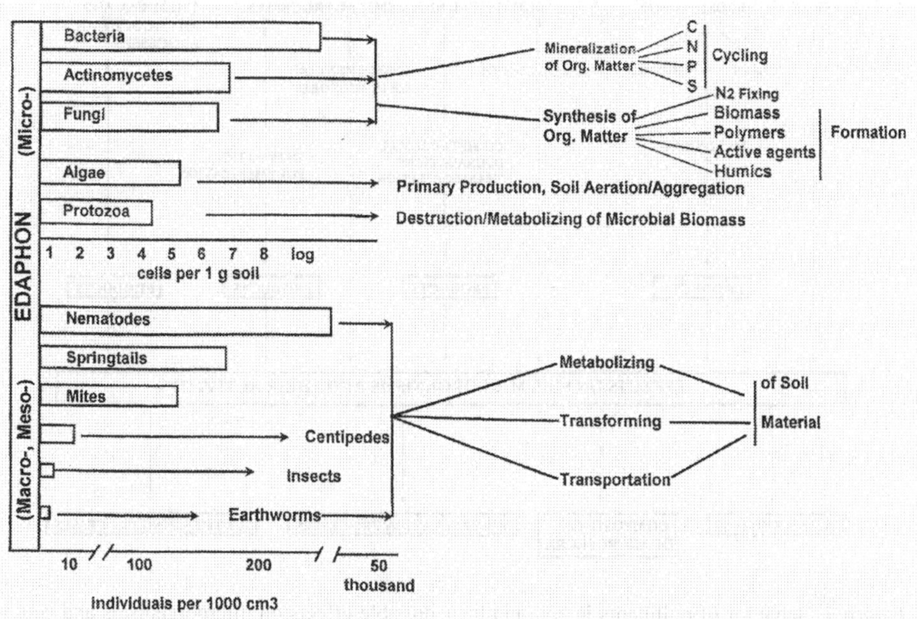

Figure 4. Main groups of soil organisms, their approximate counts, and ecologically important activities (From Filip, 2002).

As discussed in Section 2, pollutants originating mainly from anthropogenic activities may exert different detrimental effects of human population directly. Similarly, and perhaps with even more serious consequences for the entire environment, pollutants, mainly chemical ones, affect both directly and indirectly microorganisms in their natural environments such as soil. In Figure 5 the possible ways and the respective final targets of soil pollution have been summarized and shown schematically.

Evidence exists (Staley et al., 1997) that microorganisms may also influence climate. Some marine algae produce dimethyl-sulfide (DMS). This compound is volatile and escapes into the atmosphere where it is photooxidized to form sulfate. The sulfate acts as a water nucleating agent and when enough is formed clouds are produced which have three major impacts. First, they shade the ocean and, thereby, slow further algal growth and DMS production, eventually decreasing cloud formation. Second, the clouds lead to increased rainfall, and finally, because clouds are reflective of incoming sunlight, they reduce the amount of heat that reaches earth, moderating global warming.

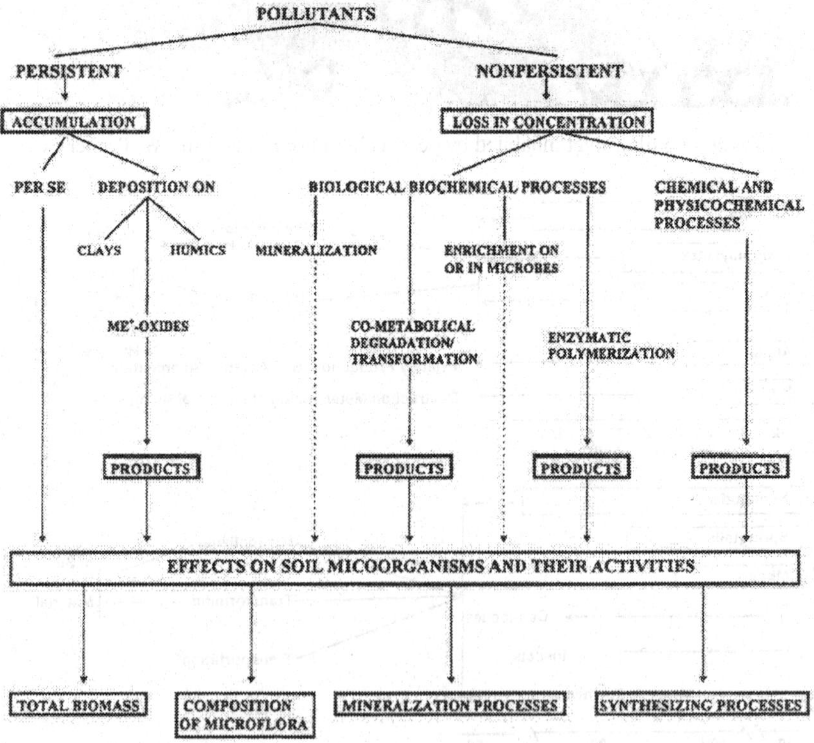

Figure 5. Behavior of pollutants in soil and their possible effects on microorganisms and ecologically important microbial processes (From Filip, 1998).

4. Societal Benefits of Microbial Exploration and Harvesting

Microorganisms have provided many environmentally important functions, and economically valuable processes and products that can be used in our all everyday life. Further exploration of microbial diversity and metabolic activities might bring about not only discovery of new unusual organisms but also novel means useful in agriculture, bioremediation, medicine, and different biotechnological industries. Table 5 lists some examples of present and predicted applications of microbial biotechnological tools and processes that were and/or could be derived from microbial exploration.

In the technologies, processes and individual utilization steps, microorganisms will not be exploited to perform industrial, agricultural, and commercial activities; they will perform these activities to sustain their own livelihood. It is solely utilization of specific microbial activities for application to the betterment of the conditions for human and environmental sustainability that is envisioned.

5. Training to meet Future Needs and Challenges in Environmental Microbiology/Biotechnology

Reviewing the facts of the previous sections: Extensive industrial and other human activities as evolved especially in the past decades have not only resulted in a general enhancement of living standards but simultaneously they substantially contributed to deterioration of terrestrial and aquatic ecosystems; consequently, they can seriously endanger the basic existence of different species on the earth, including humans. Since the functioning of ecosystems undoubtedly rely on specific activities of microorganisms, we should thoroughly preserve life-important microbial activities. Microorganisms were the first life forms, they carry out the processes that ensure clean drinking water and fertile soils, and represent important determinants of human health. They are the most genetically and biochemically diverse forms of life and are the most rapidly evolving organisms colonizing our planet. Microorganisms govern environmental cycling of elements, nutrients, and other substances necessary for the life sustainability. In light of these truths it is apparent that an effort must be undertaken to educate the youth, public, and also policy makers, in microbiology, its principles and applications, especially with respect to environmental issues and related prospects in biotechnology.

Recently, the American Academy of Microbiology, the honorific leadership group of scientists in the field, made serious recommendations on how to improve education in microbiology (AAM report, 2004). Reorienting school curricula should begin with changes in biology textbooks in which texts should be organized around a microbiology core. Microorganisms and their importance for the sustainability of life on earth should take center stage in science curricula at the elementary, middle, and high school levels. For this reason, many current teachers need to be retrained in the theory and applications associated with the modern microbial science. At the undergraduate level, education should take place with two different aspects: training future specialist in microbiology/biotechnology, and a decided contribution to the training of biologists in other fields. Even students in fields outside of the life sciences would benefit from lessons in microbiology. At a graduate level, dedicated lessons and advanced research should be performed in a broad scale, and dealing with the main future oriented topics as suggested in Table 5.

TABLE 5. Some current and anticipated benefits of microbial exploration and harvesting. (From Staley et al., 1997.)

Current successful applications of microbial biotechnology with an option of enhancement
Industrial fermentations
Industrial enzyme production (approx. value US$ 1 billion/year)
Production of bioactive compounds such as antibiotics, immune suppressors, biopesticides, vaccines, growth promoters, etc. (approx value US$ 35 billion/year)
Food and dairy industry processes and products
Pulp and paper effluent treatment
Water and wastewater treatment
Bioplastics
Anticipated products, applications, and areas expected to benefit
Novel enzymes for use in biotechnology: polymerases, ligases, alkaline phosphatases; specific robust catalysts for high and low temperatures and other extreme conditions
Novel bioactive, e.g., anti-inflammatory, compounds, and antibiotics from soil, endophytic, epiphytic, litter-inhabiting, and marine microorganisms
Sustainable agriculture and forestry
Novel biopesticides
Improvement of nitrogen fixation
Rhizoremediation; plant growth promoting rhizobacteria
Animal health and productivity promoting cultures and substances
Plant genetic engineering, e.g., with *Agrobacterium* sp.
Mycorhizae utilization
Enhanced bioremediation of hazardous waste sites, and treatment of industrial waste streams
Biorecycling of materials
Biomimetics: exopolymers, dispersants, adhesives, emulsifiers
Bioelements for computers and switches
Biorestoration and bioconservation

Green technologies
Mine reclamation
Conservation of plants and animals in endangered habitats
Human health
Diagnostics and public health issues
Drinking water and wastewater advanced treatment
Combating novel pathogenic microorganisms with new antibiotics
Improving indoor air quality
Impact on global climate (carbon dioxide, methane, dimethyl sulfide)
Microbial production of fuels (alcohol, biodiesel, hydrogen)
New discoveries that expand our understanding of the diversity of life and the impact of microorganisms on the biosphere

---

In 2004, the Czech Republic, Hungary, and Poland became full members of the European Union (EU). A few other European countries belonging to the former Eastern Block will follow, perhaps, in 2007. In addition, several other transition countries will attempt a full membership in the EU. By suffering under different problems concerning the environmental quality, the individual countries have or will have to adopt EU legislation on environmental protection including the respective quality standards, all this in favor of the entire European population. Consequently, effective measures are to be undertaken to combat the existing deterioration of natural resources in order to ensure clean air, safe water, and healthy food for the population in the individual countries. To achieve this goal, an updating and intensifying of environmental education is an important precondition. Such a necessity exists also in the Czech Republic as documented in the Section 2 of this chapter. Although in the former Czechoslovakia, valuable results have been achieved in life sciences, including general and applied microbiology/biotechnology, Mejstrik (1993) pointed on the fact that the former administrative–directive economic and social management caused a plundering manner of the utilizing of natural resources without taking care of environmental quality and human health.

The situation started to improve after democratic rules were reintroduced in the country, starting in 1989/90. Currently the Czech scientists deliver many valuable contributions in environmental microbiology/biotechnology nationally as well as internationally. The internationally renowned Department of Biochemistry and Microbiology, a part of the Institute of Chemical Technology Prague (ICTP), and chaired by K. Demnerova, may serve as an example of extensive research activities in this field. Demnerova et al. (2002, 2005), Kucerova et al. (2000), Macek et al. (2000, 2002), Mackova et al. (2002, 2005), and Pazlarova et al. (2002) contributed effectively to “a new wave” in this field. By recognizing both the high level of research work, and the existing need to develop teaching and research in the given field further, the European Commission in Brussels established a Marie Curie Chair in Environmental Microbiology/Biotechnology for 3 years at the ICTP. The aim is to extend the existing teaching program in which (i) General and Applied Biochemistry, (ii) Biochemistry and Genetics of Microorganisms, and (iii) Health Engineering, predominated until now. In Table 6 the priority topics have been summarized that are to be considered in the respective curricula.

TABLE 6. Priority scopes and individual topics for the lecturing in environmental microbiology/biotechnology as proposed for the Institute of Chemical Technology, Prague

Priority scopes	Individual topics
Biological fundamentals	Principles of general microbiology
	Microbial energetics and metabolic pathways
	Microbial population and community dynamics
	Microbial substrate utilization and nutrient cycling
	Co-metabolic utilization of refractory substances
	Enzymatic transformation of organic substrates
Ecological fundamentals	Microorganisms in terrestrial environments
	Microorganisms in aquatic environments
	Microorganisms in the air atmosphere
	Microorganisms in extreme environments
	Animals as microbial environments
	Molecular aspects of microbial ecology
Technological fundamentals	Physicochemical and technological principles of utilizing microbial activities
	Immobilization of microbial cells on natural and man-made materials
	Formation and utilization of biofilms under natural conditions, and in bioreactors
	Monitoring of microbial growth and activities under natural conditions, and in technological processes
	Current perspectives in biotechnologies
Soil and groundwater protection	Degradation/transformation/immobilization of pollutants by microorganisms
	In situ bioremediation of polluted sites
	On-site bioremediation of polluted sites
	Off-site bioremediation of contaminated materials
	Microbial formation of refractory organic compounds in soil and groundwater
	Fate of genetically engineered microorganisms in soil and groundwater
	Survival and transmission of health relevant microorganisms in soil and groundwater
	Assessment of soil and groundwater quality by microbiological and biochemical methods

Liquid and semiliquid wastes	Aerobic/anaerobic microbial degradation of organic matter in municipal and industrial effluents, sewage sludge, and landfill-leakage water
	Enzymatic conversion of organic pollutants in liquid and semiliquid wastes
	Degradation and/or transformation of organic pollutants in liquid and semi-liquid wastes using artificially designed microorganisms (e.g., immobilized cells, GEMs)
Solid wastes	Microbial degradation, transformation and/or stabilization of organic matter in solid wastes disposed of in a landfill
	Microbial processes in composting solid wastes
	Biodeterioration and transformation of plastic materials disposed of in a landfill
	Microbial leaching and/or binding of chemical pollutants (e.g., heavy metals) in solid wastes disposed of in a landfill
	Risk of dissemination of health relevant microorganisms from landfills and compost heaps
Gaseous pollutants	Microorganisms as air pollution factors
	Microbial utilization of gaseous compounds
	Cleansing of waste gases using microorganisms (biofilters)
	Microbial binding of greenhouse gases
Health risks and environmental hygiene	Detection and risk-evaluation of health-relevant microorganisms in environmental samples
	Detection and risk-evaluation of health relevant microorganisms in specific man-made environments, including health-relevant wastes
	Hygiene requirements in handling and treatment with natural and man-made microbial affected materials.

The theoretical lecturing should be accompanied by respective practical courses in the laboratory in order to make students familiar with microorganisms in a culture and with common techniques of observing and affecting microbial activities with the aim of their final utilization in environmentally oriented biotechnological processes.

According to Woese (1999), a scientific understanding of organisms has four basic components: (i) structure–function relationship, i.e., how the organisms are built and how they work; (ii) diversity, i.e., what kinds and how many of them there are, and the ways in which they are similar or different; (iii) ecology, i.e., how they interact with their environments and with one another, (iv) evolution, i.e., where they came from and how they are related to one another. Primary, these basic components should be applied to microorganisms. From a human point of view, however, we shall also ask: which kind of metabolic activities the microorganisms are capable of, and whether these could be used biotechnologically in favor of human life and social prosperity?

Indeed, a success in establishing educational curricula in environmental microbiology/biotechnology on a broad base does not depend on microbial literacy only. Far more, by completing our knowledge on microorganisms and sharing it with a broad public, we also will increase public awareness of the fact that our existence on this planet is just one single component in a life cycle, the basic processes of which are truly governed by microorganisms. Indeed, not only by respecting but also by utilizing this fact in different fields of environmental biotechnology consciously, we can enhance our chance in continuing life on earth in a sustainable way.

6. Conclusions

Due to 40 or even more years of totalitarian ruling over several countries in Central and Eastern Europe, considerable damage was developed in both humanitarian and environmental affaires. In the last decade, however, considerable improvements in environmental quality could be attained, especially in the new EU member countries. A recent Report on the Environment Quality, as issued in 2005 by the Ministry of Environment of the Czech Republic, contains much evidence in this respect. For example, the emissions of sulfur dioxide and nitrogen oxides were reduced between 1990 and 2002 from 23.5 to 3.0 t, and from 7.0 to 4.0 t/km^2, respectively. Nevertheless, many problems are still to be solved. To improve environmental quality in a sustainable way, environmental education is an urgent issue. This has been recognized by the responsible EU and Czech authorities. Even more, all European countries can count on effective support of their environmentally oriented educational activities by the European Commission as clearly evidenced, e.g., by the respective program of a Marie Curie Chair in environmental microbiology/biotechnology, as temporarily established at the ICT Prague.

Acknowledgment

The senior author (Z.F.) wishes to express his gratitude to the European Commission, Brussels, for the granting of a Marie Curie Chair in Environmental Microbiology/Biotechnology, tenable at the Institute of Chemical Technology Prague, Czech Republic.

References

AAM report, 2004, *Microbiology in the 21st Century,* American Academy of Microbiology, Washington, DC, 19 pp.

Demnerova, K., Mackova, M., Pazlarova, J., Novakova, H., Stiborova, H., Ryslava, E., 2002, PCBs – problem of their removal from the environment, in: *Abstract Book, 12th International Biodeterioration and Biodegradation Symposium*, Prague, 14–18 July 2002, p. 13.

Demnerova, K., Lovecka, P., Ryslava, E., Beranova, K., Vrchotova, B., Kochankova, L., Macek, T., Mackova, M., 2005, Complex evaluation of microbiological, chemical and ecotoxicological approaches to monitor biological degradation of PCB, in: *Proceedings of the 3rd European Bioremediation Conference*, Chania, Greece, 4–7 July 2005, p. 126.

Filip, Z., 1978, Effect of solid particles on the growth and endurance to heat stress of garbage compost microorganisms, *Europ. J. Appl. Microbiol. Biotechnol.* 6: 87–94.

Filip, Z., 1998, Soil quality assessment: an ecological attempt using microbiological and biochemical procedures, *Adv. GeoEcol.* 31: 21–27.

Filip, Z., 2002, International approach to assessing soil quality by ecologically-related biological parameters, *Agric. Ecosystem Environ.* 88: 169–174.

Filip, Z., 2004, Nachhaltige Entwicklung – unsere Sorge und Hoffnung, *Acta Univ. Agric. et Silvic. Mendel. Brun.* LII: 171–174.

Harold, F.M., 2001, Postscript to Schrödinger: so what is life? *ASM News* 12: 611–616.

Kanazawa, S., Filip, Z., 1986, Effects of trichloroethylene, tetrachloroethylene and dichloromethane on enzymatic activities in soil, *Appl. Microbiol. Biotechnol.* 25: 76–81.

Kanazawa, S., Filip, Z., 1987, Effects of trichloroethylene, tetrachloroethylene and dichlotomethane on soil biomass and microbial counts, *Zbl. Bakt. Hyg.* B 184: 24–33.

Kucerova, P., Mackova, M., Chroma, L., Burkhard, J., Triska, J., Demnerova, K., Macek, T., 2000, Metabolism of polychlorinated biphenyls by *Solanum nigrum* hairy root clone SNC-90 and analysis of transformation products, *Plant Soil* 225: 109–115.

Leadbetter, E.R., 2002, Procaryotic diversity: form, ecophysiology, and habitat, in: *Manual of Environmental Microbiology*, Hurst, Ch.J., (ed.), ASM Press, Washington, DC, pp. 19–31.

Lietzmann, K.M., Vest, G.D. (eds.), 1999, *Environmental Security in an International Context*, NATO Committee on the Challenges of Modern Society, Report No. 232, 174 pp.

Macek, T., Mackova, M., Kas, J., 2000, Exploitation of plants for the removal of organics in environmental remediation, *Biotechnol. Adv.* 18: 23–35.

Macek, T., Ryslava, E., Rezek, J., Kucerova, P., Francova, K., Chroma, L., Demnerova, K., Vosatkova, M., Rydlova, J., Mackova, M., 2002, The effect of rhizospheric colonization on PCB degradation in contaminated soil, in: *Abstract Book, 12th International Biodeterioration and Biodegrradation Symposium*, Prague, 14–18 July 2002, p. 173.

Mackova, M., Francova, K., Sura, M., Szekeres, M., Bancos, S., Silvestre, M., Demnerova, K., Macek, T., 2002, Preparation of transgenic plants with bacterial PCB degradation enzyme, in: *Abstract Book, 12th International Biodeterioration and Biodegradation Symposium*, Prague, 14–18 July 2002, p. 130.

Mackova, M., Francova, K., Vrchotova, B., Najmanova, J., Kochankova, L., Silvestre, M., Zidkova, J., Demnerova, K., Macek, T., 2005, Biotransformation of PCBs by plants and bacteria – consequences of plant-microbe interactions, in: *Proceedings of the 3rd European Bioremediation Conference*, Chania, Greece, 4–7 July 2005, p. 63.

Mejstrik, V., 1990, Ectomycorrhizas and forest decline, *Agric. Eco. Environ.* 29: 325–337.

Mejstrik, V., 1993, Air pollution and some aspects of the ecotoxicological situation in Czechoslovakia, *Sci. Total Environ.* Supplement 1993: 207–215.

Ministry of Environment of the Czech Republic, 2005, *Report on the Environment in the Czech Republic*, Prague.

Moldan, B. (ed.), 1990, *Environment in the Czech Republic*, Academia, Prague (in Czech).

Moldan, B., Billharz, S., 1997, *Sustainability Indicators*, Wiley, Chichester.

Pazlarova, J., Novakova, H., Vosahlikova, M., Mackova, M., Demnerova, K., 2002, PCB metabolism by *Pseudomonas* sp. P2, in: *Abstract Book, 12th International Biodeterioration and Biodegradation Symposium*, Prague, 14–18 July 2002, p. 28.

Smith, J.L., Paul, E.A., 1990, The significance of soil microbial biomass estimations, in: *Soil Biochemistry*, Vol. 6, Bollag, J.M., Stotzky, G. (eds.), Vol. 6, Dekker, New York, pp. 357–396.

Staley, J.T., Castenholz, R.W., Colwell, R.R., Holt, J.G., Kane, M.D., Pace, N.R., Saylers, A.A., Tiedje, J.M., 1997, *The Microbial World: Foundation of the Biosphere*, American Academy of Microbiology, Washington, DC, 32 pp.

Woese, C., 1999, The quest for Darwin's grail, *ASM News* 65: 260–263.

PART III

LESSONS LEARNED AS ILLUSTRATED VIA RESEARCH AND CASE STUDIES

14. LESSONS LEARNED FROM THE NORTH: BALANCING ECONOMIC DEVELOPMENT, CULTURAL PRESERVATION, AND ENVIRONMENTAL PROTECTION

SUSAN ALLEN-GIL*
Environmental Studies Program, Ithaca College, Ithaca, NY 14850 USA

Abstract: This paper reviews the environmental stressors and their effects on the ecosystems and cultures of the circumpolar Arctic. Transition countries face a similar set of conditions in that they are ripe for economic development, must manage severe contamination issues, and are home to many communities steeped in Old World culture facing intense modernization and globalization.

Keywords: Arctic; metals; POPs; cultural heritage

1. Introduction

The Arctic may seem an unlikely place to look for lessons learned with respect to environmental security, but in many ways the Arctic has experienced an onslaught of environmental pressures since the 1800s when whale blubber became desirable for lamps and baleen became popular for making hoop skirt frames, buggy whips, umbrella ribs, fishing rods, and mattress stuffing (Blackman, 1989). In the last 150 years, the Arctic has served as a natural resource frontier, as a repository for anthropogenic pollutants, and more recently as an early warning area for the effects of climate change. Meanwhile, the native inhabitants of the Arctic struggle to preserve their cultural heritage, which is very strongly tied to the land and its resources.

*To whom correspondence should be addressed. Susan Allen-Gil. Environmental Studies Program, Ithaca College, Ithaca, NY 14850 USA

R. N. Hull et al. (eds.), Strategies to Enhance Environmental Security in Transition Countries, 187–199.

2. Economic Development

The Arctic region is rich in natural resources, leading to extensive mineral mining, and oil and gas development. The world's largest supplier of nickel and palladium is located in Norilsk, Russia, which produces nearly 0.25 million tons of nickel and 0.5 million tons of copper annually (mining-technology.com, 2006). While over 90% of its sales are to foreign customers, this complex also supplies 90% of Russia's nickel and 55% of its copper (mining-technology. com, 2006).

The Red Dog Mine in Alaska (USA) is the largest lead–zinc mine in the world, with probable reserves of 93,696,461 t of ore (18.2% zinc and 4.6% lead) (Northern Alaska Environmental Center, 2002). Currently, the mine extracts roughly 7–10 million tons of ore per year, and is expected to continue operating until 2012, and possibly longer if nearby deposits prove to be economically viable (Northern Alaska Environmental Center, 2002). In the last few years, projects have begun to mine gold deposits in Russia and diamonds in northern Canada (All Things Arctic, 1998; CBC News, 2006). This development provides an important source of income for many Arctic residents, but also adds environmental stress to the ecosystem.

Arctic oil and gas extraction is becoming ever more widespread and provides a source of livelihood for many Arctic communities. The Prudhoe Bay and Kuparak River oil fields in northern Alaska are two of the top three oil fields in the USA in terms of production volume, accounting for nearly 10% of all US oil production (US Department of Energy – Energy Information Administration, 2005). Likewise, over 90% of current natural gas production in Russia and 90% of all Russian offshore hydrocarbon reserves are located in the Arctic (Burakova, 2005; US Department of Energy – Energy Information Administration, 2006).

3. Environmental Effects

The combination of development within the Arctic and beyond its borders has impacted the Arctic environment and its people in a variety of ways, many of which are similar to what has occurred and continues to occur in transition countries.

3.1. MINERAL EXTRACTION AND PROCESSING

Nonferrous metal production is a major source of heavy metal emissions to the atmosphere. Emissions during production accounts for the largest source of atmospheric arsenic (As; 69%), cadmium (Cd; 73%), copper (Cu; 70%), indium (In; 100%), and zinc (Zn; 72%) (Pacyna, 2002; Walsh, 2003).

The Norilsk metallurgical complex emits 5000 t of sulphur dioxides daily (Walsh, 2003). The environmental impacts of these large metal producing complexes are quite severe but predominantly local (Blais et al., 1999; Allen-Gil et al., 2003). A large area surrounding Norilsk is devoid of most lichen and moss species, and trees are dead from acidification (Vlasova et al., 1991).

The Red Dog Mine is Alaska's largest polluter, emitting over 80% of the total toxics released in the state, including 308 million pounds of zinc compounds and 123 million pounds of lead compounds annually (Northern Alaska Environmental Center, 2002). Along the Red Dog Mine haul road, concentrations of Pb, Zn, and Cd in moss species are 1.5–2.5 higher than severely polluted regions in Central Europe (Ford and Hasselbach, 2001).

Additional heavy metal emissions are transported to the Arctic from distant sources. Emissions from Asian sources continue to increase, and now account for 40% and 60% of total emissions (depending on the metal), and are expected to soon become more significant than Russian sources for delivery of heavy metals into the Arctic atmosphere (Pacyna, 2002).

3.2. GLOBAL CLIMATE CHANGE

While the Arctic region is not a major source of CO_2 emissions, it pays a large price for global fossil fuel use. The environmental effects of these enterprises are varied and complex, ranging from local vegetation impacts from heavy machinery on the tundra to pronounced global climate change (Forbes et al., 2001), (Hollister et al., 2005).

The Arctic has shown early, significant signs of warming (Lemonick, 2004; Schiermeier, 2006). In the last 30 years, the US Arctic has experienced greater changes in temperature than anywhere else on earth (Whitfield, 2003). This warming has led to many ecological changes. For example, the delayed formation of sea ice in Alaska disrupts the timing of polar bears from moving northward to hunt seals and walruses (Derocher et al., 2004; Lemonick, 2004). As the distribution of more species extends northward, Arctic species will face increasing competition for scant resources and yet have little capability to extend their own territories towards the pole (Arctic Council, 2004).

3.3. PERSISTENT ORGANIC POLLUTANTS

Transboundary movement of persistent organic pollutants (POPs) also affects the Arctic. The major source for POPs is escape into the atmosphere from sites of manufacture, use, and disposal. Depending on their physical and chemical properties, POPs can be transported through time from temperate areas of origin to the Arctic through global fractionation, resulting in accumulation at northern latitudes (Wania and Mackay, 1993; Arctic Monitoring and Assessment Program, 2002). Thus, for many anthropogenic air pollutants, the Arctic serves as the ultimate global sink.

Many lipid-soluble organochlorine pesticides (such as DDTs and HCHs) and PCBs migrate to the Arctic this way, where they bioaccumulate in the long, fat-based Arctic food web. Some evidence suggests that levels of these POPs compounds are high enough in polar bears and northern fur seals to account for observed reproductive and immune impairment (Arctic Monitoring and Assessment Program, 2002).

The extent to which global climate change will impact Arctic pollution is as yet uncertain, but there is reason to expect that it will be significant (Arctic Monitoring and Assessment Program, 2003). Clearly, climatic factors (such as temperature, precipitation, and ocean and atmospheric currents) strongly influence routes and mechanisms of long-range atmospheric transport. Warmer temperatures may also serve to increase the rates of decomposition of contaminants. As climate change alters the species composition in the Arctic, a series of potential ecological changes are possible. For example, long range biotransport from southern latitudes may become more significant as the range of more migratory fish (especially salmon) extends northward. More disease vectors may be introduced, which when combined with decreased immunocompetency, may result in greater mortalities rates among wildlife at the top of the food chain exposed to the highest concentrations of POPs.

4. Cultural Impacts

The strong connections between the people and the land in the Arctic serves to magnify the effects of ecosystem change on individuals, communities, and entire cultures by threatening human health and cultural traditions.

4.1. HUMAN HEALTH

Human health is affected by economic development and contaminant exposure in the Arctic through direct and indirect pathways. The most striking direct impacts can be seen in the city of Norilsk, "the polluted place in Russia", where the average life expectancy for factory workers is 10 years lower than in the rest of Russia due to the severe air pollution and occupational hazards associated with the metallurgical complex (Walsh, 2003).

In other communities, where reliance on fat-based subsistence foods (such as fish liver, and seal and whale blubber) can increase human exposures to these harmful contaminants (Arctic Monitoring and Assessment Program, 2002). In into account acidification of overlying deposits, mainly carbon-containing predatory species (e.g., seals and toothed whales), average daily intake rates of POPs (toxaphene and chlordanes) exceed World Health Organization (WHO) guidelines (Arctic Monitoring and Assessment Program, 2002; Van Oostdam et al., 2005). Within the Arctic, the highest exposures to POPs occur in the Inuit population of Greenland and Canada and in eastern Russia, where traditional diets depend most heavily on marine mammals (Arctic Monitoring and Assessment Program, 2002; Webster, 2004). In the Chukotka region of eastern Russia, the levels of PCBs in males were 10,000 ng/g blood lipid, among the highest ever reported anywhere in the world (Webster, 2004). There is also evidence that this elevated exposure has resulted in neurological impairment in children (Arctic Monitoring and Assessment Program, 2002).

4.2. CULTURAL HERITAGE

There are 40 minority groups identified in the Arctic (All Things Arctic, 1998), all of which share a common tradition of strong ties to the land. Most Arctic cultures are closely tied to local resources for food, clothing, and cultural identity such that subsistence activities are the "cultural glue" that holds Arctic communities together (Van Oostdam et al., 1999). As Suzy Erlich, an Inuit woman said, "Our grocery store is millions of acres wide, not just a few thousand feet, and it brings us pride" (Berger, 1985). Thus, the knowledge that local fauna carry high burdens of contaminants that can be passed onto the people has been a cause for alarm in many communities. Establishing a balance between cultural preservation and human health effects is not easy. Traditional foods are an important source of protein and essential minerals, and are associated with lower obesity, diabetes, and heart disease than supermarket alternatives (Van Oostdam et al., 2005).

Managing this delicate issue relies on the careful crafting of dietary advice that encompasses *all* of the risks and benefits associated with a traditional diet. Community involvement must be central to establishing dietary advice that protects human health, and preserves cultural and spiritual heritage and identity.

Aside from the difficulties posed by contamination, coastal Arctic communities must also confront erosion and flooding associated with global climate change. One village on an island in northwestern Alaska is no longer protected by sea ice from enormous storm surges. In 2002, the villagers voted to relocate to the mainland, leaving their ancestral homeland forever and becoming the first refugees of global warming (Drapkin, 2006).

5. Lessons Learned

5.1. THE ARCTIC AS A SOURCE OF LESSONS LEARNED

Although far removed from the historic centers of civilization, the Arctic has experience confronting issues of environmental security across a vast territory and diverse cultures. The Arctic experience includes threats from resource extraction, air pollution, and global climate change.

The shared concerns among Arctic peoples with respect to environmental and cultural preservation led first to the adoption of the Arctic Environmental Protection Strategy (AEPS) in 1991 and the subsequent formation of the Arctic Monitoring and Assessment Programme (AMAP) to implement this strategy. Today, AMAP functions as a technical advisory group and database clearinghouse for heavy metal, POPs, and radioactive environmental assessments. In 1996, the Arctic Council was formed by the eight Arctic countries and numerous indigenous communities, to incorporate AEPS and other shared concerns and challenges. One such concern has been global climate change, which led to a comprehensive examination of the state of knowledge with respect to the Arctic, known as the Arctic Climate Impact Assessment.

The nongovernmental Inuit Circumpolar Conference (ICC) also plays an important role in preserving the culture and the environment of the 150,000 Inuit living in the USA, Canada, Greenland, and northeastern Russia. Its primary goals include "to ensure and further develop Inuit culture and society for both the present and future generations; to seek full and active participation in the political, economic, and social development in our homelands, and to develop and encourage long-term policies which safeguard the Arctic environment" (Inuit Circumpolar Conference, 2006).

These frameworks and the projects that have been completed through them serve as a model for other coalitions, such as transition countries, to mold coordinated efforts for environmental security.

5.2. IMPORTANCE OF PARTNERSHIPS

The Arctic constituents of most countries, especially the USA, often feel sidelined in the political process. Forging partnerships both within and beyond national borders has strengthened the collective political clout of these communities. While many governments and institutions view indigenous peoples as "powerless victims of change", the indigenous communities of the Arctic have been surprisingly effective at shaping international environmental policy (Downie and Fenge, 2003). Perhaps the best example of the effectiveness of this approach is when circumarctic communities joined forces politically to combat transboundary air pollution. Both the ICC and the Arctic Council, as will as numerous other indigenous groups were significant players in the Stockholm Convention of 2001, which called for a ban of the 12 most harmful POPs globally (Arctic Monitoring and Assessment Program, 2002).

John Buccini, who served as the chair of many intergovernmental bodies working on the Stockholm agreement, noted important factors to the success of translating science into policy: (1) sustained involvement of stakeholders, (2) sustained publicity, (3) open and transparent communication throughout the process, and (4) strong and reliable leadership (Buccini, 2003).

5.3. CONSULTING THE CONSTITUENTS

Decisions regarding environmental security directly impact people's lives, particularly in areas where their lives are closely tied to the land and surrounding ecosystem. "Nobody is better equipped to warn of environmental changes with potentially global impact than Indigenous peoples drawing first-hand information and traditional knowledge handed down from generation to generation by hunters and Elders" (Downie and Fenge, 2003). Oftentimes, the local residents can be an invaluable source of ecological information in the planning and execution of environmental monitoring and assessment program (Ford and Martinez, 2000). In our experience, partnering with local elders helped us identify study lakes and appropriate species to include in our monitoring program. Local residents can also be an important source of quailtative information about local ecosystems and changes they have observed through time. For example, in the Arctic, subsistence fishing communities noticed that the livers of burbot (*Lota lota*) had a different texture and taste than in the past. This led researchers to investigate whether these observed changes reflected high contaminant loads or changes in the burbot's diet and/or overall health. Employing local residents as field crew is also an important action to maximize the efficiency of the field collection, increase the local understanding of the project in the community, and foster greater trust of scientific research and resultant management decisions.

In the Arctic experience, one of the areas where this is critical is consumption advisories. Without detailed information about dietary habits, including preparation techniques, such advisories can be completely inappropriate. In Canada, the primary goal of the Northern Contaminants Program is "to reduce and wherever possible eliminate contaminants in traditionally harvested foods, while providing information that assists informed decision making by individuals and communities in their food use" (Muir et al., 2005). This illustrates the role of science and governments to provide information and guidance without imposing difficult decisions on local communities regarding health and culture.

5.4. ESTABLISH A MONITORING PROGRAM EARLY

Interpretation of new data on contaminants and global climate change requires a temporal and spatial context. In our experience with AMAP, the following considerations are critical.

5.4.1. Statistical approach to site selection

If the goal of a monitoring program is characterize the level of contamination over a broad area and to identify specific areas of concern, then it is important to design the sampling protocol such that site selection is not biased by accessibility or familiarity. For the US AMAP, we used the unstratified hexagonal grid methodology developed by White et al. (1992). Within each hexagon, one sampling location was selected at random, and located on the ground by global positioning system (GPS). This provided excellent representation of the contaminant concentrations in vegetation across a large geographic area.

5.4.2. Collection of baseline data

Baseline data is absolutely necessary to determine the level of contamination for naturally occurring heavy metals and polyaromatic hydrocarbons (PAHs). This is especially important in areas such as the Arctic and transition countries where minerals and fossil fuels are naturally abundant.

It is vitally important to collect data prior to economic development to compare predevelopment and postdevelopment conditions. For many historic developments, this was not done. For example, we do not have any data on the levels of PAHs in local fauna prior to the development of the Prudhoe Bay oil fields. Determining the effects of the development then, is limited to comparisons to other areas that hopefully (but not always) have sufficiently similar geological, climatological, and ecological conditions. For this reason, the US Fish and Wildlife Service has collected extensive background data for the Arctic National Wildlife Refuge (ANWR), and oil-rich area under intense national pressure for exploitation (Snyder-Conn and Lubinski, 1993). Lake sediment cores are also very helpful in establishing a temporal baseline for heavy metals and other contaminants. The time frame in which different layers of sediment were deposited can be determined using radio-dating techniques (Gubala et al., 1995), (Allen-Gil et al., 2003). This allows comparisons of recent deposition with historical values, including preindustrial observations; this is especially useful for heavy metals in areas naturally enriched in certain metals.

5.4.3. Identifying a common set of matrices

Establishing a common set of matrices (species and tissues) to be sampled is often not a trivial task. In many cases, our knowledge of the life history and physiology of species is limited, which adds variability and uncertainty in the interpretation of data. Lichens are a good choice for atmospheric deposition, as they receive all their nourishment from the atmosphere. However, lichens often exhibit very patchy distribution, making large spatial-scale monitoring problematic. Therefore, mosses may be a more appropriate choice for large-scale assessments.

Other factors may come into play for fauna. The choice of candidate species may depend on their distributions across the sampling area. But it may be equally important to sample species with subsistence or commercial value. In the Arctic, burbot are a highly prized fish among local residents, but are not equally abundant in lakes and rivers. Thus, it was included in monitoring efforts, despite the inability to develop a strong understanding of contamination trends across a large spatial area from this data. Another consideration is whether to sample the whole individual, or just certain tissues. If the goal is to evaluate food web biomagnification, then it is important to sample whole individual (or whatever is eaten by the predator). However, if the concern is human consumption, then only the tissues consumed should be sampled (usually muscle and or liver).

5.4.4. Establishing common field methodology

If different individuals are going to be collecting samples in different locations or at different times, it is important to standardize field collection procedures. For flora, it is important to decide exactly what parts of the plant will be sampled, if adhered soil will be removed prior to sample transportation, and how best to avoid contamination at the field site (from generators or other equipment).

For fauna, it is important that procedures be consistent with respect to the sampling gear as certain types of gear introduce inherent sampling bias. For example, gill nets come in a variety of mesh sizes, and the choice of the mesh size will affect the individuals caught. Electrofishing is an alternate sampling strategy, which is very effective for small streams, but of limited value in larger rivers.

5.4.5. Collecting as much ecological and physiological information as possible

One of the difficulties in environmental assessment work is that we rarely understand the systems that we wish to evaluate well enough to avoid multiple confounding factors. This can lead to either misinterpretation of the results, or the inability to develop any meaningful conclusions. For contaminant studies, the most important confounding factors are age, sex, lipid content, reproductive status, and trophic position and life history information. Fortunately, there are established methods to collect most of this information. Age can be determined by growth rings. In fish, otoliths (inner ear bone) provide a reliable series of growth rings, particularly in temperate environments with seasonal changes in growth patterns. Sex can be determined by examination of the gonads. Lipid content can be determined in the laboratory by standard lipid extraction techniques. Trophic (food web) position can be determined by nitrogen isotope analysis (Kidd et al., 1998; Peterson, 1999). At each step of the food web, the lighter nitrogen isotope (^{14}N) is preferentially excreted while ^{15}N is incorporated into proteins. Thus, the ratio of $^{15}N/^{14}N$ increases predictably with each trophic level.

5.4.6. Establishing a common set of analytes

It is also very important to determine the endpoints of interest for chemical analysis at the outset. For heavy metals, one must consider if information on individual compounds is required, or if the total concentration of a given metal will suffice. For many metals, most notably mercury, the chemical speciation is critical in determining bioavailability. Inorganic mercury is not readily available to almost all organisms, whereas organic mercury (methyl mercury) is readily available and bioaccumulates in the food chain. Determination of metal speciation usually significantly increases the cost of chemical analyses.

The same issue applies to POPs. DDT, for example, degrades readily in the environment to several forms of dichlorodiphenyldichloroethylene (DDE) and p,p-dichlorodiphenyl Dichloroethane DDD. If one is concerned about total dichloro-diphenyl-trichloroethane (DDT), then it would be important to measure all DDT derivatives. PCBs occur as over 200 individual compounds, each of which has a different persistence in the environment and toxicity. Measuring total PCBs alone does not provide much useful information from a risk perspective; it is much better to have the concentrations of specific PCB congeners identified.

It is increasingly important to expand the list of analytes to include “new and emerging” contaminants of concern. For the Arctic, this includes polybrominated compounds (such as diphenyl ethers (PBDEs)), which are still increasing in Arctic biota while “legacy” contaminants are decreasing.

5.5. NOTHING HAPPENS IN ISOLATION

Lastly, it is important to remember that environmental systems are astoundingly complex, and that we cannot look at individual issues in isolation. Concentrating solely on local sources may ignore the role of long-range atmospheric transport. Focusing solely on individual contaminants ignores the important interactions that occur between them with respect to toxicology. Addressing pollution or global climate change as separate issues discounts the growing evidence that climate change is likely to alter current patterns of distribution and accumulation for contaminants (Arctic Monitoring and Assessment Program, 2003).

6. Conclusion

The Arctic is the proverbial "canary in the coal mine" in terms of global ecosystems. As a significant source of the world's fossil fuel and mineral resources, it has undergone extensive local economic development. Its people have among the highest levels of industrial compounds in their bodies. The effects of global climate change are expected to be sooner and more severe in the Arctic than almost any other location on our planet.

The communities and local governments in the Arctic have been very effective in organizing, gathering scientific data, and lobbying for change. Cooperation, focus, and determination have been key factors in their success. The environmental security issues and the way Arctic communities have faced these challenges provide an important model for other cultures, communities, and governments facing similar pressures.

The irony is that, as the globe warms and the Arctic ice melts giving rise to a new round of economic development opportunities (Arctic Council, 2004), (Burakova, 2005), the communities of the Arctic will need to draw on their recent experiences in addressing the complex environmental and social problems of development, contamination, and climate change just as much as any other community, culture, nation, or region of the world.

References

All Things Arctic, 1998; http://www.allthingsarctic.com/people/index.aspx

Allen-Gil, S.M., Ford, J., Lasorsa, B.K., Monetti, M., Vlasova, T., Landers, D.H., 2003, Heavy metal contamination in the Taimyr Peninsula, Siberian Arctic. *Sci. Total Environ.* 301(1–3): 119.

Arctic Council, 2004, *Arctic Climate Impact Assessment Policy Document*, Issued by the Fourth Arctic Council Ministerial Meeting, Reykjavík.

Arctic Monitoring and Assessment Program, 2002, *Arctic Pollution 2002 (Persistent organic pollutants, heavy metals, radioactivity, human health, changing pathways)*, Arctic Monitoring and Assessment Programme, Oslo, Norway, 112 pp.

Arctic Monitoring and Assessment Program, 2003, *AMAP Assessment 2002: The Influence of Global Change on Contaminant Pathways to, Within, and from the Arctic*, Arctic Monitoring and Assessment Program, Oslo, Norway, 65 pp.

Berger, T.R., 1985, *Village Journey: The Report of the Alaska Native Review Commission.* Hill & Wang, New York, 201 pp.

Blackman, M., 1989, *Sadie Brower Neakok: An Inupiaq Woman*, University of Washington Press, Seattle, WA, 274 pp.

Blais, J.M., Duff, K.E., Laing, T.E., Smol, J.P., 1999, Regional contamination in lakes from the Noril'sk region in Siberia, Russia. *Water Air Soil Pollut.* 110: 389–404.

Buccini, J., 2003, The long and winding road to Stockholm: the view from the Chair, in: *Northern Lights Against POPs: Combatting Toxic Threats in the Arctic*. Downie, D., Fenge, T. (eds.), McGill-Queen's University Press, Montreal, Canada, pp. 224-255.

Burakova, I., 2005, Development of Arctic areas to bring trillions dollars of profit to Russia, Pravda. Moscow, Russia; http://english.pravda.ru/main/18/89/357/15328_arctic.html

CBC News, 3 July 2006, Canada's diamond rush; http://www.cbc.ca/news/background/diamonds/

Derocher, A.E., Lunn, N.J., Stirling, I., 2004, Polar bears in a warming climate. *Integr. Comp. Biol.* 44: 163–176.

Downie, D., Fenge, T. (eds.), 2003, *Northern Lights Against POPs: Combatting Toxic Threats in the Arctic*, McGill-Queen's University Press, Montreal, Canada.

Drapkin, J., 2006, A struggle to stay afloat: in the chilling reality of a warming Arctic, Alaskan villagers are fighting a battle to save their community and way of life. *Smithsonian* 37: 48–50.

Forbes, B.C., Ebersole, J.J., Strandberg, B., 2001, Anthropogenic disturbance and patch dynamics in circumpolar Arctic ecosystems, *Conserv. Biol.* 15: 954–969.

Ford, J., Hasselbach, L., 2001, *Heavy Metals in Mosses and Soils on Six Transects Along the Red Dog Mine Haul Road*, U. N. P. Service, Alaska, 73 pp.

Ford, J., Martinez, D., 2000, Traditional ecological knowledge, ecosystem science, and environmental management, *Ecol. Appl.* 10: 1249–1250.

Gubala, C.P., Landers, D.H., Monetti, M., Heit, M., Wade, T., Lasorsa, B., Allen-Gil, S., 1995, The rates of accumulation and chronologies of atmospherically derived pollutants in the Arctic Alaska, USA, *Sci. Total Environ.* 160/161: 347–361.

Hollister, R.D., Webber, P.J., Tweedie, C.E., 2005, The response of Alaskan arctic tundra to experimental warming: differences between short- and long-term responses, *Global Change Biol.* 11: 525–536.

Inuit Circumpolar Conference, 12 July 2006, General Information, ICC; http://www.inuit.org/index.asp?lang=eng&num=2

Kidd, K.A., Hesslein, R.H., Ross, B.J., Koczanski, K., Stephens, G.R., Muir, D.C.G., 1998, Bioaccumulation of organochlorines through a remote freshwater food web in the Canadian Arctic, *Environ. Pollut.* 102: 91–103.

Lemonick, M.D., 2004. Meltdown! The Arctic is warming up even faster than scientists feared. That could spell doom for the polar bear, *Time* 164: 72.

mining-technology.com, 3 July 2006, Norilsk Mining Centre Nickel, Palladium and Copper Production Facility, Russia; http://www.mining-technology.com/projects/norilsk/

Muir, D.C.G., Shearer, R.G., Oostdam, J.V., Donaldson, S.G., Furgal, C., 2005, Contaminants in Canadian arctic biota and implications for human health: Conclusions and knowledge gaps, *Sci. Total Environ.* 351-352: 539.

Northern Alaska Environmental Center. Mining, 29 June 2006; http://www.northern.org/artman/publish/mining.shtml

Pacyna, J.M., 2002, *Sources and Emissions: Heavy Metals in the Arctic*, AMAP Assessment 2002, AMAP, Oslo, Norway.

Peterson, B.J., 1999, Stable isotopes as tracers of organic matter input and transfer in benthic food webs: a review. *Acta Oecologica* 20: 487.

Schiermeier, Q., 2006, Arctic ecology: on thin ice, *Nature* 441: 146.

Snyder-Conn, E., Lubinski, M., 1993, Contaminant and water quality baseline data for the Arctic National Wildlife Refuge, Alaska, 1988–1989.

US Department of Energy – Energy Information Administration, 2005, *US Crude Oil, Natural Gas, and Natural Gas Liquids Reserves 2004*, Annual report.

US Department of Energy – Energy Information Administration, 2006, Country analysis briefs, Russia.

Van Oostdam, J., Gilman, A., Dewailly, E., Usher, P., Wheatley, B., Kuhnlein, H., Neve, S., Walker, J., Tracy, B. Feeley, M., 1999, Human health implications of environmental contaminants in Arctic Canada: a review. *Sci. Total Environ.* 230: 1.

Van Oostdam, J., Donaldson, S.G., Feeley, M., Arnold, D., Ayotte, P., Bondy, G., Chan, L., Dewaily, E., Furgal, C.M., Kuhnlein, H., 2005, Human health implications of environmental contaminants in Arctic Canada: A review. *Sci. Total Environ.* 351–352: 165.

Vlasova, T.M., Kovalev, B.I., Filipchuk, A.N., 1991, Effects of point source atmospheric pollution on boreal forest vegetation of northwestern Siberia, in: Weller, G., Wilson, C.L., Severin, B.A.B. (eds.), Proceedings of a Conference on the Role of Polar Regions in Global Change, 11–15 June 1990, University of Alaska Fairbanks, Vol. II, Fairbanks, pp. 423–428.

Walsh, N.P., 18 April 2003, Hell on earth. *The Guardian*; http://www.guardian.co.uk/g2/story/0,939043,00.html

Wania, F., Mackay, D., 1993, Global fractionation and cold condensation of low volatility organochlorine compounds in polar regions, *Ambio* 22: 10–18.

Webster, P., 2004, Study finds heavy contamination across vast Russian Arctic, *Science* 306: 1875.

White, D.A., Kimerling, J., Overton, W.S., 1992, Cartographic and geometric components of a global sampling design or environmental monitoring. *Cartogr. Geogr. Inf. Sys.* 19: 5–22.

Whitfield, J., 2003, Alaska's climate: too hot to handle, *Nature* 425: 338.

15. A PROCESS FOR FOCUSING CLEANUP ACTIONS AT CONTAMINATED SITES: LESSONS LEARNED FROM REMOTE NORTHERN SITES IN CANADA

STELLA M. SWANSON*
Golder Associates Ltd., Calgary, AB, Canada

Abstract: Contaminated sites in remote areas are challenging to assess and to remediate because of practical difficulties with access, high cost, the general lack of data on the natural environment or human populations in these areas, and the lack of remediation techniques designed for cold climates, poor soils, and slow-growing plants. Risk-based remediation at these sites must be carefully focused in order to manage costs and uncertainty, while truly achieving risk reduction. The focusing process starts with identification of links between contaminant sources, migration pathways, and the most highly exposed, sensitive receptors. A qualitative risk-ranking exercise follows, which identifies the most important sources and pathways that contribute to human health or ecological risk. The qualitative analysis includes an evaluation of uncertainty and the identification of any critical data gaps. However, data gaps are only deemed to be critical if risk management decisions cannot be made without additional information. In some cases, remediation can proceed at the end of the qualitative risk analysis because there is sufficient confidence in the identification of sources and pathways that drive the risk. In other cases, quantitative risk assessment is required in order to refine the understanding of how much cleanup is required (and when the cleanup can stop). Key lessons learned about risk-based cleanup at remote sites are: (1) start with the obvious sources of risk; (2) do not forget nonchemical effects such as changes in water flow or water level; (3) interact with government and community representatives as much as possible; (4) learn about the site's human and ecological characteristics from local people; (5) realize that a cleanup decision can be made at any step in the process; (6) cleanup actions should promote "self-healing" by initiating or enhancing natural processes such as biological degradation and soil development; and (7) remediated sites should require very little maintenance or monitoring.

*To whom correspondence should be addressed. Stella M. Swanson, Golder Associates Ltd., Calgary, AB

R. N. Hull et al. (eds.), Strategies to Enhance Environmental Security in Transition Countries, 201–213.

Keywords: risk-based cleanup; remote sites; lessons learned

1. Introduction

There are hundreds of contaminated sites requiring cleanup in northern Canada, ranging from former military bases to abandoned mines. These sites are usually in very remote locations, with access by road only during a very limited winter ice road season or with no road access at all (Figure 1). The sites may be within towns or villages, but can be hundreds of kilometers away from the nearest human settlement. However, even if the sites are far away from settlements, they may still be within traditional hunting and fishing territories.

Figure 1. Abandoned mine tailings in Northern Canada.

Cleanup of remote, northern contaminated sites according to environmental quality criteria or standards can be extremely difficult. In many cases, there are no locations for placement of contaminated material for hundreds or thousands of kilometers. The construction of a landfill, or the covering of contaminated sediments with clean material may cause more harm than the original contamination if the material required for lining the landfill or covering the sediments has to be obtained from new (and often widely scattered) borrow pits. There are no well-established techniques for cold-climate remediation of contaminated soils, and the science of revegetation using native subarctic or arctic plants is in its infancy. Long-abandoned sites may have some natural revegetation occurring, which would be destroyed if regulatory soil criteria are not met and soil has to be removed (Figure 2). Given the length of time required for plants to become reestablished in arctic or subarctic environments, destruction of a revegetated area to meet conservative criteria may not be the best option.

The alternative to the use of environmental quality criteria or standards is to develop risk-based cleanup options. The advantage of risk-based remediation is that the cleanup action can be matched to the actual risk (Swanson and Gerein, 2006). "Matching the effort with the risk" ensures that money and time spent at these sites truly results in a net reduction in risk. The risk assessment method also requires active consultation with regulators and the public, thus ensuring that the site is left in a state that meets government requirements and public expectations.

Figure 2. Natural revegetation occurring at a subarctic contaminated site.

Although risk-based remediation has clear advantages at these remote sites, its use creates challenges because of the lack of data on either human uses of resources or the characteristics of plant and animal communities in the vicinity of the sites. In many cases, no scientist has ever been near the site, and there is no information on key variables that drive the linkages between the sources of contaminants and biota.

There is often some urgency attached to the cleanup of these remote sites, creating even more challenges. With very little time, almost no data, difficult (and very expensive) access, how is a credible risk-based remediation plan possible? This paper provides some ideas for how to achieve an acceptable risk-based cleanup when time and/or money are in short supply.

2. Getting Started: Scoping the Problem

2.1. WHAT IS KNOWN ABOUT THE SITE?

Relevant information about the site can be obtained from a wide variety of sources. These sources include local, regional, and national government records of activities at the site, old aerial photographs, interviews with local residents or former workers, and any published information in newspapers or magazines (such as mining trade magazines). In addition, there may be data on soils, groundwater, surface water, plants, or animals in the region, if not at the actual site. Information on the site is then assembled to create a basic understanding of the primary source–pathway–receptor linkages (Figure 3). This understanding is required to conduct a risk assessment according to standard guidance, such as that provided by the United States Environmental Protection Agency (USEPA) (1998).

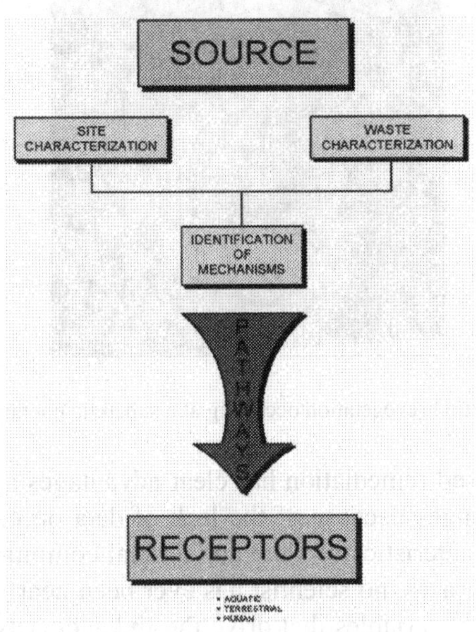

Figure 3. Source–pathway–receptor linkages at contaminated sites.

The key information required to develop the source–pathway–receptor understanding of the site is outlined below.

2.1.1. Source

1. The location of all contaminant sources; e.g., waste dumps, effluent outfalls, underground storage tanks, aboveground storage tanks, pipelines, drainage ditches, and sumps

2. The nature of the contamination; i.e., the specific chemical or physical stressors associated with the site, such as metals, acids, suspended solids, pesticides, solvents, diesel, or gasoline
3. Other nonchemical effects of the site on the local or regional ecosystem; e.g., blockage or diversion of water flow, dust deposition, erosion, land subsidence, and slumping of banks, physical hazards created by large debris, abandoned buildings, or old wells and shafts

2.1.2. Pathway

1. Air pathway
   - Dust generation and prevalence of dusty conditions
   - The presence of vapors (e.g., from old storage tanks or waste dumps)
   - The presence of any active air emissions (e.g., from incineration or spontaneous combustion of materials at the site)
2. Groundwater pathway
   - The depth to groundwater
   - The nature of soil and rock overlying the groundwater (in order to evaluate the hydraulic conductivity of these materials)
   - Groundwater flow direction
3. Surface water pathway
   - Identification of any streams, ponds, wetlands, or lakes on or adjacent to the site
   - Stream flow (preferably seasonal data)
   - Lake or pond depth and surface area
   - Drainage patterns leading off the site
   - Sediment characteristics (texture, organic content)
4. Soil pathway
   - Soil texture and moisture content
   - Presence or absence of an organic, litter layer
   - Soil depth

2.1.3. Receptor

1. Current human use of the site
2. Potential or planned future human use of the site

3. The location of important plant or wildlife habitat on or near the site
4. Plant and animal presence on or near the site

Not all of the above information will be available; however, some educated guesses will usually be possible by taking advantage of information on regional climate, geology, and biology. In addition, a great deal of information can be obtained from local people, especially regarding the history of the area, past and current presence of plant and animal life, and past and current human use of the site (Golder, 1999) (Figure 4).

Figure 4. Interviewing native elders regarding locations of critical habitat for marine life.

If possible, a site visit should be conducted. The site visit can reveal many key variables to the experienced team, such as evidence of past spills, surficial geology, surface water drainage patterns, likely direction of groundwater flow, and dominant plant species. Representatives of each key specialty should be involved; i.e., soil scientists, hydrogeologists, and biologists. In our experience, the time and money spent sending a qualified and experienced team to the site has saved much more time and money later. For example, a site visit showed that deep groundwater flow to a stream containing an endangered fish species was not the key issue (as had been assumed); rather, shallow groundwater contaminated by the site emerged as surface springs within 100 m of the site in the middle of a cattle pasture (Golder, 1995). This finding greatly altered the scope and nature of the risk assessment and the resulting mitigation.

Chemical contamination should not be the only focus for the investigation of the site. Physical alterations to habitat caused by the past site activities or current conditions can be equally or more important sources of stress to ecosystems (and to human land use). For example, dams across streams may have been constructed to divert water for use at the site. These dams may have altered or eliminated important fish habitat. Erosion and slumping of stream banks caused by clearing of forest and riparian vegetation is another prime cause of degradation of fish habitat. The building of access roads to a previously remote area may have been the largest source of stress to the environment because the roads provided access to people and, thus, to greater exploitation of fish, game, or marginal agricultural land.

2.2. IDENTIFYING THE KEY SOURCE–PATHWAY–RECEPTOR LINKAGES

We have found that a combination of knowledge about the site and professional judgment can be used to identify the key source–pathway–receptor linkages (Swanson and Gerien, 2006). Some linkages can be screened out based upon the physical or chemical form of chemicals in the environment; e.g., if none of the contaminants are volatile, then the vapur pathway can be eliminated. Other linkages can be screened out because the pathway is not operable under site conditions; e.g., leaching to groundwater is highly unlikely because of clay–silt layers between the source of contamination and groundwater. Linkages to certain ecological receptors can be eliminated if it is known that those plants or animals do not live on or near the site. Linkages to humans can be eliminated on the basis of resource use; e.g., people may not use the groundwater for drinking water.

Key linkages can be shown using simple diagrams, where the most important pathways are illustrated by the thickest arrows (Figure 5). Such diagrams are useful communication tools during discussions with regulators. In particular, these diagrams help explain why the next steps in the assessment of risks at the site is focused on specific pathways.

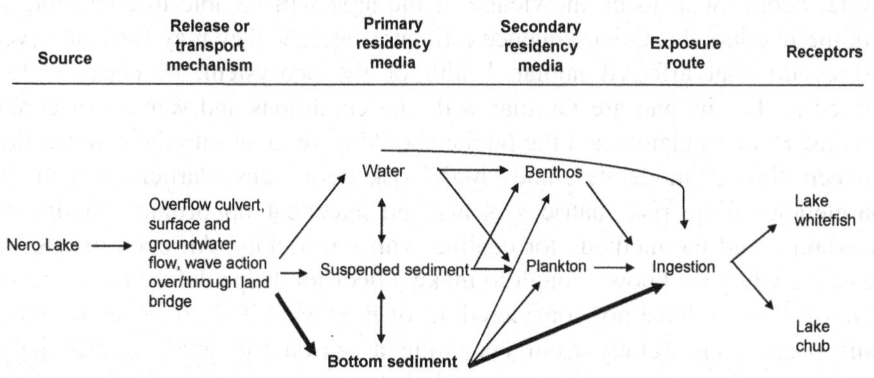

Figure 5. Example of a source–pathway–receptor diagram with key linkages shown by the thicker arrows.

3. What are the Relative Risks from the Key Linkages?

It is usually possible to complete a qualitative risk analysis of the key linkages using simple risk matrices that combine our understanding of the likelihood of an event and the consequences of that event (Figure 6). Such matrices can be used to identify the "risk scenarios" that would result in low, moderate, and high risk. The focus for remediation and other risk management actions (e.g., restrictions on land use) would be those risk scenarios that receive a "high" risk rank.

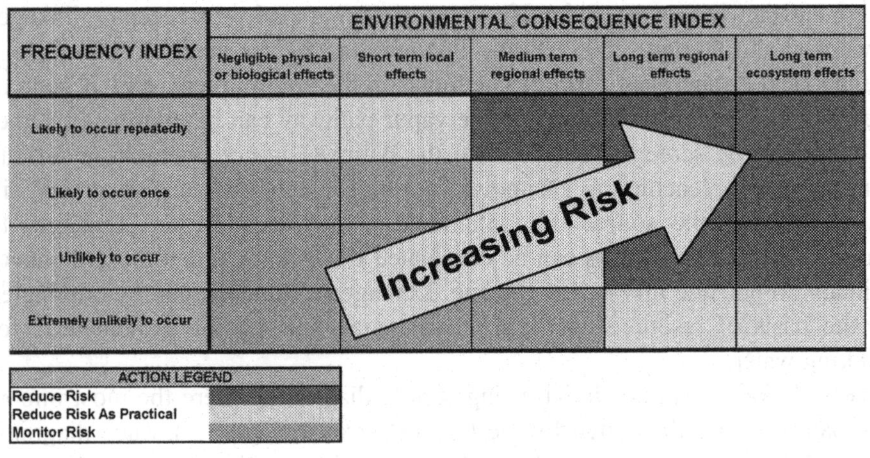

Figure 6. Risk matrix.

Risk matrices should be completed in consultation with regulators and the public. People with local knowledge of the area will be able to contribute to both the likelihood and consequence estimates because they may have observed past events that affected human health or the ecosystem, or because they worked at the site and are familiar with site conditions and sources of contaminants. Both regulators and the public should have input into the demarcation between "low", "moderate", and "high" risk definitions. Participation in the completion of the risk matrices is also an excellent opportunity to discuss uncertainty and the methods for dealing with uncertainty. An open discussion regarding when we know enough to make a decision helps defuse the common criticism that we have not considered all of the "what if's". In other words, it helps prevent uncertainty from becoming a reason for delay. It also helps prevent unnecessary conservatism and the inappropriate application of the precautionary principle.

Examples of risk scenarios that may receive a "high" risk rank are:

- Seepage of acid drainage through sandy soil to a stream that flows through the site and that contains sensitive fish species
- Tailings dam that is subject to breaching in a once in 25 years storm event with downstream human uses that include agriculture and drinking of surface water
- Corroding barrels containing pesticides located in an unlined landfill immediately above the water table where the near-surface groundwater is used for irrigation

All of the above examples include a likely release mechanism (e.g., seepage or release because of corrosion), an active pathway (e.g., surface water flow), and a sensitive receptor (e.g., humans drinking surface water). When both the likelihood of a release and transport along a pathway and the consequences of that release and transport are high, then the risk is high.

Examples of risk scenarios that may receive a "low" risk rank are:

- Low-volume spill to clay soil with no nearby surface water and deep groundwater overlain with impervious material with no human use of the site and only sporadic use by animals
- Localized sediment contamination with concentrations of chemicals that are only marginally greater than environmental quality criteria and where the sediments have a high contaminant binding capacity
- Minor habitat loss caused by very localized land clearing with good natural revegetation potential

The examples of risk scenarios with a "low" risk rank have low volumes or concentrations of chemicals or minor physical effects on habitat, very low potential for transport along pathways, and no or very little contact with receptors.

The most difficult risk scenarios to deal with are those that fall into the "moderate" category. The demarcation between "low" and "moderate" or "moderate" and "high" can be quite arbitrary, and is often affected the most by uncertainty in our knowledge and by data gaps. Examples of risk scenarios with a "moderate" risk rank are:

- Continuous, but low-volume seepage containing high concentrations of salts and low concentrations of metals into a small stream that drains into a wetland
- Periodic slumping of river banks along 5 km of the river resulting in sedimentation; however, no critical fish spawning habitat in the area

- Leaching of metals into groundwater used as drinking water by humans; however, zone of increased metal concentration very small and does not coincide with any known drinking water wells

The examples of "moderate" risk scenarios include situations where additional knowledge could move the risk rank in either direction; thus, uncertainty plays an important role. If we discover that the low-volume seepage with high salts drains into a wetland with endangered species that cannot tolerate elevated salts, the risk shifts to "high". If historic information reveals that fish spawning used to occur in the area of bank slumping, then it is likely that the slumping and sedimentation have eliminated spawning habitat and the risk shifts to "high". If more data on groundwater shows that metal concentrations are all less than drinking water guidelines, then the risk shifts to "low".

4. Can Remediation Proceed?

The key question to ask at this point is "Are we sure enough about the risks to make a decision?" In some cases, it is not necessary to proceed with any further assessment of risk after key linkages have been identified and a qualitative ranking of risks has been completed. This is because either (a) it is obvious that there are significant chemical and/or physical stressors that are affecting receptors on and/or near the site and uncertainty is low or (b) it is clear that chemical and/or physical stressors are not present in significant quantities or that they are not being transported to receptors and uncertainty is low. For example, if the site is an acid-generating mine tailings area that is discharging directly to a receiving stream which has little dilution capacity and acid-sensitive species, then remediation should proceed. Further risk assessment is not necessary to demonstrate the need for mitigation of acid drainage because we know a lot about the effects of acid on aquatic environments. If, however, we are not sure about the actual extent of release and transport of chemicals, or we are not sure that receptors would actually come into contact with the chemicals, then additional data will increase our confidence in the decision to either proceed with remediation or not.

The identification of critical data gaps must be in the context of the key linkages and the relative risk matrix. As noted above, a discussion of uncertainty and the likelihood that additional data would change the risk rank (and thus the decision to remediate or not to remediate) should include regulators and the public.

In some cases, uncertainty will be too great to support a confident decision to proceed with remediation. Additional data collection and a quantitative risk assessment will be required. Quantitative risk assessment goes beyond the risk matrix methodology and proceeds to the development of a risk estimate based upon modeling and (if sufficient field data are available) a weight of evidence regarding the current level of impacts from the contamination and physical stressors at the site. Methods for quantitative risk assessment are widely available from sources such as the US EPA (1998) and Suter (2006) and will not be described in detail here.

The quantitative risk assessment usually proceeds from screening-level modeling using conservative assumptions (e.g., use of the maximum or 95th percentile upper confidence limit of contaminant concentrations) (Suter, 2006). If the screening-level modeling shows acceptable risks, then no further action is required. If screening-level modeling shows risks above the acceptable threshold, then further assessment proceeds until there is sufficient confidence in the risk analysis to support remediation plans.

A full-scale quantitative risk assessment usually is not required at remote contaminated sites unless the site is very large and complex. In most cases, confident decisions regarding the nature and extent of remediation requirements can be made using knowledge of key linkages, relative risks from those linkages, and extensive consultation with regulators and the public.

5. Take Advantage of Natural Processes in Order to Minimize Long-Term Active Remediation Requirements

Remediation that requires extensive engineering works or long-term active intervention often is not practical at remote contaminated sites. Furthermore, engineered solutions at remote sites in cold climates have uncertain outcomes because of the lack of long-term experience in such environments. The alternative is to take advantage of natural processes.

The natural processes that can contribute to the remediation of a site include:

- Adsorption to organic materials such as peat
- Adsorption to fine particulates such as silts and clays
- Uptake into wetland plants such as cattails with subsequent sequestration in highly organic sediments composed of decaying plant material
- Neutralization of acidic drainage through naturally occurring limestone deposits
- Bacterial and fungal degradation
- Photodegradation
- Filtering of particulates through sand deposits;

- Settling of particulates in wetlands
- Soil development in the lee of large surface boulders or in depressions downslope of natural seasonal surface water drainage areas

In many cases, these natural processes must be assisted with an initial engineering effort. For example, to take advantage of the natural filtering and adsorptive capacity of a wetland, site drainage may have to be diverted towards the wetland. Acidic drainage may have to be collected and diverted through limestone materials. Initial water clarification of ponds with high suspended solids will be required before photodegradation can be effective. The objective of the engineering effort should be to create a situation where natural processes can take over, with little to no further intervention required.

An analysis of the net benefit of all remediation efforts should be conducted prior to any active engineering works proceeding at the site (Swanson and Gerein, 2006). For example, if a large volume of borrow material is required to construct berms for diversion of surface water, the impact of obtaining the borrow material on wildlife habitat must be balanced against the benefit of water diversion. Caution is required to ensure that the site does not become an "attractive nuisance"; i.e., a site that is now attractive to wildlife, but still contains sufficient contamination to pose a risk via food chain uptake or direct contact with soils or water. In some cases, measures to control or divert wildlife movement may be required until it can be confirmed that contamination levels and physical hazards no longer pose an unacceptable risk to wildlife. These measures can include fencing and flagging that startles and repels animals. Such measures have been required at northern Canadian mine sites to divert migrating caribou away from tailings dams and waste rock piles.

An additional objective for the remediation design of remote contaminated sites is to minimize the requirement for long-term monitoring. A monitoring design that focuses on key indicators of the success (or failure) of the remediation plan should be developed, with a clear definition of when it will be acceptable to stop monitoring. This plan will require input from regulators and the public. Public participation in the monitoring can be effective and economical, since local residents will be familiar with the site and can be trained to perform the required sampling and observations. This avoids the expense and practical difficulties of transporting people and equipment great distances to conduct the monitoring program. It also provides direct feedback to local communities via their own community members regarding the performance of the remediation plan at the site.

6. Summary

Risk-based remediation at remote contaminated sites offers a practical alternative to cleanup based upon conservative environmental quality criteria or standards. Risk assessment at these sites should proceed through a series of steps that provide focus and an understanding of the relative risks of key source–pathway–receptor linkages. A check to see if a confident remediation decision can be made should be performed at regular intervals within the process.

Key lessons learned about risk-based cleanup at remote sites are: (1) start with the obvious sources of risk; (2) do not forget nonchemical effects such as changes in water flow or water level; (3) interact with government and community representatives as much as possible; (4) learn about the site's human and ecological characteristics from local people; (5) realize that a cleanup decision can be made at any step in the process; (6) cleanup actions should promote "self-healing" by initiating or enhancing natural processes such as biological degradation and soil development; and (7) remediated sites should require very little maintenance or monitoring.

References

Golder, 1995, Ecological risk assessment for the Quirk Creek gas plant area, Prepared for Imperial Oil Resources Ltd., by Golder Associates Ltd., Calgary, AB.

Golder, 1999, Application of a risk-based approach to environmental management at 27 power stations operated by Northwest Territories Power Corporation. Prepared for Northwest Territories Power Corporation, Hay River, NWT, by Golder Associates Ltd., Calgary, AB.

Swanson, S., Gerein, K., 2006, Risk-based closure of uranium mill tailings – how to get from theory to practice – 1: risk assessment in support of closure planning, in: Uranium production and raw materials for the nuclear fuel cycle – supply and demand, economics, the environment and energy security, *Proceedings of international Symposium of the International Atomic Energy Agency*, Vienna, 20–24 June 2005, IAEA, Vienna STI/PUB/1259, pp. 293–301.

Suter, G.W. II, 2006, Ecological risk assessment, 2nd edn., CRC Press, Boca Raton, FL.

USEPA (United States Environmental Protection Agency), 1998, Guidelines for ecological risk assessment, EPA/630/R-95/002F, April 1998, Risk Assessment Forum, Washington, DC.

16. DECOUPLING ENVIRONMENTAL PROBLEMS FROM THE OVERALL ASPECTS OF POLITICAL DISPUTES: IS IT POSSIBLE? THE CASE OF THE PM2.5 PROJECT

SHMUEL BRENNER*, CLIVE LIPCHIN, ALLYSON AMSTER
The Arava Institute for Environmental Studies, D.N. Hevel Eilot, 88840 Israel

Abstract: More and more evidence is being accumulated regarding the health hazards from massive exposure to particulate matter, especially fine particles with a very small diameter. Transboundary effects of air pollution, especially particulate matter resulting from anthropogenic activities may result in negative impacts on the fragile stability existing between Israel and its neighbors. The Arava Institute for Environmental Studies (AIES) initiated a comprehensive study of PM2.5 to be conducted jointly between AIES (Israel), The Jordan Society for Sustainable Development (Jordan), and Al-Quds University (Palestinian Authority [PA]). It is believed that the cooperation during the study and the direct and open flow of information will bring about a clear, objective picture of the PM2.5 situation in the area and will lead to the establishment and enforcement of appropriate emission standards wherever required in the region. It is hoped that the study will also play its role in contributing to lessening of the ongoing tension. This paper analyzes the role that the environment has in bilateral agreements and the interplay between bilateral relations and environmental problem solving. Bilateral agreements are most often used to resolve a conflict, but the soft approach of compliance relies on good relations between States. Joint environmental projects offer a tool to move countries that do not have good relations towards common ground, which may lead to effective bilateral agreements.

Keywords: particulate matter; air pollution; transboundary; Israel; Palestinian Authority; Jordan; bilateral agreements

*To whom correspondence should be addressed. Shmuel Brenner, Kibbutz Ketura – AIES, D.N. Hevel Eilot, 88840, Israel; e-mail: aies22@ketura.ardom.co.il

R. N. Hull et al. (eds.), Strategies to Enhance Environmental Security in Transition Countries, 215–224.

1. Introduction

Environmental problems play a large role in the Israeli–Palestinian dispute. Yet, their role in the peace process is minimal compared to major issues of dispute such as political borders and the right of return (Brenner, 2001). Of all the environmental issues, water has received the greatest attention in peace talks over the years, and as a result is the most studied and has led to many joint Palestinian and Israeli projects. That other environmental problems are ignored is troublesome as without due recognition, they have the potential to become irreversible and cause tremendous ecological and social harm (Brenner, 2001).

One of the main difficult issues concerning the Israeli–Arab conflict is the question of whether there is a chance to separate, or decouple, the environmental problems from the "big" overall sources of the dispute. Until now, we can see that the conflict is mostly governed by political, ethnic, religious, and economic subjects with only minimal contribution of environmental aspects other than water. We know that cooperation is possible as other fields (most notably, those with promising economic rewards such as communication) have experienced successful partnerships (Brenner, 2001). How to successfully tackle environmental problems in the region is still an ongoing debate. Suggested solutions range from joint advisory boards to involving the "private sector and the auspices of foreign stakeholders in each of the relevant projects" (Brenner, 2001). Both the Israeli peace agreement with Jordan (1994) and the Oslo Accord with the Palestinian Authority (PA) (1995) contain specific paragraphs related to water and environment emphasizing the importance of these subjects in the integrated scheme of the agreements. Even though air quality issues are considered less influential than water, still transboundary movement of air pollutants play an important role in the interrelations between nations sharing a common border. And in fact, this subject is clearly mentioned in both agreements. Unfortunately, little has been done to counter the problem of transboundary air pollution in the region, and to date, there is no environmental agreement between Israel and the PA, and the environmental agreements between Israel and Jordan are largely restricted to issues of agriculture.

Scientists and environmental advocates in the region recognize the need for additional scientific studies to determine the sources and effects of pollution. This paper will examine the role of bilateral environmental agreements (BEAs) in North America and Eastern Europe within the context of cooperation and compliance techniques. The paper then looks at how environmental issues have been treated in bilateral agreements between Israel and the PA and Israel and Jordan. We will highlight the development and implementation failures of these components. We then review issues of transboundary air pollution and its health risks, including an overview of the current PM2.5 research project. This section highlights some of the projects barriers and more importantly, the motivation and goodwill behind the project's continuation. We conclude with questions about the feasibility of constructing strong transboundary environmental agreements in a region ripe with political conflict.

2. Bilateral Environmental Agreements

BEAs date as far back as 1351 with a fisheries agreement between England and Castile and now number around 1500 total agreements.[1] In the second half of the 20th century, many states have found bilateral agreements to be a successful tool in environmental problem solving, one that is simpler to implement than multilateral environmental agreements merely because fewer parties means fewer conflicts (British Columbia Guide to Watershed Planning, 2006; Downs and Jones, 2002; Zaelke et al., 2005). Yet, according to the International Environmental Agreement Database, between Jordan, Israel, and the PA there is only one agreement, the Agricultural Agreement Between The Government of Jordan And Israel (1995). We looked at three BEAs – the Agreement Between the United States of America and the United Mexican States on the Cooperation for the Protection and Improvement of the Environment in the Border Area (the La Paz Agreement) (1983); the Great Lakes Water Quality Agreement between the USA and Canada (1987); and the Szentendre Agreement between Bosnia and Herzegovina (BiH) and Srpska Republic (1998). While all three are different, they all lack common standards and enforcement capabilities.

The La Paz Agreement provided a general framework for environmental cooperation in the region, but was dependent on the adoption of later annexes to provide specific action regarding environmental regulations, which has occurred (American University, 2006). The Great Lakes Water Quality Agreement between Canada and the USA was created to control pollution and clean up the Great Lakes. The agreement is overseen by the International Joint Commission, which was established in 1909. Since 1972, the agreement was reviewed and resigned one time (1978) and has had one additional protocol (1987) added. The treaty works with both a water quality and science advisory board, which oversee the monitoring of pollution to the lakes. More recently, the Memorandum of Understanding between Bosnia and Herzegovina and Republika Srpska is an example of an environmental agreement in a region suffering from violent conflict. This agreement follows only 3 years after the 1995 peace

---

[1]This information is based on the International Environmental Agreements Database, which itself acknowledges difficulties in quantifying environmental bilateral agreements. Ronald B. Mitchell, 2003, *International Environmental Agreements Website*. Available at: http://www.uoregon.edu/~iea/accessed 19 July 2006.

agreement between the two entities, and establishes an Environmental Steering Commission to ensure a continuation of environmental regulations. Whatever criticism there may be about these treaties, what exists is a continuation and progression of the original treaties, suggesting some level commitment by both parties. This is in stark contrast to Israel, the PA, and Jordan where only one environmental agreement exits.

2.1. THE ENVIRONMENTAL COMPONENT IN THE AGREEMENTS BETWEEN ISRAEL AND JORDAN AND ISRAEL AND THE PALESTINIAN AUTHORITY

While Israel, Jordan, and the PA do not have any BEAs (with the exception of Israel–Jordan Agricultural Agreement), the two peace agreements between Israel and the PA and Israel and Jordan contain environmental components: Annex IV of the Treaty of Peace Between the State of Israel and the Hashemite Kingdom of Jordan (1994) and Annex III of The Israeli–Palestinian Interim Agreement on the West Bank and the Gaza Strip (The Oslo Peace Accords) (1995).

Annex IV of the Treaty of Peace Between the State of Israel and the Hashemite Kingdom of Jordan lists ten environmental subjects which are further divided into four geographical areas (FOEME, 2006):

1. Protection of nature, natural resources, and biodiversity, including cooperation in planning and management of adjacent protected areas along the common border, and protection of endangered species and migratory birds.
2. Air quality control, including general standards, criteria and all types of man-made hazardous radiations, fumes, and gases.
3. Marine environment and coastal resources management.
4. Waste management including hazardous wastes.
5. Pest control including house flies and mosquitoes, and prevention of diseases transferred by pests, such as malaria and leishmaniosis.
6. Abatement and control of pollution, contamination, and other manmade hazards to the environment.
7. Desertification: combating desertification, exchange of information and research knowledge, and the implementation of suitable technologies.
8. Public awareness and environmental education, encouraging the exchange of knowledge, information, study materials, education programs and training through public actions and awareness campaigns.

9. Noise: reducing noise pollution through regulation, licensing, and enforcement based on agreed standards.
10. Potential cooperation in case of natural disasters.
11. It is important to understand that the treaty does not include any steps to carry out the above environmental component of the treaty, rather they merely recognize the shared responsibility to the environment. This is very similar to the Oslo Accords signed in 1995 by Israel and the PA (PLO at the time of signing).

Article 12 of Annex III to the Oslo Accords is Environmental Protection. Here, air pollution is mentioned three times: Articles 12(A1) and 12(B4) and once again in Article 12(B5). Article 12(A1) lists air pollution as one of the environmental issues that the two sides agree to address, while section B is "Cooperation and Understandings". Article 12(B4) states, "Each side shall act for the protection of the environment and the prevention of environmental risks, hazards and nuisances including all kinds of soil, water and air pollution", while Article 12(B5) goes further to state:

> Both sides shall respectively adopt, apply and ensure compliance with internationally recognized standards concerning the following: levels of pollutants discharged through emissions and effluents; acceptable levels of treatment of solid and liquid wastes, and agreed ways and means for disposal of such wastes; the use, handling and transportation (in accordance with the provisions of Article 38 (Transportation)) and storage of hazardous substances and wastes (including pesticides, insecticides and herbicides); and standards for the prevention and abatement of noise, odor, pests and other nuisances, which may affect the other side.

This is the strongest language in this article as it calls for the adoption and compliance with international standards. The remainder of the environmental protection article is equally weak as it fails to outline any goals or requirements of either side (aside for agreements on environmental impact statements). However, no time lines were set, and little environmental progress between the two sides was made. Where some success did occur as a result from this agreement is with regard to water. Here, the treaty calls for the establishment of the Joint Water Committee, which while highly criticized for its weakness, it was in fact established, a clear sign that without clear stages and goals, little progress will occur after the first signing.

It is true that little has come from the environmental components of the two peace treaties, but it would be unwise to dismiss the significance of the environmental components of the two agreements. That they were included speaks to the recognition of transboundary environmental problems and the need for joint cooperation by scientists, something that exists in the three BEAs mentioned above.

3. Air Quality

3.1. AIR QUALITY AS A TRANSBOUNDARY ISSUE

The Convention on Long-Range Transboundary Air Pollution (1979) defines long-range transboundary air pollution as:

> air pollution whose physical origin is situated wholly or in part within the area under the national jurisdiction of one State and which has adverse effects in the area under the jurisdiction of another State at such a distance that it is not generally possible to distinguish the contribution of individual emission sources or groups of sources. (UNECE, 2006)

The treaty's origins lie in the United Nation's Economic Commission of Europe (UNECE), of which the USA and Canada are both members, concern regarding sulfur emissions and the acidification of water bodies. However, it now includes eight protocols, which address persistent organic pollutants (POPs), heavy metals, volatile organic compounds, and nitrogen oxides. The main topics of the convention are: air quality management, research and development, exchange of information, implementation and further development of the cooperative program for the monitoring and evaluation of the long-range transmission of air pollutants in Europe; and the settlement of disputes. This convention and its subsequent protocols reveal the change in understanding of the impacts of air pollution on human health. While it was created to address ecological problems (acid rain), we now understand the impact of air pollution on human health and the need to address air pollution as a transboundary pollutant in order to address issues of morbidity and mortality as a result of air pollution. It is now well established that the air pollution's biggest risk to human health is with fine particulate matter, PM2.5 (Vallius, 2005).

3.2. FINE PARTICULATE MATTER

Particulate matter is the suspended solid and liquid particles in air (Vallius, 2005). Most often classified as either fine or course particles, particulate matter is most often referred to as PM10 (course) and PM2.5 (fine). Fine particulate matter, PM2.5, is particles with a diameter smaller than 2.5 μm. The main sources of PM2.5 are industrial and residential combustion; vehicle exhaust;

particles resulting from the chemical transformation of combustion gases such as sulfur dioxide and nitrogen oxides. PM2.5 impacts the respiratory and cardiovascular systems. Respiratory problems include chronic bronchitis, asthma attacks, respiratory symptoms, decreased lung function, and airway inflamemation. Cardiovascular problems include heart attacks, cardiac arrhythmias, congestive heart failure, and peripheral vascular and cerebrovascular disease.

3.3. THE RELATIVE IMPACT OF AIR QUALITY ON NATIONAL WELFARE (ISRAEL)

In Israel, environmental problems have remained a low priority of the government their severity and public awareness masked by the political situation. And yet, air pollution poses a major hazard in Israel, claiming the lives of more people than car accidents and the conflict itself. According to Israel's Ministry of Transport, in 2003, there were 486 road casualties, and in 2004 and 2005 there were 518 and 471, respectively (Ministry of Transport). The Israel Defense Forces' (IDF) own record on casualties from hostilities in Israel reveal an average of 200 casualties per year since 2000 (figure includes both civilians and security officials) (IDF, 2006). While these numbers sound high, an Israeli air pollution study from 1995 to 1999, revealed higher numbers of mortality and morbidity resulting from urban pollution.

In the mid-1990s, the Israel Union for Environmental Defense (IUED), the Israel Ministry of Environment, and the United States Environmental Protection Agency (US EPA) partnered (along with the Tel Aviv Municipality and the Ashdod–Yavne Regional Association of Towns for Environmental Protection) to conduct a study of the public health risks due to air pollution. Their study revealed that air pollution is responsible for hundreds to thousands of deaths and hospital visits in Tel Aviv an Ashdod each year. They also found that the two pollutants most responsible were ozone and particulate matter. They were able to estimate that PM2.5 was responsible for 900 additional deaths in 1997 in Greater Tel Aviv and Ashdod (Israel Ministry of Environment et al., 2003). While Israel has proposed adopting the USA PM2.5 standard of 15 $\mu g/m^3$, all sites exceed the standard (Israel Ministry of the Environment, 2005), and there is no knowledge of sources and no indication as to how to regulate. Furthermore, there is no detailed PM2.5 monitoring currently being conducted in the PA and Jordan. This is what drives a study such as ours, so that the region can best mitigate the effects of air pollution, determine the sources, and avoid conflict from transboundary air pollution.

4. The PM2.5 "MERC" Project: Israel, Jordan, and the Palestinian Authority

In 2005 The Arava Institute for Environmental Studies (AIES) launched an air pollution project with two other research teams, one from Al-Quds University in the PA and another from the Jordan Society for Sustainable Development in Amman, Jordan.

The project has four goals:

- To conduct simultaneous monitoring of ambient PM2.5 in Israel, Jordan, and PA
- To characterize the chemical composition of ambient PM2.5 and identify sources
- To compare results from localities within the same airsheds with different urban activities (e.g., Jerusalem vs. East Jerusalem, Aqaba vs. Eilat)
- To produce relevant recommendations for policy makers in order to reduce harmful impacts.

and asks four key questions:

- What are the specific sources, components or PM attributes linked to adverse effects?
- How do we know what sources to control to reduce public health risks?
- What is the association between ambient PM and actual personal exposure? What about specific PM components?
- Who is susceptible and why?

There are 11 monitoring sites. In Israel: Tel Aviv, Haifa, West Jerusalem, and Eilat. In the PA: Nablus, East Jerusalem, and Hebron. In Jordan: Amman, Aqaba, Zarqa, and Rahma. Filters from the sites will be collected every 6 days, and chemical analysis will be conducted in all four countries (Israel, Jordan, the PA, and USA). This study is unique because it will be the first time that such detailed PM sampling and analysis is conducted for such a long duration in this region.

There are to be two outcomes of this study: first will be the elucidation of PM2.5 composition, sources, and distribution; and second is capacity building for advanced air quality or health effects research and analysis in Middle East. In addition, at the project's completion, we will be able to answer four questions:

1. What is the relative impact of regional sources and local sources of particulate matter in these regions?

2. What are the seasonal variations in sources and composition of PM in these regions?
3. How does the composition and concentrations of PM2.5 compare across sites? To PM2.5 in Europe and North America?
4. What are the sources of high particulate matter episodes in this region of the world and what is the association of these particulate matter episodes with other pollutants such as ozone and carbon monoxide?

In May, the project coordinated visits to all 11 sites and held a 2-day conference in East Jerusalem. There is no doubt that the political situation complicates joint research teams. Logistically, travel restrictions on Palestinians and Jordanians can make simple communication between partners difficult. The potential for travel restrictions and new political developments that can affect travel through the region are difficult to predict, which can have impacts on a scientific study, and in fact this occurred in May, preventing certain site visits and shortening the 2-day conference. It is important for all parties to understand the emotional toll that these problems can have on different partners. Clearly, the complications faced by scientific projects in this region must be considered in all stages of project development.

5. Discussion

International agreements rely on goodwill for success. Few have within them true mechanisms for compliance; that is, they lack punitive measures for countries that fail to comply. Rather, diplomatic measures are used to entice noncompliant countries to come back into the fold. The bilateral agreements of Israel and the PA were constructed similarly to international law, and therefore relied on the goodwill of the parties to fully comply with the agreements. However, can international agreements be constructed using international norms and theory of reputational consequences when reputation may not matter? Noncompliance between Israel and the PA should not be viewed as a reputational failure, but rather a likely consequence of a bad relationship. If the two parties in a bilateral agreement lack goodwill from the outset, ambiguous language will not forward the agenda. The one area that the Oslo Accords had some (albeit minimal) success was with water. Here, the treaty outlined steps for future activities, and, in fact, a joint water commission was created. One component that bilateral agreements all have in common is the creation of a joint scientific board. The importance of joint monitoring cannot be stressed enough.

In a region of such volatility, the potential for conflict surrounding pollution exists already. The need for joint research projects is two-fold: first, continued relations between different sectors is the best way to achieve political stability; second, given the strained relations between the different parties, the potential for bias or methodological tampering could undermine any project. The PM2.5 project is one such step in recognizing the need to address environmental problems regardless of the political situation, but also be acutely aware of the impacts that politics can have on science.

References

American University, 2006, US-Mexico Lapaz agreement on waste export (LAPAZ), Trade and Environment Database Projects; http://www.american.edu/TED/LAPAZ.HTM 19 July 2006.

Brenner, S., 2001, Environmental trustees, position paper.

British Columbia Guide to Watershed Planning (July 10, 2006); http://www.bcwatersheds.org/issues/water/bcgwlp/s4.shtml

Downs, G.W., Jones, M.A., 2002, Reputation, compliance, and international law, *J Legal Studies* 31: S95–S114.

FOEME (July 19, 2006); http://www.foeme.org/regional.php

IDF (July 25, 2006); http://www1.idf.il/SIP_STORAGE/DOVER/files/7/21827.doc

Israel Ministry of Environment, Israel Union for Environmental Defense, Tel Aviv Municipality, Ashdod–Yavne Regional Association of Towns for Environmental Protection, USEPA, 2003, A comparative assessment of air pollution public health risks in two Israeli metropolitan areas, 1995–1999, pp. i–F-5.

Israel Ministry of the Environment (January 2005); http://www.sviva.gov.il/Enviroment/Static/Binaries/odotHamisrad/jan05_1.pdf. (pp. 18–21).

UNECE (July 18, 2006); http://www.unece.org/env/lrtap/

Vallius, M., 2005, Characteristics and sources of fine particulate matter in urban air, Academic dissertation, National Public Health Institute, Kuopio, pp. 13–23.

Zaelke, D., Stilwell, M., Young, O., 2005, Compliance, rule of law, & good governance: what reason demands, making work for sustainable development, in: *Making Law Work: Environmental Compliance & Sustainable Development*, Zaelke, D., Kaniaru, D., Kružíková, E. (eds.), International Law Publishers, London, pp. 29–51.

17. WATERSHEDS MANAGEMENT (TRANSYLVANIA/ROMANIA): IMPLICATIONS, RISKS, SOLUTIONS

ANGELA CURTEAN-BĂNĂDUC*, DORU BĂNĂDUC, CORNELIU BUCŞA
"Lucian Blaga" University of Sibiu, Faculty of Sciences, Ecology and Environmental Protection Department, Sibiu, Romania

Abstract: This work intends to establish ecological rehabilitation, conservation, and management measures for the Olt and Mureş Rivers and their main tributaries. These measures will be based on scientific ecological assessment and monitoring of the structure of the river communities (benthic macroinvertebrates and fish) and of the physical and chemical water parameters over the last 6 years. The major categories of human impact on these rivers are highlighted (water and sediment pollution; hydrotechnical works – dams; river channels, marshes and floodplain drainages, cut of meanders, river banks reshaping and embanking, tributary deviations; riverbed mineral exploitation; and riverine land exploitation), and also their consequences on the rivers. Specific ecological rehabilitation measures and optimum conservation measures are described for these rivers.

Keywords: running water; ecological assessment; monitoring; rehabilitation and conservation; Olt River Watershed; Mureş River Watershed; Romania

1. Introduction

The impetus for ecological assessment and rehabilitation has arisen from the major concern of hydrobiologists over the declining quality of lotic systems as complex resources, on the Romanian territory, an accentuated phenomenon over the last seven decades (Antonescu, 1934, 1947; Bănăduc, 1999, 2000,

*To whom correspondence should be addressed. Angela Curtean-Bănăduc, "Lucian Blaga" University of Sibiu, Faculty of Sciences, Ecology and Environmental Protection Department, Sibiu, Romania; e-mail: banaduc@yahoo.com

R. N. Hull et al. (eds.), Strategies to Enhance Environmental Security in Transition Countries, 225–238.

2001, 2005, 2006; Bănărescu, 1964, 1969, 1994; Curtean 1999, 2004, 2005a, b; Curtean-Bănăduc, 2000, 2001, 2004, 2005; Curtean-Bănăduc and Bănăduc, 2001, 2004a, b, 2006; Curtean et al., 1999; Curtean-Bănăduc et al., 2001).

In any historical period and in almost all geographic regions, water was a priceless resource, but always was handled by people with divergent interests, different methods, and with significant different spatial and temporal effects. (Antipa, 1909; Antonescu, 1953; Băcescu, 1947; Bunn and Arthington, 2002; Collier et al., 1996; Curtean-Bănăduc and Ilie, 2001; Gore and Schields, 1995; Moyle and Leidy, 1992).

The Mureş and Olt are located mainly in the Romanian Carpathians arch, and together with their tributaries, vary substantially along their courses in climate, geology, relief, hydrology (Posea et al., 1983), and anthropogenic impact.

Although the macroinvertebrate communities and the fish communities may have a high degree of natural variability, they can be useful indicators of aquatic ecosystem status or health. Also, it is recommended that macroinvertebrates and fish be given consideration in biological water-quality surveys of lotic systems because they generally are discerned by the public to be ecologically relevant, and they are in direct relation to legislative mandates because of human health and endangered species concerns. Analyses of these taxonomic groups are essential for establishment of lotic system management.

The dimensions of these rivers, and the number and dimensions of their tributaries, their natural and historic economic importance, as well as the variable and aggressive types of human impact, justify such a study for this specific watershed area (Figure 1).

Early in time (prehistory), these river basins constituted a zone of local high potential for human activities and therefore these rivers were used intensively. Historically, important human impact on the river started in the 1200s (water and sediment pollution; hydrotechnical works – dams; river channels, marshes and floodplain drainage, cut of meanders, river banks reshaping and embanking, tributary deviations; riverbed mineral exploitation; riverine land exploitation and both sewage waste and industrial pollutant discharges) and continued for more than seven centuries until the present day.

Especially in the last five decades, human impact was heavily increased, causing visible disruption in the ecological functioning of the rivers. Research was needed to assess the observed ecological effects, to find new management actions for the new situation and to predict some aspects of the evolution of the future lotic system.

These two major river watersheds were in different degrees, in different historical periods, under long-term human impact pressure that threatened their

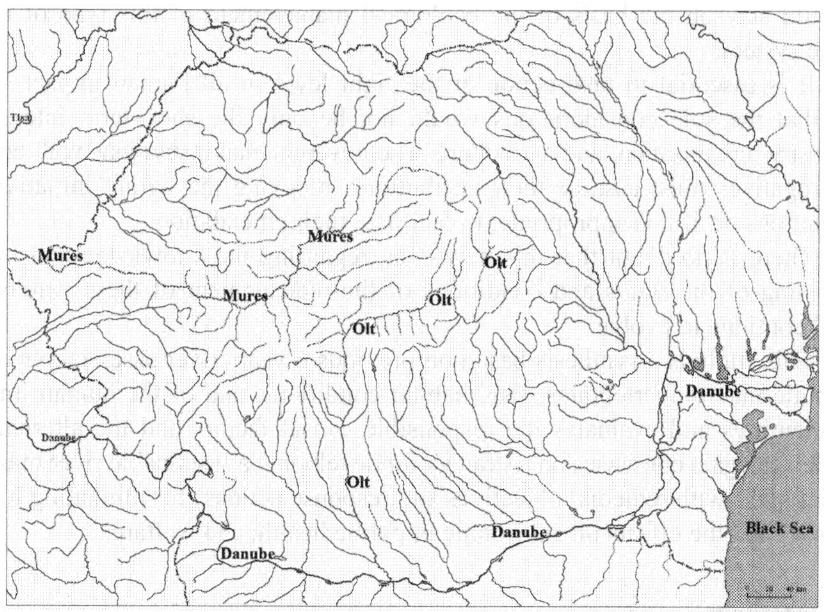

Figure 1. The Olt and Mureş Rivers watersheds study unit location.

structure, functions, and capabilities to offer priceless services to the riverine human communities. As a direct consequence of human impact, the river system habitats have become in some sectors severely modified and/or degraded, aquatic organisms contaminated with various pollutants and river and riverine biological communities changed.

Although there is a considerable number of damaging practices and activities affecting the Mureş and Olt River basins resources, the potential for their recovery is substantial, if an ecological rehabilitation strategy is implemented. Past and present response actions, carried out or planned, do not or will not sufficiently remedy the injury to natural resources without further integrated actions.

The anthropogenic impact consequences on these two major Romanian Carpathians river systems shaped the patterns for the needed local and regional specific ecological rehabilitation measures and for the needed optimum conservation measures for these rivers.

Finding the causes of the degradation of the natural equilibrium and the rehabilitation and conservation measures, finally highlighted the elements which should stay on the basis of the ecological management of this type of major lotic system.

It is essential to take action at the right level in all human impact cases, so that the sole considerations would not be only the short-term interest of industry, urbanization, and agriculture. The governmental institutions with control prerogatives must assume their attributions, blocking the wrong initiatives or intentions or, if it is appropriate, to facilitate their emendation.

There is no doubt that this new data, regarding the knowledge and understanding of the status and conditions of the management of these watersheds will continue to evolve.

This study is a synthesis based on previous 6 years river assessments of the benthic macroinvertebrates, fish, habitat conditions, and of the human impact. It identifies and evaluates the responsible causal factors and the alternatives, which can gain consensus on a strategy for developing a sustainable river management plan (with remedial objectives and response actions) for mitigating human impact and the effects on environment, public health, and welfare.

2. Methods

The presence and the effects of human impact relative to a reference ecological state was analyzed in terms of relative biologic integrity, biodiversity indexes (Margalef, Shannon-Wiener, Simpson, equitability, Belgian Biotic Index, Hilsenhoff Biotic Index, Carpathian Fish – Index of Biotic Integrity (Bănăduc and Curtean-Bănăduc, 2002), based on a variety of benthic macroinvertebrates [plecopterans, trichopterans, and ephemeropterans] and fish), and on physical and chemical water parameters (temperature, pH, total hardness (TH), dissolved oxygen (DO), biochemical oxygen demand (BOD_5), chemical oxygen demand (COD-Mn), Cl^-, SO_4^{2-}, NO^-_3, PO_4^{3-}, total N, total P, Pb, Zn, Cu, Cd, and Mn), which were used as ecological indicators of the state of the river.

The 124 river sectors were chosen according to: (1) the valley morphology, (2) the type of river substratum, (3) the confluence with the main tributaries, (4) the relative undisturbed areas, and (5) to the human impact presence bias (riverine land use, hydrotechnical works, urban and industrial pollution sources).

Generated data during this phase filled previously existing data gaps in order to provide a comprehensive understanding upon which to evaluate possible remedial alternatives for the human and ecological risk situations. The biological monitoring was used to characterize the response of the aquatic ecosystems to multiple disturbances, considering that the integrity of the biota inhabiting the river ecosystems provides a direct and integrated measure of the integrity or health of the river.

In the end, with these offered alternatives, a specific case proposal is developed concerning natural resources restoration, conservation, and management of the Olt and Mureş Rivers.

3. Significant Contributors to River Disturbance

In examining the river biotic and abiotic characteristics, research has determined that primary contributors to the disturbance of natural conditions are river water pollution, hydrotechnical works, and riverbed mineral over-exploitation.

3.1. WATER POLLUTION IMPACT

3.1.1. Habitat, water as direct resource

A wide range of natural resources and natural resource services are affected or potentially affected by the release of different pollutants.

Historically, these rivers received both sewage waste and industrial pollutants for more than seven centuries; accidental and permanent release of hazardous substances has at least in the last five decades, the anthropogenic impact being in this period drastically higher.

The main sources of river contamination include inactive and active hazardous waste disposal sites and technical installations, combined sewer overflows, sewage effluent, and tributaries entering the rivers. The main released pollutants (organic substances in different degrees of decomposition, phytosanitation products, chemical zootechnics disinfectants, suspensions, nitrites, nitrates, detergents, oil products, sodium chlorite products, organic solvents, synthetic resins, SO_2, SO_3, NO_2, CO, Cr, Pb, Zn, Cu, and Cd) are over the legal limit concentrations, discharging accidentally and/or permanently from many important riverine sources (steel, machines, textile, leather, wood, food, zootechnical, metallurgy, ceramic, enamel, glass, building materials, silk, wool, glue, carbide, sulphur acid, plastics, etc. industries) of some of the main localities.

It has to be highlighted that not all riverine main localities (especially the smaller localities) have water treatment plants and sewage systems, not enough sewage systems and/or deteriorated sewage systems; a lot of water treatment plants work under the necessary standards; a lot of illegal waste deposits exist on the river banks.

The quantity and concentration of the hazardous substances and the frequency of the releases are sufficient to cause injury to natural resources (sediment, water, biota), resulting in various long-term ecological effects in the Olt and Mureş rivers. The affected resources with their sundry ecological and human services (habitat, water as a direct resource, recreational activities, etc.) are described further in the context of contamination impact.

The rivers provide habitat for macroinvertebrate and fish communities, including feeding, breeding, and nursery services. Bioassessment using these groups of organisms reveals that many sectors are considerably affected from this point of view by the elevated hazardous substance concentrations, particularly damaging aquatic organism populations. Concerning the timescale, important pollution levels were registered through the local biocoenosis reaction starting around 1960 and the rivers have undergone a progressive loss of biodiversity as a result of habitat degradation or loss.

These rivers represent the cheapest high-quality direct water resource in the area only in their upper third part sectors. The significant modifications of the analyzed invertebrate and fish associations reflect that, downstream of these upper sectors, these rivers are no longer appropriate to be used as reliable and cheap sources of water. Beginning with these hot spots, water pollution in these rivers acts as a constraint on the economic and social development of the downstream localities. The improved raw water quality for the water supply can improve the local economy efficiency and public health standards.

Numerous sectors of these rivers, except their mountainous part, and their environment are less appropriate to support both consumptive and non-consumptive recreational activities such as recreational fishing, with no health hazards coming from fish consumption, swimming, boating, and wildlife viewing, activities with (for the moment) a theoretical high touristic potential in the area.

3.2. HYDROTECHNICAL WORKS IMPACTS

Hydrotechnical works impacts include river channels, marshes and floodplain drainages, cut of meanders, river bank reshaping and embanking, and tributary deviations.

River water and sediments serve as media for the transport of energy and nutrients, and as habitat for various aquatic biota. Both the services for transport of energy and nutrients and as habitat were affected due to hydrotechnical works impact.

The long-term negative impact is higher then the short- or medium-term positive "economic" one, due to: (1) the economic overevaluation on a long timescale of arable and grazing land, and the extensions of arable or grazing land through elimination of wetlands which are considered "unprofitable land"; (2) the wrong idea of unhealthiness (fog, mosquitoes) generally attached to wetlands; and (3) the widely held opinion that wetlands are useless, owing to a lack of understanding concerning their complex role in maintaining fundamental ecological balances and exchanges between terrestrial and aquatic ecosystems.

Historically, the first important human impact presence on these rivers was due to the building and later industrial development and urbanization of main localities on their banks. Many drainage works in the river floodplains, marshes and, secondary channels and tributary deviations were initiated, to build and develop towns starting especially in the 12th century. Today, these rivers, drastically affected by such human activities, in both terms of biotope and biocoenosis modifications, pass through these anthropogenically affected sectors.

In the modern period, activities like cutting meanders, embanking of some river sectors and some ponds and wetlands connected with the river drainage, were promoted in order to ensure increased arable land surfaces.

The natural resources (sediment, water, biota) have been adversely affected by construction and management of all these hydrotechnical works. These human activities have had major negative consequences for aquatic biocoenosis, causing the loss of many natural ecosystems, loss of biodiversity, as well as the disturbance of the natural ecological balance within the aquatic and terrestrial ecosystems.

Actually, the hydrotechnical works in the Olt and Mureş river basins area include major river banks reshaping and embankments, floodplain and marshes drainages, and tributary deviations.

The river continuum, both in structure and function is also disturbed in the present in numerous sectors by riverbed "cleaning" activities, realized by the local administrations and Romanian Waters Administration, actions which affect the assimilative and self-cleaning capabilities of the river biocoenosis, and by the clear cutting of the trees and bushes which formed in the past a continuous "functional green corridor" on the river banks. These ex-riverine green corridors with no more buffer function are missing for long distances, also (mainly) due to the arable land extensions in proximity to the river.

Dam construction and management (water abstraction and flow regulation) modify the natural hydrologic regime of these rivers, and also cause riverbed and channel modifications. These major disturbances induced on a spatial and temporal scale an anthropogenic variability of habitats and associated biota, more accentuated in the proximity of dams.

The variable rainfall pattern and the occurrence of sporadic droughts, exacerbate the impact of water abstraction on instream and riparian habitats. The resulting cessation of surface flow in a naturally perennial river during the dry season and during droughts with loss of fast flowing instream habitats in the main stem of the river had detrimental consequences for the associated biota.

Habitat integrity changes are particularly severe for aquatic organisms (shrinking or destroying the spawning territory and the breeding success of the local or valuable characteristic aquatic fauna) inducing biocoenosis changes.

Not only the major instream hydrotechnical constructions have important influences on these river habitats and associated biota. For example, the small concrete hydrotechnical works in the riverbed induced a decrease in the local biotic integrity.

3.3. MINERAL RESOURCE OVEREXPLOITATION IMPACT

Environmental impacts also arise from gravel mining activity along these rivers. Important riverbed rocks or boulders and sand exploitation started in the 12th century to supply construction materials to the building of the important medieval city of the area. These activities have continued, and riverbed exploitation activities have increased significantly in the last five decades.

Natural resources (sediment, water, biota) have been adversely affected by mineral riverbed overexploitation. The exploitations increase banks and riverbed erosion resulting in excessive downstream siltation and a decrease in water quality.

If we consider that restoration of aquatic ecosystems can be defined as "return of an ecosystem to a close approximation of its condition prior to disturbance", the term restoration meaning the reestablishment of predisturbed aquatic functions and related physical, chemical, and biological characteristics, a holistic process which is not the sum of simply creation or reallocation or enhancement activities, it is obvious that this cannot be achieved through isolated manipulation of individual existing elements.

It is necessary that ecological restoration be approached only under the conditions of proper management of ecological integrity. Ecological integrity includes a critical range of variability in biodiversity, ecological processes and structures, regional and historical context, and sustainable cultural practices.

In terms of budgetary constraints, the single reasonable approach in the short and medium term, is a sectoral one (including pollution, hydrotechnical works, and riverbed mining overexploitation). This approach attempts to restore these rivers to their ecological state from the middle of the 18th century. This period is considered as appropriate because it was a time of aggressive human impact. In addition, comparable data exist about this period due to the activity of different societies for natural sciences in the area.

Integrated studies all over the world show that old hydrotechnical works contribute to continuously increasing stresses on aquatic ecosystems. These studies also show increasing financial efforts to exploit river resources and a decreasing quality of human life for present and future generations. We must accept the need for amelioration to already-constructed dams, and consider decommissioning some of them and also the river embankments.

Due to the present unsustainable mentality and "technical" education of the decision makers, we will exclude here the radical but the most efficient actions which are able to change the present situation. However, the potential for the recovery of natural resource damages of these rivers is substantial even in the actual technical administration "hostile" environment, if the minimum five amelioration actions levels described in Section 4 are implemented.

4. Remedies for River Disturbances

Remedies are proposed for four main riverine disturbances, as detailed in Sections 4.1–4.4.

4.1. WATER AND SEDIMENT POLLUTION REMEDIES

It is well known that there is a general lack of compliance with Romanian and international regulations governing wastewater management. This, along with a lack of adequate effluent dilution, appears to be the major cause of degradation in these rivers. Action is needed when: (1) the rivers or river sectors do not have an abundant supply of water for dilution; (2) the effluent treatment conditions are not sufficient; (3) past and present response actions have not sufficiently restored the injured river natural resources; and (4) there is actual or potential continued release of hazardous substances. Actions could include:

- Unconditional respect of Romanian and international environmental laws by everybody.
- Increasing water consumption efficiency through the use of general contour meters and a reliable pipe transport system.
- Keeping hazardous waste sites inactive.
- Creation of a hazardous waste site evaluation unit. This unit should include biologists to represent the interests of wildlife. This unit would be involved in the process of identification and cleanup of inactive and active hazardous waste sites.
- Developing a potential resource damage claim against the major polluters.
- Changing the inadequate physical and chemical standards used to characterize and manage wastewater treatment in Romania to protect the downstream environment in the vicinity below wastewater works.
- Countermeasures against accidental oil spills.

- Ensuring the protected and semiprotected river sectors are large and dense enough to allow the river self-cleaning capacity to be active enough to face human impact pressure.
- Managing the healthy river biocoenosis as biological capital whose interest is collected through reducing the expenses for water cleaning technologies.

4.2. HYDROTECHNICAL WORKS IMPACT REMEDIES

Remedies for impacts arising from hydrotechnical works include the following:

- Activities for restoring the river assimilative capacity, including land acquisition and wetlands restoration.
- The dam management strategy must be based on the equitable allocation of water resources. It must include the minimum water flow necessary for downstream users (domestic, industrial, agricultural, etc.), known as the servitude discharge, as well as the minimum discharge required in downstream watercourse sectors, to provide the natural conditions for the existing aquatic ecosystems, known as the sanitation discharge.
- Revitalization of the best traditions for land protection and use (ecological systematization of riverbanks, protection of lands near the river channel).
- A strategy of "no net loss" of wetlands. As long as wetlands may legally be destroyed, these impacts can be mitigated and compensated through restoration, creation, or enhancement of other wetlands.
- Protect and restore sectors of typical local ecosystems.

4.3. RIVERBED MINERAL AND RIVERINE LAND EXPLOITATION REMEDIES

Remedies for impacts arising from exploitation of river land and the riverbed include the following:

- Rational gravel mining activity, based on riverbed exploitation quantities less than the annual riverbed regeneration rate, and filtering the used industrial water for sand or gravel washing
- Determination of incentive policies for cultivation of multiyear cultures (vineyards, orchards, forests)
- Rehabilitation of riverine forest corridors, with interdiction of arable land extension within the minimum 10 m riverine corridor along the river banks
- Prohibiting access to the upper parts of catchment areas (limiting damage from water erosion) so that spontaneous perennial vegetation could regenerate in the best conditions

- Rotating rational silviculture and grazing activities, having regard to seasonal conditions, especially on the river banks

4.4. REMEDIES FOR PROTECTED SPECIES AND AREAS

Remedies for protected species and areas include the following:

- Protected habitats are needed which are sheltered from all human or man-related aggression. These would make excellent observation areas and serve as reference areas for monitoring and comparing to other more deteriorated environments. The larger a protected area is, the better it could fulfill its conservation function, to offer diverse habitats and enable species to live in sufficient numbers to maintain healthy populations.
- A properly conceived protection plan is a useful aid to direct and indirect economic development of rural communities.
- Complementarity should exist between local development and conservation, a demonstration that could be achieved in the context of rational use of wildlife.
- Protection of key, rare and/or endangered species.

5. Technical Aspects Related to Management Plan Implementation

The implementation of a management plan should be done with the participation of associations, groups of farmers, nongovernmental organizations (NGOs), and groups of specialists of different disciplines including those who work for improvement of forests, pastures, improvement of practices for use of terrain, control of erosion, etc. This would be facilitated by the creation of a functional interdisciplinary professional watersheds council. The management plan should include the criteria and indicators required for monitoring the relevant structures of the ecosystem. This management plan should be reviewed periodically in concordance with the EU Water Framework Directive.

6. Conclusions

The "argument" with which the local and/or national authorities will reject as "impossible" the implementation of such an ecological rehabilitation will be the lack of necessary funding. In fact, the basic problems are their incorrect understanding of the problem as a whole, their indecision and the fact that they did not use sufficient or integrated professional advisory services.

The situation of these rivers is, in general, a common one among Romanian rivers. The Romanian specialists involved in water management are astonishingly determined to follow the same wrong wetland management strategies which were "successfully" collapsed or are changing in western countries. There are "good examples" of destroying wetlands resources, overexploitation of Western European, North American, and ex-communist countries, and destructive and long-term unprofitable effects. There also are good examples, from the last two decades, of the reshaping of the human–wetlands relationship (which includes even the destroying of some dams and embankments!), in developed countries. If our "specialists" have no capacity to learn from the mistakes and experiences of others, claiming again and again false economic considerations, we will learn on our own, but at our expense.

Based on the disappointing last 60 years of experience in Romania, it is idealistic to think that ecological restoration of these rivers can be complete especially in our lifetime even if this complex action will start now. The long-term human impact effects, the decision makers "understanding" concerning nature, and economic and social interrelations, will make ecological restoration possible only in the actual European conditions of high standards and observance of regulations in both the environment and economic fields.

References

Antipa, G., 1909, Fauna ihtiologică a României, Edit, Academiei, Bucureşti, 294 p.

Antonescu, C.S., 1934, Peştii apelor interioare din România, Imprimeriile Monitorului Oficial.

Antonescu, C.S., 1947, Peştii din apele României, Colecţia Îndrumări Nr. 5, Institutul de Cercetări Piscicole al României, Imprimeria Naţională, Bucureşti.

Antonescu, C.S. şi colab., 1953, Valorificarea economică şi piscicolă a râului Sâmbăta (din Munţii Făgăraş).

Băcescu, M., 1947, Peştii aşa cum îi vede ţăranul pescar român, *Publ. Inst. Cerc. Pisc. Rom., col. Monographia* nr. 3.

Bănăduc, D., 1999, Data concerning the human impact on the ichthyofauna of the upper and middle sectors of the Olt River, *Transylv. Rev. Syst. Ecol. Res.*, Sibiu, 1: 157–164.

Bănăduc, D., 2000, Ichthyofaunistic criteria for Cibin River (Transylvania, România) human impact assessment, *Trav. Mus. Hist. Nat. "Grigore Antipa"*, Bucureşti, Vol. XLII: 365–372.

Bănăduc, D., 2001, *Valea Lotrioarei* – capitolul Ihtiofauna râului Lotrioara, Editura "Mira Design", pp. 48–54.

Bănăduc, D., 2005a, Speciile şi subspeciile din România ale genului Gobio (Gobioninae, Cyprinidae, Pisces) *Brukenthal Acta Musei I.*, 3: 125–144.

Bănăduc, D., 2005b, Fish associations – habitats quality relation in the Târnave Rivers (Transylvania, Romania) ecological assessment, *Transylv. Rev. Syst. Ecol. Res.*, 2: 123–136.

Bănăduc, D., 2006, Râul Mare Ichthyofauna, *Transylv. Rev. Syst. Ecol. Res.*, 3: 202.

Bănăduc, D., Curtean-Bănăduc, A., 2002, A biotic integrity index adaptation for a Carpathian (first–second order) river assessment, *Acta oecologica*, Sibiu, Vol. IX, 1–2; 5–23.

Bănărescu, P., 1964, Pisces-Osteichthyes, Fauna R.P.R., Vol. 13, Edit, Academiei, Bucureşti, 961 p.

Bănărescu, P., 1969, Cyclostomta şi Chondrichthyes, Fauna R.S.R., Vol. 12 (1), Edit, Academiei, Bucureşti.

Bănărescu, P.M., 1994, The present – day conservation status of the freshwater fish fauna of Romania, *Ocrotirea naturii şi a mediului înconjurător*, Bucureşti, 38 (1): 5–20.

Bunn, S.E., Arthington, 2002, Basic principles and ecological consequences of altered flow regimes for aquatic biodiversity. *Environ. Manage.* 30: 492–507.

Collier, M., Webb, R., Schmidt, J., 1996, *Dams and Rivers; Primer on the Downstream Effects of Dams*. Circular 1126. Denver, CO, US Geological Survey.

Curtean A., Sîrbu, I., Drăgulescu, C., Bănăduc, D., 1999a, *Impactul antropic asupra biodiversităţii zonelor umede din bazinul superior şi mijlociu al Oltului*, Editura Universităţii "Lucian Blaga" din Sibiu, p. 102, ISBN 973-651-003-4.

Curtean, A., Morariu, A., Făcălău, S., Lazăr, B., Chişu, S., 1999b, Data concerning the benthic communities of the Cibin River (Olt River Basin), *Transylv. Rev. Syst. Ecol. Res.*, Vol. 1, Editura Universităţii "Lucian Blaga" din Sibiu, pp. 99–111, ISSN 1841-7051, ISBN 973-9410-69-3.

Curtean-Bănăduc, A., 2000, Cibin River (Transylvania, Romania) ecological assessment, based on the benthic macroinvertebrates communities, *Acta oecologica*, Vol. VII, Nr. 1–2, Editura Universităţii "Lucian Blaga" din Sibiu, pp. 97–109, ISSN 1221-5015.

Curtean-Bănăduc, A., 2001, Aspects concerning Cibin River (Transylvania, Romania) stonefly (Insecta, Plecoptera) larvae associations, Analele Universităţii "Ovidius" Constanţa, *Seria Biologie – Ecologie*, Vol. 5, Ovidius University Press, pp. 1–8, ISSN 1453–1267.

Curtean-Bănăduc, A., 2004, *Cibin River macroinvertebrates communities diversity analyze*, Analele Universităţii din Oradea, Fascicula Biologie, TOM XI, Editura Universităţii din Oradea, pp. 21–32, ISSN 1224-5119.

Curtean-Bănăduc, A., 2005a, Târnava Mare River (Transylvania, Romania) ecological assessment, based on the benthic macroinvertebrates communities, *Transylv. Rev. Syst. Ecol. Res.*, Vol. 2, Editura Universităţii "Lucian Blaga" din Sibiu, pp. 109–122, ISSN 1841-7051, ISBN 973-739-141-1.

Curtean-Bănăduc, A., 2005b, Aspects concerning Târnava Mare and Târnava Mică rivers (Transylvania, Romania) stonefly (Insecta. Plecoptera) larvae communities, *Transylv. Rev. Syst. Ecol. Res.*, 2, Editura Universităţii "Lucian Blaga" din Sibiu, pp. 75–84, ISSN 1841-7051, ISBN 973-739-141-1.

Curtean-Bănăduc, A., 2005c, *Râul Cibin – caracterizare ecologică*, Editura Universităţii "Lucian Blaga" din Sibiu, p. 240, ISBN 973-739-195-0.

Curtean-Bănăduc, A., 2005d, Contributions to the study of Cibin River Odonata (Insecta) larvae communities, Studii şi Comunicări, Muzeul Brukenthal Sibiu, Ştiinţele Naturii, Vol. 30, pp. 115–124, ISSN 1454-4784.

Curtean-Bănăduc, A., Ilie, D., 2001, *Aspecte privind starea ecologică a râului* Lotrioara, Valea Lotrioarei, Ghid de ecologie montană, Editura Mira Design Sibiu, pp. 45–48, ISBN 973-8232-25-2.

Curtean-Bănăduc, A., şi Bănăduc, D., 2001, Cibin River (Transylvania, Romania) management, scientific foundation proposal, *Acta oecologica*, VIII: 1–2.

Curtean-Bănăduc, A., şi Bănăduc, D., 2004a, Cibin River fish communities structural and functional aspects, Studii şi Cercetări Ştiinţifice – Seria Biologie, Universitatea din Bacău, Vol. 9, pp. 93–102, ISSN 122-919-X.

Curtean-Bănăduc, A., Bănăduc, D., 2004b, Aspecte privind dinamica faunei râului Cibin în ultimii 150 de ani, Studii şi Comunicări, Muzeul Brukenthal Sibiu, Ştiinţele Naturii, Vol. 29, pp. 205–214, ISSN 1454-4784.

Curtean-Bănăduc, A., Bănăduc, D., 2006, Târnave rivers (Transylvania, Romania) ecological management proposal, Studia Universitatis Vasile Goldiş Arad, seria Ştiinţele Vieţii, Vol. 16., Vasile Goldiş University Press ISSN 1584-2363, pp. 153-162.

Curtean-Bănăduc, A., Bănăduc, D., Sîrbu, I., 2001, *Oameni şi râuri împreună. Impactul antropic asupra Târnavelor şi Ampoiului*, Editura Mira Design Sibiu, pp. 15–67, ISBN 973-8232-32-5.

Gore, J., Schields, D. jr, 1995, Can Large Rivers Be Restored? *BioScience* 45: 142–152.

Moyle, P., Leidy, R., 1992, Loss of biodiversity in aquatic ecosystems: evidence from fish faunas. In: *Conservation Biology: The Theoty and Practice of Nature Conservation, Preservation and Management*, Fiedler, P.L., Jain, S.K. (eds.), Chapman & Hall, New York.

Posea, G., şi colab., 1983, Enciclopedia Geografică a României, Editura Ştiinţifică şi Enciclopedică, Bucureşti.

18. WATER SUPPLY EMERGENCY FOR LAKE SEVAN

GAGIK TOROSYAN*
State Engineering University of Armenia 105 Teryan, 375009 Yerevan, Republic of Armenia

Abstract: This paper presents results of research in the field of use of natural materials such as zeolites (mordenite, clinoptilolite, or zeolite containing tufas) as sorbents[2] and as promising materials for raising the water quality at Lake Sevan in Armenia. The Lake Sevan ecosystem is increasingly in a nonequilibrium state, and the water level of the lake dropped by approximately 19 m. At the eutrophic biolevel, equilibrium was lost and organic matter began to accumulate, entailing a positive balance of biogeochemical circulation. All this brought enhanced autotrophic subsystem activity and a subsequent sharp decline of water quality. The advantages of natural zeolites in comparison with other sorbents, such as technological stability, low cost, availability, and filtering properties are determined first for use in the Sevan region.

Keywords: Lake Sevan; eutrophication; natural zeolites; mordenite; clinoptilolite; adsorbent; wastewater treatment; ammonia; organic pollutants (benzene, toluene, ethylbenzene, xylenes, phenol, aniline, and their derivatives; water-soluble oil products)

1. Introduction

The role of Lake Sevan was, and still is, very important in the national economy of Armenia. The lake water is used for hydroelectric power generation and agricultural land irrigation.

The Republic of Armenia is located in the South Caucasus. It is a highland country, with about 90% of its territory situated at an elevation of over 1000 m

*To whom correspondence should be addressed. Gagik Torosyan, Chemical Technology & Environmental Engineering, Department of State Engineering, University of Armenia, 105 Teryan, 375 009 Yerevan, Republic of Armenia; e-mail: gtorosyan@seua.am or gagiktorosyan@seua.am

R. N. Hull et al. (eds.), Strategies to Enhance Environmental Security in Transition Countries, 239–247.

and 40% over 2000 m above sea level (a.s.l.). Lake Sevan is in the central part of the Republic of Armenia, in the Geghama mountain chain, and is one of the highest freshwater lakes in the world, at an elevation of 1897 m a.s.l. It is the main source of freshwater of the Transcaucasian region. Lake Sevan also plays an important role in regulating water quantity in Armenia as well as in the Transcaucasian water balance. The area of the lake is 4.89 km^2 or 16% of the whole territory of Armenia.

Armenian water resources generally originate within its own territory and add about 7.8 km^3 annually, which includes 4.7 km^3 of surface water and 3.1 km^3 of groundwater. There are more than 300 rivers over 10 km in length and 9500 small drains up to 10 km long. About 30 small rivers and drains empty into Lake Sevan, adding about 700 million (mln.) cubic meters per year to the lake, with only the Razdan River draining from it.

Up until 1933, Lake Sevan contained 58.5 billion cubic meters of water or 80% of Armenia's water resources. During the period 1933–1981, water of the lake was intensively used in the agricultural and energy sectors (six hydropower stations were constructed on the Razdan River) resulting in the lake level dropping by 19 m, and reducing the volume of the lake by 25 billion cubic meters (42%). The lowering of the lake water level started in 1933, but intense water discharges and a drastic drop in the lake water level has occurred mainly since 1949 when the annual discharge reached 1.000–1.700 million cubic meters. In 1962, when the water level had dropped by 15.7 m, the "water bloom" phenomena were observed in Lake Sevan.

2. Ecological Problems

The ecological balance of Lake Sevan was greatly disturbed and has led to degradation of the whole ecosystem and changes in various processes (biodegradation, sedimentation, etc.). The Lake Sevan ecosystem is increasingly in a nonequilibrium state now, and the changes currently taking place within a 2–3-year time frame, would have taken from 50 to 120 years before the water level of the lake dropped by approximately 19 m. With this drop in water level, the concentration of ammonia and organic pollutants began to increase in the lake.

At the eutrophic biolevel, equilibrium was lost and organic matter began to accumulate, entailing a positive balance of biogeochemical circulation. All this brought enhanced autotrophic subsystem activity and a subsequent sharp decline of water quality.

The catchment basin influences the lake system mainly through river flow. Twenty-eight rivers flow into Lake Sevan; of them, only 10 have permanent flow. Since 1981, the waters of the Arpa River have been flowing into the lake. Among large rivers, only the Gavaraget River flows into Minor Sevan. The total annual flow of tributaries to Lake Sevan amounts to 1.0165 million cubic meters, which is 3% of the volume of the lake.

The main cause of the anthropogenic eutrophication of the lake is the concurrent influence of two effects: the water level drop of 19 m, and the vast quantities of biogenic material and toxic substances flowing into the lake.

The problem of removal of hydrocarbons from the Sevan basin is also an urgent task. Many manufacturing processes form organic pollutants (benzene, toluene, ethylbenzene, xylenes, phenol, aniline, and their derivatives), which are toxic substances. Organic toxic substances are the main sources of anthropogenic pollution and come from agriculture (stockbreeding, farming, mineral fertilizers, poisonous chemicals), industry, and household sewage wastewater.

The intensification of economic activity in the Lake Sevan catchment basin is the cause of essential changes in the hydrocarbon content of the rivers. The amount of dissolved organic matter in the water has increased up to 3 mg/L; the content of chlorides and sulfates has increased several fold; the concentration of nitrogen nitrate and ammonia forms have increased ten-fold; the total mineralization has grown from 130 to 190 mg/L. Noticeable changes took place in the concentration of mineral nitrogen which is the consequence of mineral fertilizers and stockbreeding development.

The most polluted rivers are: the Gavaraget, Masrik, Makenis, Argichi, Vardenik, and Tsakkar; the Lichk River is the cleanest one.

Ammonia concentrations have been determined for Lake Sevan. The rivers Gavaraget, Vardenik, and Masrik are most polluted with ammonia and heavy metals; the cleanest one in this respect is the river Arpa.

The estimation of water quality compared to average values does not reflect the alterations in the ecosystem of Lake Sevan during the process of eutrophication. The growth of the tropic level in many lakes of the world is dependent on the increased load of organic elements, particularly nitrogen and phosphorus, from the catchment basin. However, the cause of eutrophication of Lake Sevan differs from those of other lakes of the world. Ecological changes in Lake Sevan in the early 1990s determined its ecosystem and, caused perceptible structural changes in fish symbioses (biological and population indicators have changed).

Lake Sevan eutrophication is accompanied by profound structural changes in phytoplankton. In response to the problem of eutrophication, the decision was made to increase inflow to Lake Sevan from the neighboring Arpa River. Since 1982, about 250 million cubic meters of water annually has been carried through the 48 km long Arpa–Sevan tunnel. From 1982 until 1990, the lake level increased by 1.2 m. A number of bioproduction, physical, biochemical, and chemical processes had stabilized, resulting in significant decrease in the levels of eutrophication. Since establishment of the National park in 1981, efforts have been made to preserve the huge water reservoir, in general addressing proper organization of services at recreation areas and minimizing pollution resulting from agricultural activity. Tree planting was also an important step in that direction and the number of forest areas was increased.

3. Improvement of Water Quality

The communities and food, fish, and other factories are situated mainly along the lake shore. Under above-mentioned conditions of water supply, wastewater treatment in the Lake Sevan catchment basin should be effective after a corresponding pretreatment of raw water from the rivers using a treatment technology. This paper suggests a method of sorption treatment. For a sorption process, the use of natural zeolite was proposed. The increasing demand for ion-exchange materials to solve ecological problems stimulated the intensive study of natural and modified zeolites because they are considered to be cheap modified sorbents (Kallo, 1995). The technological stability of zeolites as sorbents is determined by such characteristics as mineral and chemical composition, sorption ability, mechanical and physical properties, and then filtering properties (Breck, 1974).

Natural zeolites are used in the following fields of wastewater cleaning (Collela, 1999):

1. Removal and recovery of NH_4
2. Removal and storage of radionuclides
3. Removal and storage of heavy metals
4. Removal of organics

The advantages of zeolites in comparison with other sorbents are their reserves in Armenia, a unique complex of technological properties (sorptional, molecular sieving) their natural origin, possibilities of their modification in various ways, and regeneration and utilization properties (Torosyan et al., 2002). The application of the Armenian natural zeolites in the processes of water preparation has been scientifically approved according to the all-round evaluation of mechanical, physical, physicochemical, and technological properties of developed zeolites. It has been established that natural zeolites in combination with other systems are necessary to remove metal ions from water. There are established extraction processes for ions of iron, magnesium, calcium, zinc, copper, nickel, cobalt, lead, and ammonium from water. The efficiency and mechanism of sorption of these components from water, filtering parameters, length of contact, liquid and solid phase ratios, and other factors are obtained. The chemical stability and mechanical strength of natural Armenian zeolites such as mordenite and clinopilolite meet the requirements of filtering materials.

4. Technological Decision

The creation of a protective system is proposed at a confluence of river water and Lake Sevan. It will consist of a primary treatment system, and also secondary treatment where protective walls filled with zeolite are established. Some walls with forward pools are designed to create an opportunity for replacement of a sorbent.

4.1. PRIMARY TREATMENT

This treatment involves settling clarifier pools, where organic materials settle out and are pumped away, while oil floats to the top and is skimmed off.

4.2. SECONDARY TREATMENT

After primary treatment river water is subjected to secondary treatment. This process includes adsorption of organic pollutants on sorbents. This treatment uses zeolites in powder form (10–25 mm size) in the form of a bed.

Results of research are presented for the application of processed zeolites as sorbents for ammonia, phenol, and other aromatics in wastewater for further application in the Sevan basin. Mordenite, clinoptilolite, H-form of mordenite, and clinoptilolite, modified by Catamine AB clinoptilolite have been used as adsorbents. The linear dependence between the concentration of phenol in a water solution and the appropriate molar refraction is preset at 20°C. The measurements were carried out within concentration limits from 0.05 up to 0.3 mol/L. It was earlier established that the sorption within these limits increases and has a linear dependence on the molar refraction. The amount of adsorbed phenol is determined from graphic dependence. It is necessary to note that partial sorption of water (1–2 mL from 10 mL of a solution after 4 h sorption of a solution onto sorbents) takes place. A similar result was experienced with aniline and BTEX (benzene, toluene, ethylbenzene, xylene).

The increasing demand for ion-exchange materials to address ecological problems stimulated the intensive study of natural and modified zeolites because they are considered to be cheap modified sorbents.

5. River Water Treatment of Ammonia by Zeolites

Wastewater is often dumped directly into rivers. This fact is the primary cause of river water pollution in Lake Sevan. Various attempts are now being made on a local scale to purify such river wastewater using simple methods. One of the methods is using natural zeolites for removing ammonia from wastewater (Dyer, 1988; Lahav and Green, 1998).

At first the ability of mordenite to purify river wastewater is compared to normal river gravel. The tested zeolite was superior to gravel in the oxidation of ammonium ions through nitrite ions into nitrate ions. The decrease in the amount of ammonium ions resulted not only in exchange with Ca^{2+} mainly in the zeolites, but also microbial oxidation on the surface of the zeolites (Table 1). Thus, a long-term decrease in the amount of ammonium ions may be possible. This observation suggests that mordenite seems to promote purification with the help of microbial oxidation, with frequent exchange between Ca^{2+} and NH_4^+ at the zeolite surface.

TABLE 1. The adsorption of NH_3 [g] on the zeolites, 10 g

No.	Zeolite	1 day	3 days	5 days	7 days	10 days
1.	Mordenite	10.38	10.51	10.62	10.70	10.76
2.	Mordenite Treated by 0.5 N $CaCl_2$	10.86	11.02	11.15	11.22	11.38
3.	Mordenite Treated by 1 N $CaCl_2$	11.14	11.71	12.13	12.46	12.98
4.	Clinoptilolite modified by Catamine AB	11.86	12.16	12.49	12.97	12.98

Higher ammonia sorption is also evident in material treated by ammonium salt–Catamine AB clinoptilolite.

6. Wastewater Treatment of Organic Impurities by Zeolites

In the present work, results of research are submitted as field solutions and wastewater application to natural zeolites as sorbents for the removal of organic substances from water.

Water polluted by hydrocarbons passes through a column filled with adsorbents. Hydrocarbons are taken from the water, remaining in the adsorption column. The cleaned water leaves the column for direct use or further treatment.

The higher adsorptivity shows natural clinoptilolite modified by quaternary ammonium salt at about 100% extraction of phenol from 0.01 M solutions, and also rather high sorption activity of benzene, toluene, xylene from water solutions from 40% up to 50%. The quantitative adsorption takes place in the H-form natural mordenite.

The linear dependence between the concentration of phenol in a water solution and the appropriate molar refraction is preset at 20°C. The measurements were carried out within concentration limits from 0.05 up to 0.3 mol/L. It was earlier established that the sorption within these limits increases and has a linear dependence on the molar refraction. The amount of adsorbed phenol is determined from graphic dependence (Torosyan et al., 2000). The results are given in Table 2. It is seen from the tables that with an increase in concentration of the solutions, the amount of absorbed phenol is increased. Most active is modifided clinoptilolite by Catamine AB – the mixture of dimethylbenzyl C_{12}–C_{18} alkyl ammonium chlorides. The H-form of mordenite shows activity where, in all probability, the formation of hydrogen bonds takes place. The sorption characteristics of water-soluble oil products are compared between different samples of zeolites – zeolites tuff, treating zeolites, and modified zeolites. Zeolites used were clinoptilolite containing tuff from the Idjevan deposits of Armenia. They are modified by Catamine AB.

The content of oil products in water before and after the sorption was estimated by gravimetric method, followed by a correction by liquid chromatography. The results are presented in Tables 3 and 4.

TABLE 2. The sorption of phenol from a water solution on sorbents

Concentration of phenol in water solution	N_D^{20} initial	N_D^{20} after sorption on clinoptilolite (g phenol/g sorption)	N_D^{20} after sorption on mordenite (g phenol/g sorption)	N_D^{20} after sorption on H-mordenite (g phenol/g sorption)	N_D^{20} after sorption on on clinoptilolite modified (g phenol/g sorption)
0.05	1.3324	1.3320/0.0236	–	Full sorption	Full sorption
0.10	1.3328	1.3320/0.0470	–	1.3314/0.0752	Full sorption
0.15	1.3342	1.3325/0.0710	1.3338/0.0019	1.3322/0.0846	Full sorption
0.20	1.3355	1.3335/0.0940	1.3346/0.0047	1.3329/0.1034	Full sorption
0.30	1.3371	1.3350/0.1034	1.3361/0.0063	1.3345/0.1175	–

Temperature 20°C, duration 4 h

TABLE 3. Distribution coefficient of the oil products K_a (mg/g) during the sorption on clinoptilolite

Sorbent	K_a/light organic[a]	K_a/heavy organic
Clinoptilolite-tuff	50	300
Clinoptilolite-treating	160	540
Clinopti lolite-modified	420	740

[a]Conditions $t = 2$ days, concentration of oil product in water 70 mg/L

TABLE 4. Sorption of the oil from water solutions on clinoptilolite

Sorbent	Weight of sorpted oil product (mg)[a]	Percentage of sorpted oil product
Clinoptilolite-tuff	20	30
Clinoptilolite-treating	32	45
Clinoptilolite-modified	70	85

[a]Oil product concentration in water 70 mg/L, 100 g sorbent, V water 1000 mL

Phenol and other aromatic sorption from water solutions has also been investigated. Phenols and aromatics such as benzene, toluene, xylenes, and others are discharged in open reservoirs, where they destroy the microflora and have a negative effect on human health. The major way to diminish the discharge of dissolved phenols is strong purification and reuse.

Methods are presented for successful sorption of phenols and other aromatics from wastewater using natural and modified zeolites. The given method can be applied at rather low initial concentrations of phenols and aromatics. The oil product content in water before and after sorption was estimated by gravimetric method, followed by a correction by liquid chromatography.

7. Experimental Materials and Methods

7.1. MATERIALS

Natural zeolite deposits of sedimentary origin are widespread in the Idjevan (clinopilolite) and Shirak (mordenite) regions of Armenia. The natural zeolite – clinoptilolite and mordenite – were dehydrated at 110°C.

In this study clinoptilolite and mordenite were prepared by the heated pile method, using 0.5 and 1 N solution of $CaCl_2$. Clinoptilolite modified by Catamine AB was prepared according to Torosyan et al. (2000).

The ammonia adsorption by zeolites was estimated using the gravimetric method (Table 1), and was also determined by IR spectroscopy. The sorption data were plotted after sorption of ammonia vapor produced by a 1 N NH_3 solution by natural zeolites and its treated forms for 10 days at room temperature.

Liquid chromatography is passed on higher-effective liquid chromatography (HELCh), detector Waters 486, controller Warers 600S, Pump, Waters 626, colon 250 × 4 mm^2, Si-100 C 18, P 150 Bar, V 1 mL/m, mobile phase acetonitryl–water (50:50), detector UV-254).

UV spectrometry is passed on UV-Specord spectrometer.

7.2. WASTEWATER TREATMENT OF ORGANIC IMPURITIES BY ZEOLITES

10 mL of a phenol in water solution were added to 1 g of sorbent. The mix was carefully shaken for 4 h. Measurements of the molar refraction of the solution were carried out before and after sorption. The difference in concentration in the organic solution was the expected amount of adsorbed phenol. The results were checked by liquid chromatography and UV spectrum data also.

The removal of organic substances is carried out as follows. Precisely weighed portions of sorbents are added to certain volumes of organic substances in water, where initial concentrations vary. The mix is carefully shaken for 6 h, and allowed to settle. The quantity of the adsorbed substance on the zeolites is determined by the precipitated organic fraction in the filtered solution, using UV spectroscopy, highly effective liquid chromatography and refractometry analytical methods.

8. Conclusion

It is advantageous to continue the research in ammonia and organic pollutant sorption using Armenian natural zeolites. This offers a convenient method for the successful sorption of ammonia and organic pollutants from water. The method can be applied to river water treatment processes.

References

Breck, D.W., 1974, *Zeolite Molecular Sieves: Structure, Chemistry & Use*, Wiley, New York/London/Sydney/Toronto, p. 782.

Collela, C., 1999, Natural zeolites in environmentally friendly processes and applications, in: Porous materials in environmentally friendly processes, Kiricsi, I., Pál-Borbély, G., Nagy, J.B., Karge, H.G. (eds.), *Studies in Surface Science and Catalysis*, Vol. 125, Elsevier, Amsterdam, pp. 641–655.

Dyer, A., 1988, *An Introduction to Zeolite Molecular Sieves*, Wiley, Chichester, p. 83.

Kallo, D., 1995, *International Committee on Natural Zeolites*, Brockport, New York, p. 341.

Lahav, O., Green, M., 1998, Ammonia removal using ion exchange and biological regeneration, *Water Res.* 32(7): 2019–2028.

Torosyan, G., Sargsyan, S., Grigoryan, A., Harutjunyan, S., 2000, The phenol sorption on the zeolites, *Bulletin of Armenian Constructors* 2(18): 30–32.

Torosyan, G.D., Hovhannisyan, S. Shahinyan, H., 2002, Armenian zeolites & its possibilities in industrial & municipal waste water cleaning. *Ecol. J. Armenia*, 1(1): 93–96.

19. SEDIMENT BIOBARRIERS FOR CHLORINATED ALIPHATIC HYDROCARBONS IN GROUNDWATER REACHING SURFACE WATER (SEDBARCAH PROJECT)

KELLY HAMONTS, ANNEMIE RYNGAERT, MIRANDA MAESEN, JOHAN VOS, DANIEL WILZCEK, JAN BRONDERS, LUDO DIELS, WINNIE DEJONGHE*
Flemish Institute for Technological Research, Mol, Belgium

JOHN DIJK, DIRK SPRINGAEL
Catholic University of Leuven, Heverlee, Belgium

MARK STURME, HAUKE SMIDT
Wageningen University, Wageningen, The Netherlands

JAN KUKLIK, PETR KOZUBEK
AQUATEST, Prague, Czech Republic

THOMAS KUHN, RAINER MECKENSTOCK
Forschungszentrum für Umwelt und Gesundheit, Neuherberg, Germany

ANDRE RIEGER, THOMAS LANGE
Consulting und Engineering GmbH, Chemnitz, Germany

HARALD KALKA, NIELS-HOLGER PETERS
Umwelt- und Ingenieurtechnik GmbH, Dresden, Germany

JÖRG PERNER, LUTZ ECKARDT
Eurofins-AUA GmbH Jena, Germany

Abstract: Polluted groundwater in urban and industrial areas often represents a continuous source of diffuse contamination of surface waters. However, the fate of the infiltrating groundwater pollutants might be influenced by the sediment in eutrophic water bodies since they might possess characteristic biological and physicochemical properties influencing pollutant degradation. In a part of the eutrophic river Zenne (Vilvoorde, Belgium) we studied the intrinsic capacity of river sediment microbial communities to degrade chlorinated aliphatic hydrocarbons

*To whom correspondence should be addressed. Dejonghe W. Flemish, Institute for Technological Research, Mol, Belgium

R. N. Hull et al. (eds.), Strategies to Enhance Environmental Security in Transition Countries, 249–261.

(CAHs). The CAH concentrations determined in groundwater sampled at screenpoints and using piston sampler equipment indicate that the Zenne locally drains groundwater polluted with vinyl chloride (VC) and *cis*-1,2-dichloroethene (*cis*-DCE). To further monitor the CAH influx zone into the Zenne, Teflon pore water samplers were installed at certain positions and at a depth of 20, 60, and 120 cm in the sediments of the Zenne. The CAH concentrations in the interstitial water in the sediments will in the future be compared with CAH concentrations in groundwater sampled from monitoring wells positioned 5 and 20 m away from the Zenne. Anaerobic microcosm tests with sediment material from the Zenne showed a rapid microbial dehalogenation of both VC and *cis*-DCE to nontoxic ethene (1 mg/L in 19 days). In congruence, chloroethene degrading *Dehalococcoides* sp. were present in the sediment as shown by detection of their 16S rRNA genes. Furthermore, the VC reductive dehalogenases, *vcr*A and *bvc*A, from respectively *Dehalococcoides* sp. strain VS and BAV1 were detected in the river sediment by polymerase chain reaction (PCR) analysis. Both *Dehalococcoides* sp. can grow on VC and *cis*-DCE as a sole energy source and are probably responsible for the observed degradation of these groundwater pollutants in the microcosms and *in situ* in the riverbed. Our results indicate that the interface between groundwater and surface water harbors a unique microbial community structure that changes with depth and that is capable of degrading CAHs. However, at specific locations in the studied test area with high groundwater influx rates, VC and *cis*-DCE seem to reach the surface water and discharge into the river. At these locations, the observed microbial degradation potential of the river sediment apparently is inadequate to prevent the CAHs from reaching the surface water. Future research will therefore focus on how to stimulate the degradation of the CAHs, both in the river sediment and in the aquifer material upstream of the river Zenne.

Keywords: CAH degradation; chlorinated aliphatic hydrocarbons; monitoring; reductive dehalogenation

1. Introduction

Although groundwater and surface water are usually evaluated as separate water masses, they are connected by the groundwater or surface water transition zone in an hydrologic continuum. Understanding contaminant fate and transport in this zone is important as many contaminated industrial, urban, and agricultural sites are located within 1 km of a surface water body and more than 50% of the contaminated sites have impacted surface water (EPA, 2000). In this way, contaminated groundwater reaching surface waters such as rivers and lakes is considered as an important source of continuous pollution of these surface water bodies. However, the fate of the infiltrating groundwater pollutants might be influenced by the sediment zone in eutrophic water bodies since such sediments possess characteristic biological and physicochemical degradation properties (Bretschko and Moser, 1993; Fraser and Williams, 1998; Naegeli and Uehlinger, 1997; Push et al., 1998). Considering the generally high abundance of bacteria in sediments and the high bacterial activity therein, these microorganisms could play a major role in the removal of both organic (Conant et al., 2004) and inorganic contaminants (Feris et al., 2003; Seuntjens, 2002) from groundwater. Knowledge on natural attenuation of passing pollutants and the potential to stimulate and sustain occurring degradation processes in the sediment zone are however scarce. This is especially due to the lack of appropriate monitoring devices and tools to measure *in situ* mass balances of pollutants and other reactants or products.

In the SEDBARCAH project, we investigate the boundaries of the sediment zone as a barrier against the infiltration of chlorinated aliphatic hydrocarbons (CAH) into surface water and how this zone can be diverted into a sustainable and efficient (stimulated) biobarrier technology for protection of surface waters from groundwater contamination. We (i) determine the role of the microbial community present in sediments in the biodegradation of groundwater pollutants infiltrating a riverbed; (ii) explore the boundary conditions and the possibility to increase and sustain removal activities in the sediment zone; and (iii) select tools to follow such removal activities *in situ*. Therefore, a thorough study both in the field and in the laboratory of the physicochemical and microbial processes occurring in these sediments is performed and coupled to the CAH degradation potential present in the sediment interface of the Zenne (Vilvoorde, Belgium) and the Bělá River (Pelhřimov, Czech Republic). The final goal of SEDBARCAH is to define the potential of these (stimulated) sediment biobarriers as a groundwater remediation technology and a surface water pollution and risk prevention technology.

In this paper, the results from the Zenne site available until now are presented.

2. Study of the Sediments of the River Zenne (Belgium) as a Biobarrier Against the Discharge of CAH-Contaminated Groundwater into an Eutrophic River

2.1. DETERMINATION OF THE CAH-INFLUX ZONES IN THE ZENNE RIVERBED

In Vilvoorde (Belgium), a 1.2 km wide CAH groundwater plume, originating from former industrial activities, extends downgradient of multiple source zones to the river Zenne. Previous research indicated that perchloroethene (PCE), that was spilled at the sources, is successively converted to trichloroethene (TCE), dichloroethene (DCE), and mainly vinyl chloride (VC) in the aquifer while the groundwater is flowing towards the Zenne.

To determine the zones where the highest CAH concentrations are reaching the river, a screenpoint measurement campaign was organized. During this campaign, temporary monitoring wells (screenpoints) were installed along the side of the Zenne at a distance of ~5 m from the river and 125 m from each other. Water samples were taken at a depth of 7 and 10 m below ground surface (bgs) at these screenpoints and in selected monitoring wells. The obtained water samples were analyzed, determining the concentration of CAH, ethene, and ethane in the laboratory by respectively gas chromatography mass spectrometry (GC-MS) and gas chromatography flame ionization detector (GC-FID) analyses.

Figure 1 represents the positions of the screenpoints and monitoring wells along the Zenne that were sampled. Figure 2 indicates the measured concentration of CAH and ethene in the obtained water samples.

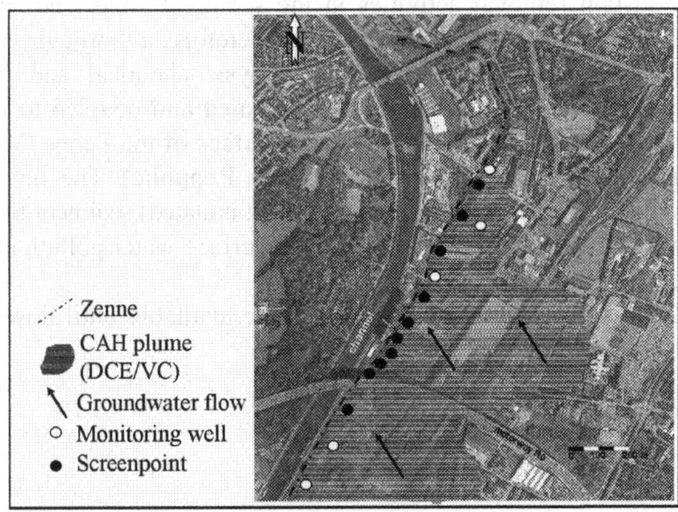

Figure 1. Site map showing the CAH groundwater plume that is discharging into the river Zenne.

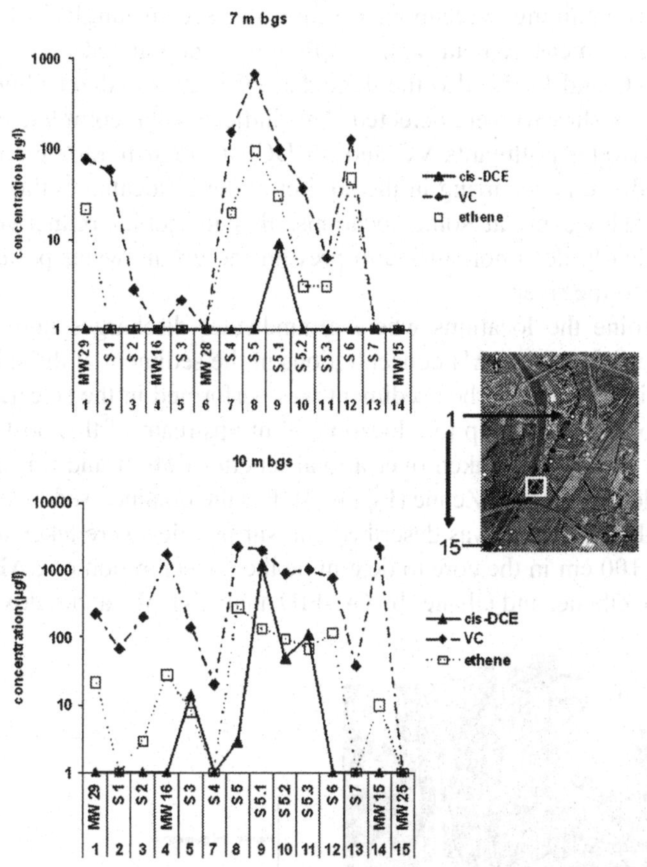

Figure 2. Representation of the measured VC, *cis*-DCE, and ethene concentrations in the water sampled in the screenpoints (S) and monitoring wells (MW) in the Zenne area. The rectangle indicates the selected 50 m long test field.

The results indicate that at a depth of 7 m bgs, mainly VC and to a smaller extent DCE are flowing into the Zenne (Figure 2). The highest concentrations of VC were measured in screenpoint 4 (154 µg/L), screenpoints 5 (677 µg/L), and 6 (127 µg/L). Samples obtained at a depth of 10 m bgs contained next to VC also a high concentration of DCE (Figure 2). These compounds were again mainly found in screenpoints no. 5 (VC: 2212 µg/L; DCE: 150 µg/L), and 6 (VC: 743 µg/L; DCE: 0 µg/L). From these results we concluded that the VC plume is mainly flowing into the Zenne in the area between screenpoints 5 and 6. To narrow the area of investigation, three new screenpoints were drilled (screenpoints 5.1, 5.2, and 5.3, ~40 m apart and located between screenpoints 5 and 6). Based on the analysis results we selected a 50 m long test area in the river Zenne (containing screenpoints 5 and 5.1; see rectangle in Figure 2) at which all further measurements and sampling were conducted.

Next to VC and DCE, also the degradation products ethene (Figure 2) and ethane (data not shown) were detected. This indicates that complete degradation of the groundwater pollutants VC and *cis*-DCE to their nontoxic end products ethene and ethane is occurring in the aquifer material located at the edge of the river Zenne. However, at some locations, this microbial reductive dechlorination rate clearly does not suffice to prevent the groundwater pollutants from infiltrating into the river.

To determine the locations where groundwater discharges into the surface water and the respective CAH concentrations in the sediment at these locations, a piston sampling of the riverbed sediment was performed in the selected test area, stretching from "post 26" up to a location 50 m upstream of this post (Figure 3). Every 5 m a sample was taken over a total length of 50 m and this at the right, middle, and left side of the Zenne (Figure 3). For the obtained sediment cores, the structure of the river core was described and subsamples were taken at depths of ~10, 50, and 100 cm in the core to determine the concentration of CAHs (by GC-MS analysis), ethene, and ethane (by GC-FID analysis), pH, and redox potential.

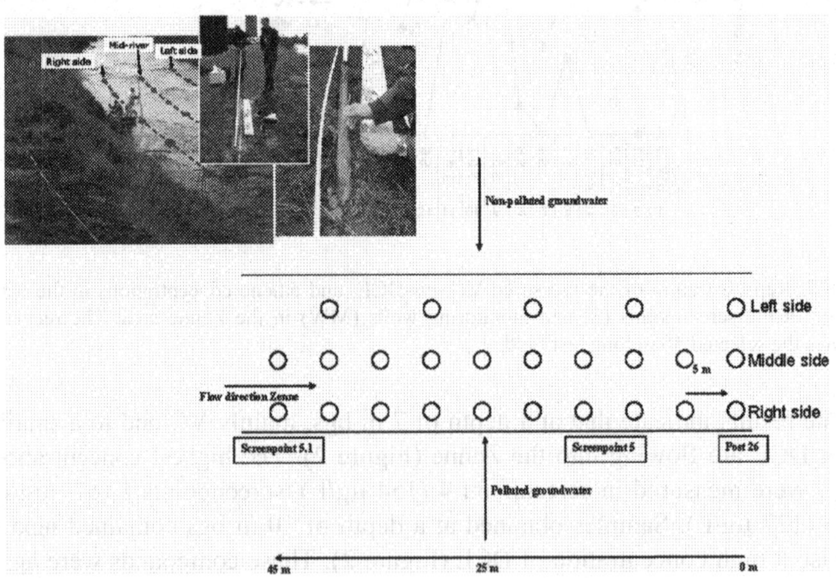

Figure 3. Piston sampling at multiple locations in the streambed of the river Zenne.

Apparently, the concentrations of the CAHs that are flowing into the Zenne depend on the sampling place in the river. The spatial distribution of the concentration of VC is shown in Figure 4. At post 26, no VC is flowing into the river. Analysis of the texture of the samples that were taken in this last area indicated that they contain layers with very fine sand that could prevent the infiltration of VC into the river. In the area from 15 to 45 m upstream of post 26, VC is flowing into the Zenne and the highest concentrations are detected between 25 and 30 m upstream of this post (285 μg/kg dry weight [dw]). While VC is detected throughout the studied Zenne area, *cis*-DCE is only present between 30 and 40 m upstream of post 26 at a concentration of maximum 19 μg/kg dw (data not shown).

Depending on the position in the Zenne, the VC concentration at the top of the piston samples (~10 cm beneath the surface water) is lower, equal, or higher than in the middle and the bottom of the cores (Figure 4). However, at one location a VC concentration of 285 μg/kg dw is detected in the top of the sediment layer. This indicates that VC is discharging into the surface water of the river Zenne. At other locations, e.g., closer to post 26, much lower VC concentrations are present. The concentrations of ethene and ethane that are detected (data not shown) indicate that full dechlorination of VC to ethene and ethane is occurring, but apparently not efficiently enough to prevent VC from flowing into the river.

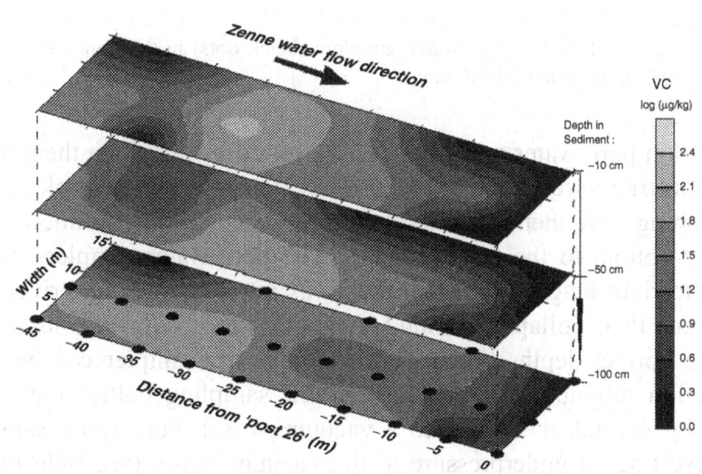

Figure 4. Measured interpolated VC concentrations in the sampled sediment of the river Zenne at the three sampling lines (*right*, *middle*, and *left*) and depths (10, 50, and 100 cm).

It can be concluded that the groundwater discharge spatial distribution is highly heterogeneous, mostly depending on local sediment permeability. At some locations (e.g., at post 26) full dechlorination was observed, while VC and *cis*-DCE reach the surface water and discharge into the river at spots with a high-velocity groundwater influx.

To further monitor the zone of CAH influx into the Zenne, Teflon pore water samplers were installed at certain positions and at depths of 20, 60, and 120 cm in the sediments of the Zenne. According to the results from the piston sediment sampling, we installed these samplers upstream and at a distance of 40, 24.5, and 13.5 m from post 26 at the right side of the river (Figure 5).

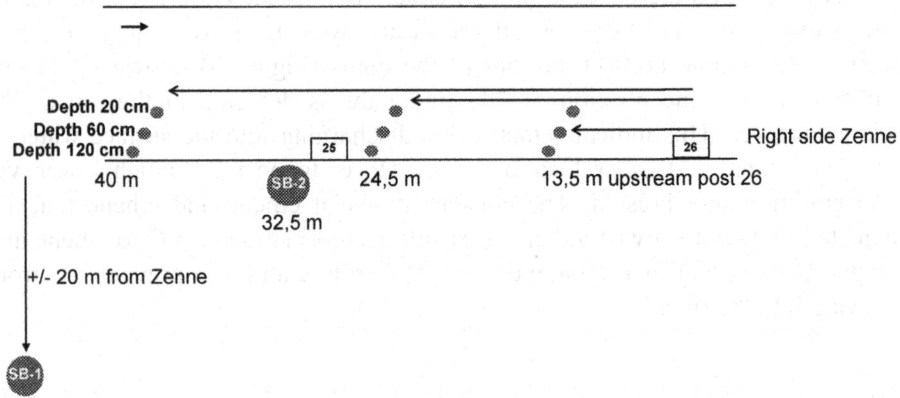

Figure 5. Installation of Teflon pore water samplers (black dots) in the Zenne and monitoring wells SB1 and SB2 at the side of the Zenne.

The Teflon pore water samplers were installed by attaching them to a "lost point" with a string (see Figure 6). All three Teflon pore water samplers, attached to Teflon tubing were then inserted into a steel tube that was hammered into the Zenne river bottom to the desired depth. The pore water samplers were then pulled to the right height. The steel tube was retracted from the river bottom. The sediment then collapsed around the pore water samplers, immobilizing them at the correct depths. Each Teflon pore water sampler can be sampled via the Teflon tubing, connected to 100 mL sampling bottles that are each connected by secondary tubings to a vacuum vessel. Pore water samples are taken by exerting an underpressure to the vacuum vessel (see right picture of Figure 6). Before detaching and closing each sample bottle, the volume in each sample bottle was refreshed two times. The samples were analyzed for CAHs and methane, ethane, and ethene.

Figure 6. Installation of Teflon pore water samplers in the sediments of the Zenne to monitor the interstitial water.

With these samplers interstitial water will be collected and the concentration of CAHs will be determined. The concentration will then be compared with the concentration of the groundwater obtained from a borehole that is positioned 5 and 20 m away from the Zenne (Figure 5). In this way we can determine which amount of CAH is degraded in the aquifer and in the sediments.

2.2. DETERMINATION OF THE DEGRADATION POTENTIAL OF THE EUTROPHIC RIVER SEDIMENT MICROBIAL COMMUNITY BY ANAEROBIC DEGRADATION TESTS AND DETECTION OF 16S RRNA AND CATABOLIC GENES OF DEHALOGENATING BACTERIA

To determine the degradation potential of the eutrophic river sediment microbial community, anaerobic microcosms were constructed using sediment material obtained from different positions in the interface of the CAH-polluted test area of the river Zenne. The actual anaerobic degradation of VC by the eutrophic river sediment microbial community was studied in anaerobic microcosms with fresh Zenne sediment obtained from the top (5–10 cm) or the bottom (100 cm) of undisturbed sediment cores that were taken in the test area. Sediment material (37.5 g) obtained from the right, middle, or left side of the Zenne was suspended in 70 mL of an anaerobic medium to which 2 ppm of VC was added. This was done with material collected at a location with high CAH concentrations (25 m upstream from post 26) and with material collected at a spot with low CAH levels (post 26). The degradation of VC was followed as a function of time by headspace GC-FID analysis. The results of these batch degradation tests (Figure 7) indicate that more than 1 ppm of VC was reduced to nontoxic ethene and ethane in less than 19 days, both in the top and the bottom river sediment microcosms.

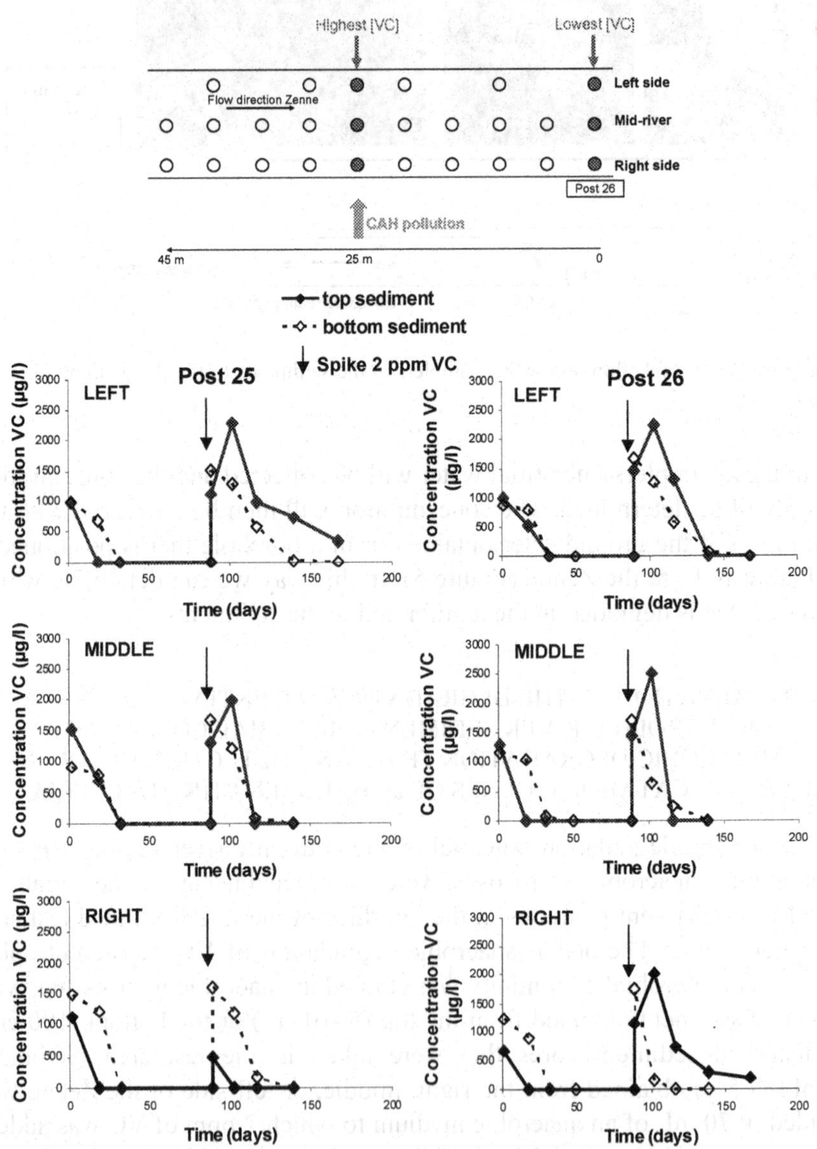

Figure 7. Anaerobic degradation of VC in batch degradation tests containing river sediment material obtained at a depth of 10 (*black line*) or 100 (*dotted line*) cm depth at the right, middle, and left side of the Zenne at a nonpolluted (post 26) and polluted (post 25) part of the Zenne.

In addition, these high dechlorination rates were observed in all microcosms, regardless of their sampling location (right, middle, or left side) for both the CAH-polluted and nonpolluted place. When 2000 µg/L of VC was readded to the microcosms (see arrow in Figure 7), it was again completely reduced to ethene within 50 days.

The CAH degradation potential that was detected by the *in situ* sampling and batch degradation tests was sustained by the polymerase chain reaction (PCR) detection of different genes that are involved in the degradation of CAHs (Figure 8B–D). First of all, the 16S rRNA genes of Dehalococcoides sp. (Löffler et al., 2000), a species that is often involved in the degradation of CAHs, was detected in piston sediment samples obtained throughout the 50 m long test area (Figure 8B). Quantitative PCR (Smits et al., 2004) however indicated that these species were most abundant at the right side, where the CAH plume flows into the Zenne, and in the top 10 cm of the sediments, where the highest amounts of organic material are present (Figure 8B). Secondly, also the catabolic genes vcrA (Müller et al., 2004) (Figure 8C) and bvcA (Krajmalnik-Brown et al., 2004) (Figure 8D) were detected which indicate that probably the species Dehalococcoides strain VS and strain BAV1 are responsible for the observed CAH-degradation in the Zenne sediments.

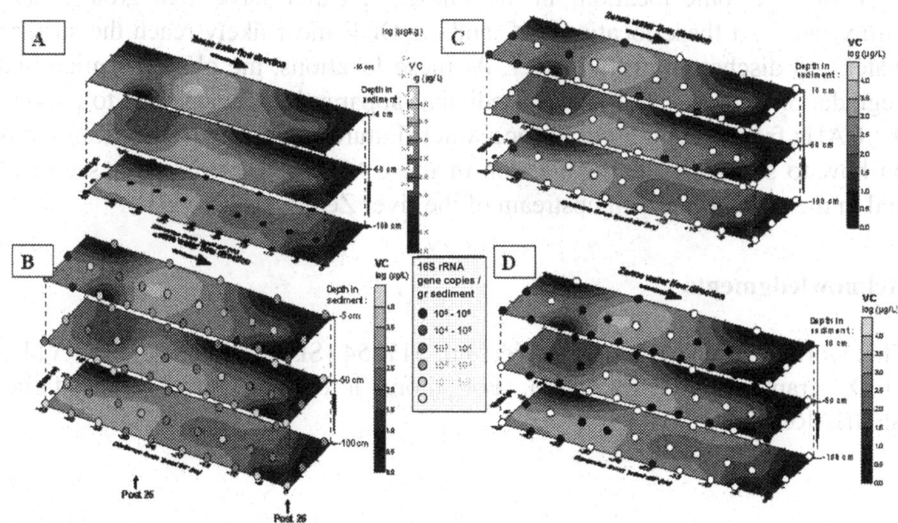

Figure 8. Measured interpolated VC concentrations (A) and detection and quantification of the 16S rRNA gene (B) and catabolic genes vcrA (C, black dots), and bvcA (D, black dots) of dehalogenating bacteria in the sampled sediment of the river Zenne at the three sampling lines (*right*, *middle*, and *left*) and depths (10, 50, and 100 cm).

3. Conclusions

The hydrogeology of the groundwater or surface water transition zone strongly influences the spatial and temporal distribution of both aerobic and anaerobic microbial processes as well as the chemical form and concentration of nutrients, trace metals, and contaminants in surface and groundwaters. Determining the location and magnitude of contaminant discharges to surface waters from groundwater plumes is a complex hydrogeological and biogeochemical problem.

By the sampling of groundwater in boreholes and screenpoints, the CAH influx zone of a 1.2 km wide CAH plume into the Zenne was determined. It can be concluded that the groundwater discharge spatial distribution is highly heterogeneous, mostly depending on local sediment permeability.

The sampling of interstitial water in Teflon pore water samplers installed in the sediments of the Zenne, the performance of microcosm tests with sediments of the Zenne, and the PCR detection of the 16S rRNA and catabolic genes of dehalogenating bacteria, indicated that a high microbial CAH degradation potential was detected in the river sediment of the Zenne in Belgium. By degrading VC and *cis*-DCE, the sediments act as a natural biobarrier for the CAHs present in the groundwater that is passing through the sediment zone, hereby reducing the risk of surface water contamination.

However, some locations in the studied test area have high groundwater influx rates. At these locations VC and *cis*-DCE most likely reach the surface water and discharge into the river. At these locations, the observed microbial degradation potential of the river sediment apparently is inadequate to prevent the CAHs from reaching the surface water. Future research will therefore focus on how to stimulate the degradation of the CAHs, both in the river sediment and in the aquifer material upstream of the river Zenne.

Acknowledgments

This work was funded by EC-project no. 511254 (SEDBARCAH) and a VITO Ph.D. grant (to K. Hamonts). For more information see www.vito.be/SEDBARCAH/.

References

Bretschko, G., Moser, H., 1993, Transport and retention of matter in Riparian ecotones, *Hydrobiologica* 251(1–3): 95–102.

Conant, B. Jr., Cherry, J.A., Gillham, R.W., 2004, A PCE groundwater plume discharging into a river: influence of the streambed and near-river zone on contaminant distributions, *J. Contam. Hydrol.* 73(1–4): 249–279.

EPA, 2000, *Proceedings of the Ground-Water/Surface-Water Interactions Workshop*, 2000, EPA/542/R-00/007, United States Environmental Protection Agency, Washington, DC, p. 2.

Feris, K., Ramsey, P., Frazar, C., Moore, J.N., Gannon, J.E., Holben, W.E., 2003, Differences in hyporheic-zone microbial community structure along a heavy-metal contamination gradient, *Appl. Environ. Microbiol.* 69: 5563–5573.

Fraser, B.G., Williams, D.D., 1998, Seasonal boundary dynamics of a groundwater/surface water ecotone, *Ecology* 79(6): 2019–2031.

Krajmalnik-Brown, R., Hölscher, T., Thomson, I.N., Saunders, F.M., Ritalahti, K.M., Löffler, F.E., 2004, Genetic identification of a putative vinyl chloride reductase in Dehalococcoides sp. strain BAV1, *Appl. Environ. Microbiol.* 70(10): 6347–6351.

Löffler, F.E., Sun, Q., Li, J., Tiedje, J.M., 2000, 16S rRNA gene-based detection of tetrachloroethene-dechlorinating desulfuromonas and Dehalococcoides Species, *Appl. Environ. Microbiol.* 66(4): 1369–1374.

Müller, J.A., Rosner, B.M., von Abendroth, G., Meshulam-Simon, G., McCarty, P.L., Spormann, A.M., 2004, Molecular identification of the catabolic vinyl chloride reductase from Dehalococcoides sp. strain VS and its environmental distribution, *Appl. Environ. Microbiol.* 70(8): 4880–4888.

Naegeli, M.W., Uehlinger, U., 1997, Contribution of the hyporheic zone to ecosystem metabolism in a prealpine gravel-bed river, *J. N. Am. Benthol. Soc.* 16(4): 794–804.

Push, M., Fiebig, D., Brettar, I., Eisenmann, H., Ellis, B.K., Kaplan, L.A., Lock, M.A., Naegeli, M.W., Traunspurger, W., 1998, The role of micro-organisms in the ecological connectivity of running waters, *Freshwater Biology* 40(3): 453–495.

Seuntjens, P., 2002, Field-scale cadmium transport in a heterogeneous layered soil, *Water, Air Soil Pollut.* 140(1–4): 401–423.

Smits, T.H., Devenoges, C., Szynalski, K., Maillard, J., Holliger, C., 2004, Development of a real-time PCR method for quantification of the three genera Dehalobacter, Dehalococcoides, and Desulfitobacterium in microbial communities, *J. Microbiol. Methods* 57: 369–378.

20. BULGARIAN NUCLEAR PLANTS' STRATEGY AND ENVIRONMENTAL SECURITY ON THE BALKANS

PLAMEN GRAMATIKOV*
South-Western University "Neofit Rilski", Ivan Mihailov 66, 2700 Blagoevgrad, Republic of Bulgaria

Abstract: The Bulgarian energy sector covers at the moment ~45% of the total energy deficit of the Balkan region. The policy of the Republic of Bulgaria in the area of nuclear energy utilization, legislative, and regulatory framework, and development of the regulatory basis for nuclear power plant (NPP) management and for environmental protection, assessment, and verification of safety are presented in this paper.

Keywords: energy demand; nuclear energy; regional environmental security

1. Introduction

The expected exponential growth in demand for energy services, in particular in developing regions, and the global, regional, and local environmental impacts resulting from the supply and use of energy will pose in the future the question – How can we supply energy for the inhabitants of the Earth, sufficient to meet every ones needs, without causing serious, irreversible damage to the environment?

This is timely as we face the global challenges of addressing climate change, providing a secure and reliable supply of energy and the depletion of oil. Fossil fuels provide cheap and convenient sources of energy and no single solution can replace them. But unless we change course, developing alternatives to fossil fuel sources of energy and address how we use it, the next generations could face dangerous climate change and major restrictions to their lifestyles and economic development with the potential for conflicts over energy supplies. Our

*To whom correspondence should be addressed. Plamen S. Gramatikov, Physics Dept., South-Western University "Neofit Rilski", Ivan Mihailov 66, 2700 Blagoevgrad, Republic of Bulgaria; e-mail: psgramat@yahoo.com

R. N. Hull et al. (eds.), Strategies to Enhance Environmental Security in Transition Countries, 263–277.

generation – rightly – will be blamed for knowingly squandering the planet's resources. The input for the change will be from leading scientists and engineers representing a whole spectrum of possible energy solutions.

The future prospect of all forms of energy depends most critically today on two factors: (1) environmental policies, especially with respect to greenhouse gas emissions and (2) future commercial fossil fuel prices, mainly oil and gas. Secure nuclear energy utilization is one of the modern, ecological, and real opportunities to satisfy the growth in energy needs of developing areas, such as the Balkans, and to export electricity to neighboring regions.

2. Regional Demand and Supply of Energy in the Balkans, Outlook 2003–2012

2.1. ELECTRICITY SUPPLY AND DEMAND IN THE BALKAN COUNTRIES

Four different kinds of countries at different stages of development and transition to market economics exist on the Balkan Peninsula: European Union (EU) members (Greece); two accession countries in the EU (Bulgaria and Romania); EU applicant countries (Turkey and Croatia); and others. The total population in the region is 55.7 million and the average value of gross domestic product (GDP) per capita is $1765 (Figure 1).

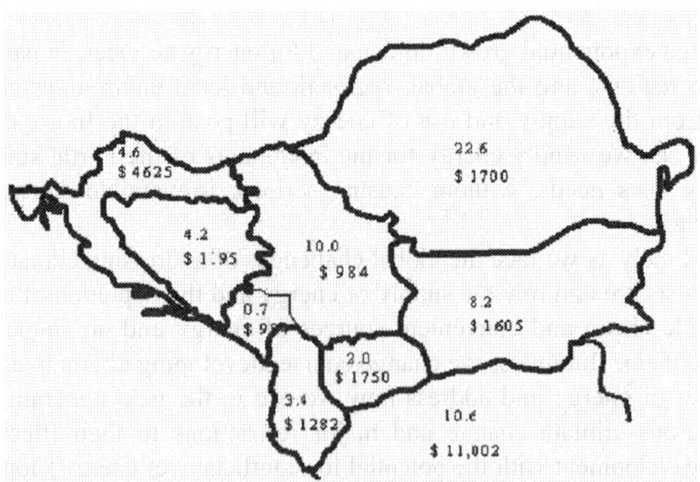

Figure 1. Average population (million) and value of GDP per capita (US dollars) of Southeast European countries.

Basic principles of the "Global scenarios of energy policy till 2050" investigation of the World Energy Council are:

- *Accessibility* – the extent to which people have access to modern energy
- *Availability* – the reliability and security of energy supply systems, once access has been achieved
- *Acceptability* – the environmental sustainability of energy supply and use

All scenarios predict great economic development and, as a result, growth of the regional GDP and power demand in the near future.

At the moment, total installed power capacity is about 49.5 gigawatts (GW). A peak load increase of 2.2% per annum (p.a.) is expected for the period 2002–2012, from 31.4 to 38.2 GW. Regional public utilities expect their electricity demand only to grow at a rate of 2.3% p.a. for the period 2002–2012, from 171 to 214 terawatt hours (TWh), hence the region plans to add about 4.5 GW through 2012 to meet demand. Expectations are that this tendency will continue beyond 2012 (Figure 2).

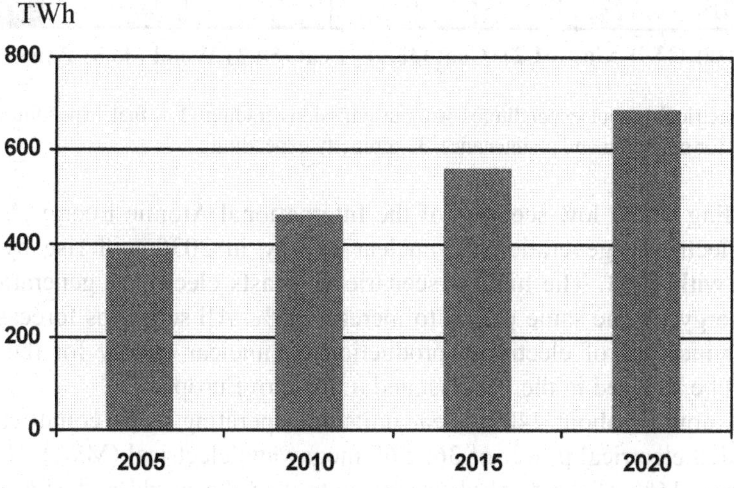

Figure 2. Forecast for total electricity consumption of the Balkan region for the period 2002–2020 in TWh (Anastassov, 2006).

Rehabilitation of about 4000 megawatts (MW) of existing capacity would be required to meet this demand. Without investments in new generation, the region may require the rehabilitation of up to 6500 MW.

2.2. NUCLEAR ENERGY PLACE

Today, nuclear energy is planned for use all over the world. The USA plans to create 110 new units in nuclear power plants (NPPs) and five of them are already licensed. Russia intends to build in total 100 nuclear units by 2030 – 40 on its territory and 60 abroad. The projects for new NPP exist also in Finland, China, India, Iran, France, Czech Republic, Slovakia, Hungary, Turkey, Vietnam, and Australia. One of the reasons is that expenditures for electrical output by nuclear energy are lower than other energy sources (Figure 3) (Tarjanne and Luostarinen, 2004).

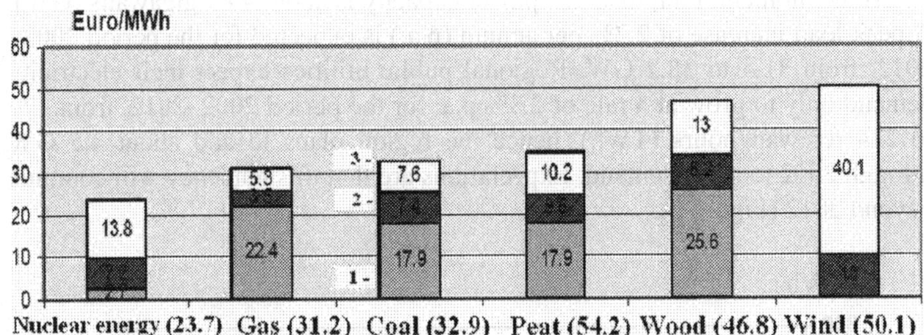

Figure 3. Electrical output expenditures without emissions trading: 1 – fuel expenditures, 2 – expenditures for operation and maintenance, 3 – property expenditure.

According to the low scenario of the International Atomic Energy Agency (IAEA), electricity generation by nuclear energy in 2030 will rise by 34% compared with 2003. The highest scenario forecasts electricity generation by nuclear energy for the same period to increase 86%. All scenarios forecast that the largest increase of electricity production by nuclear energy for the same period will be realized in the Far East and in Eastern Europe.

At the moment about 440 nuclear units are operating in 31 countries with total installed electrical power of 365,560 megawatts electrical (MW_e). Nuclear energy forms 16% of the total electricity output in the world and 35% of the total electricity output in Europe. At the end of 2004, 26 new nuclear units with a total capacity of 21,276 MW_e were in construction and 17 of them are in China, Southern and Northern Korea, Japan, and India. The Balkan regional primary energy balance is presented in Figure 4.

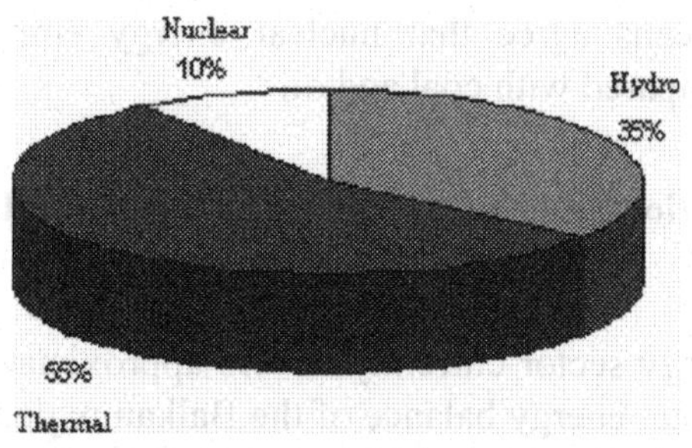

Figure 4. Regional primary energy balance.

2.3. NUCLEAR SECURITY CULTURE

In 2001, IAEA identified security culture as one of the 12 principles underlying fissile material security (IAEA, 2001). The IAEA provisionally defines *nuclear security culture* as "that assembly of characteristics, attitudes and behaviours in individuals, organizations and institutions, which support the objectives of nuclear security and ensures that it receives the attention warranted by its significance". In July 2005, a series of amendments to the "Convention on the Physical Protection of Nuclear Material" was approved elevating the status of security culture to that of a treaty obligation. Yet, a common definition of this concept has proven elusive. Unless the international community can reach a common understanding of security culture, cross-national comparison and evaluation will prove difficult, if not impossible, and governments will find it difficult to discharge their international obligations.

Ever changing environments impose new needs on the management of NPPs. In order to achieve operational excellence and optimal safety under these changing environments, the nuclear industry is adopting organizational learning approaches. NPPs continuously undergo changes as dictated by the existing political, social, economical, technological, regulatory, and other conditions each time. Implementing these changes could potentially cause minor problems that could lead to a chain reaction of events resulting in a deterioration of safety and/or public trust in the safety standards.

An investigation of the European and Transport Forum on the European citizens' opinion about nuclear power (Eurobarometer, 2005) shows that:

- 60% of Europeans believe that nuclear energy gives opportunity for diversification of energy sources
- 61% of Europeans believe that nuclear energy helps decrease European dependence on gas

- 62% of Europeans agree that nuclear energy emits less CO_2 into the atmosphere compared with coal and oil

3. Present Condition and Future Development of the Bulgarian Nuclear Energy Sector

The Bulgarian energy sector currently covers approximately 45% of the constant deficit in the total energy balance of the Balkan region, and in the 2003 hot dry summer, up to 100%. Since 1993, Bulgaria has become a net electricity exporter and plays a key role in regional stability (Gramatikov, 2005).

Six units having a total electrical capacity of 3760 MW_e have been installed in the Kozloduy NPP (Table 1) (Gramatikov, 2002). During the last several years, the first Bulgarian NPP "Kozloduy" has provided more than 44% of the total electricity generation in the country. After disconnection of units 1 and 2 from the national grid, the percentage decreased to 40.6 in 2003.

TABLE 1. Installed nuclear capacities in Bulgaria

Unit	Type	Year of start-up
Kozloduy 1	VVER-440/230	1974 (stopped in 2003)
Kozloduy 2	VVER-440/230	1975 (stopped in 2003)
Kozloduy 3	VVER-440/230	1980
Kozloduy 4	VVER-440/230	1982
Kozloduy 5	VVER-1000	1988
Kozloduy 6	VVER-1000	1993
Belene 1 and 2	VVER-1000	Construction frozen in 1999

Bulgaria also agreed with the EU request to shut down units 3 and 4 as of 1 January 2007. The shutdown will result in no export of electricity for 2 years at least. Therefore, the problem of the future existence and operation of the NPP Kozloduy and creation of the second Bulgarian NPP in Belene is extremely important not only for Bulgaria, but also for the entire Balkan region. Direct losses to Bulgaria because of the prescheduled stop of the first four units of the NPP "Kozloduy" are estimated at about €1.617 billion.

Possible scenarios for future development of the Bulgarian energy sector are shown below (Figure 5):

1. Short-term low growth – energy saving, decentralized sources
2. Long-term low growth – renewable energy sources, energy efficiency

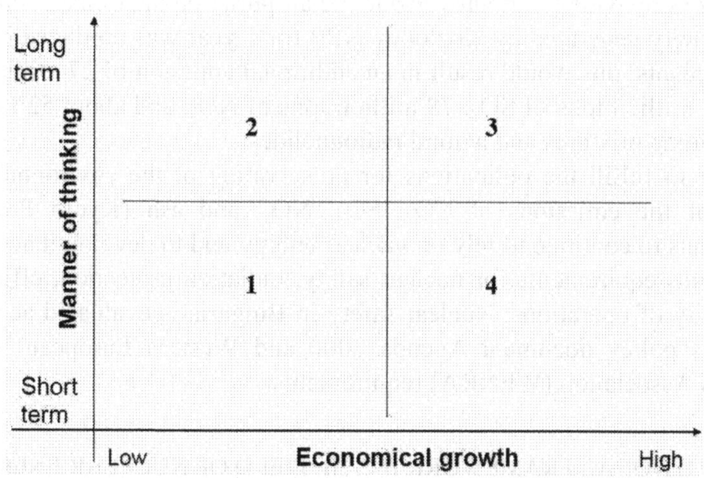

Figure 5. Possible scenarios for future development of the Bulgarian energetics.

3. Long-term high growth – nuclear energy, coal clean technologies, energy efficiency, new technologies, renewable sources
4. Short-term high growth – energy efficiency, research and development (R&D) activities, coal technologies

4. Nuclear Environmental Security in Bulgaria

4.1. GENERAL SITUATION

Nuclear hazardous waste generated in Bulgaria in 2003 was 73% of that generated in 1999. The expenditure on protection and restoration of the environment was 570 million Bulgarian Levs in 2003, which was 17% more than in 2002. The relative share of the total expenditure on protection and restoration of the environment was 1.7% of GDP in 2003 and it increased by 0.2 points compared to the previous year (Catalogue of Statistical Publications, 2005).

The main areas with the highest relative share of expenditures in 2003 were: protection of the water resources – 31.2% (25.9% in 2002), air protection – 22.8% (17% in 2002), and detoxification of waste – 20.1% (23.8% in 2002). The relative share for protection of soil and forests was respectively 3.2% and 1.8% of the total expenditure in 2003. At the end of 2003, the tangible fixed assets were 1133 million Levs (926 million Levs in 2002), an increase of about 30%.

The Kozloduy NPP contributes also to the air purity on the Balkan Peninsula. If the electricity generated by Kozloduy NPP for 1 year was produced by thermal power plants, this would result in an additional emission of 27 million tons of CO_2, 1.2 million tons of SO_2, 78 million tons of NO_x, and about 52 t of dust containing toxic mixtures and natural radionuclides.

In order to fulfill the obligations for preservation of the environment and reduction of the emissions of CO_2, SO_2, NO_x, and ash (Kyoto Protocol), Bulgaria plans to continue to rely on nuclear energy and to develop it according to the current requirements for nuclear safety, radiation protection, efficiency, and reliability of operations. Nuclear safety in Bulgaria is evaluated according to the EU's policy document Agenda 2000 and Western European Nuclear Regulator's Association (WENRA) requirements.

4.2. INSTITUTIONAL FRAMEWORK IN THE FIELD OF NUCLEAR ENERGY AND SAFETY IN BULGARIA

The Republic of Bulgaria is amongst the contracting parties that ratified the Convention on Nuclear Safety (in force for Bulgaria since 24 October 1996). With this Act the country confirmed the national policy for maintaining a high level of nuclear safety, assurance of necessary transparency, and the application of the highest standards.

With the enforcement of the new Act on Safe Use of Nuclear Energy (2002) and the set of Regulations for applying it (in force since September 2002) the regulatory basis is established for communicating to the public in the area of safe use of nuclear energy. Further development of the regulatory basis in the area is connected with the legislation of the EU as far as Bulgaria took the obligation to incorporate the European Directives. The improvement of the documents issued by international organizations like IAEA and WENRA is also used as a basis for both updating the existing and development of new regulatory documents.

According to the old Act, the body with the executive power was a collective body known as the Committee on Use of Atomic Energy in Peaceful Purposes. Included in the Committee were representatives of other bodies and organizations, some of them having direct responsibilities for utilization of nuclear energy and NPP operation.

In the new Act, the independence of the regulatory authority is supported by the fact that the function of development support for nuclear energy utilization is not included in its functions, as it was in the old Act. According to Article 4 of the new Act, the state regulation of the safe use of nuclear energy and ionizing radiation and the safety of radioactive waste management and spent fuel management is implemented by the Chairman of the Nuclear Regulatory

Agency (NRA). In the Act the idea for independence of the regulatory authority is systematically introduced in several areas:

- Political
- Financial
- Organizational

This represents a significant change in the status of the regulatory authority in regard to the superseded old Act. The decree for approval regulates the transfer of responsibilities, rights, and obligations from the former Committee on Use of Atomic Energy in Peaceful Purposes to the new Agency.

The Organizational Statute of the Agency rules the total number of administered staff, the functions of the administrative departments of the Agency and appoints the chairman of the Agency as a primary administrator of budget loans. In his work for fulfillment of the authorities under the Act, the NRA Chairman is assisted by an administrative organization in the NRA. The administration of the NRA is organized in one main directorate and four departments, split into general and specialized administrations. The overall management of the administration is performed by the Chief Secretary. The organizational and managerial flow chart of the Agency is shown in Figure 6.

The deputies of the NRA shall be designated by a decision of the Council of Ministers on a motion by the NRA Chairman. This is an additional element of the independence of the regulatory body and gives the chairman a possibility for recruitment of an effective team of well-qualified experts. The total number of administered staff is 102, including 37 inspectors of safety on nuclear facilities. Six of them are located permanently on Kozloduy NPP site. 95% of the inspectors have a university degree and 60% of them have over 15 years experience in the field of nuclear energy utilization.

As an independent regulatory body in the system of executive power, the Chairman of NRA reports directly to the Chairman of the Council of Ministers (the Prime Minister). In addition, the Chairman of the NRA informs the National Assembly about the matters of nuclear safety and radiation protection by participating at the sessions of the National Assembly and its commissions, when it is appropriate. An important aspect of the political independence of the regulatory body is the appointment of its Chairman for a 5-year mandate. This is a guarantee for its independence against the specific composition of the Council of Ministers, which appoints him and ensures continuity of the policy of the NRA. The new Act sets specific requirements for the person who may be appointed as NRA in regard to the education, experience in the nuclear field, etc.

The new Act creates conditions for financial independence of the regulatory authority. In accordance with the Act, the Agency operations are financed by the national budget and by income from the fees collected under the Act

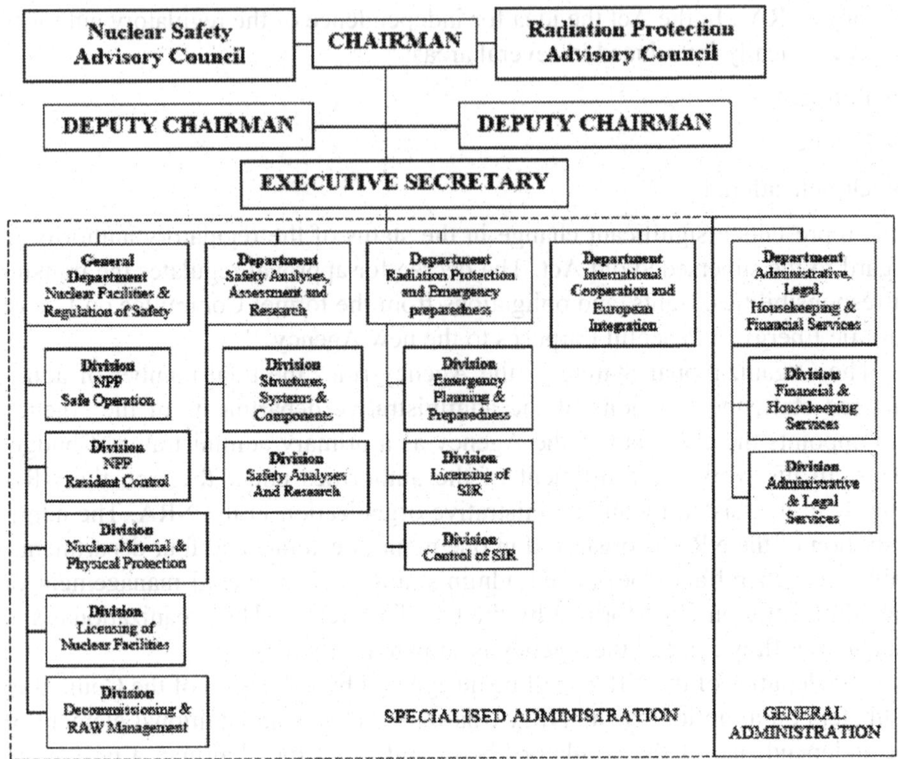

Figure 6. Organizational statute of the Nuclear Regulatory Agency (NRA).

provisions. The order of priority for expenditures of Agency financial resources is the following:

- Financing of studies, analyses, and expertise connected with the assessment of nuclear safety and radiation protection and financing of regulatory activities under the Act
- Capital expenditures on development of the Agency infrastructure
- Training and qualification of Agency staff
- Additional financial motivation of the Agency personnel

Figure 7 presents the increase of the regulatory authority budget for the last several years.

The new Act mandates the establishment of the Advisory Council on Nuclear Safety and the Advisory Council on Radiation Protection. The NRA Chairman approves the composition of the Advisory Councils. Included in the Advisory Councils are prominent scientists and experts in the field of nuclear energy and ionizing radiation, radioactive waste management, and spent fuel management.

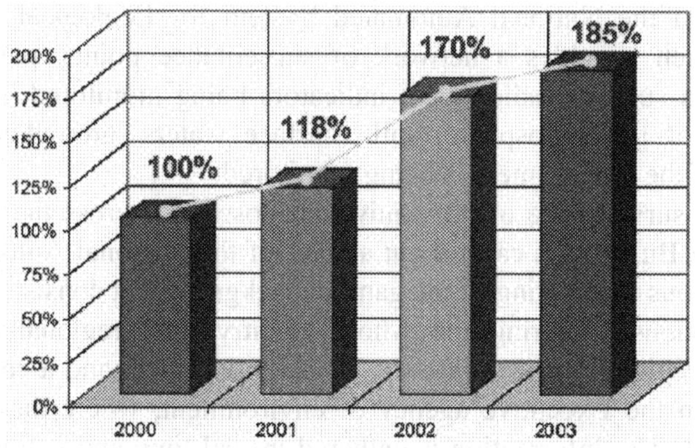

Figure 7. The regulatory authority budget for 2000–2003.

The Advisory Councils shall assist the Chairman by giving expert advice on the scientific aspects of nuclear safety and radiation protection.

The established regulatory system was highly evaluated by the IAEA International Regulatory Review Team (IRRT) mission held in June 2003 (IRRT, 2003). It was also positively assessed in the conclusions of the peer review, held in November 2003 by an expert team of Atomic Questions Group of the European Commission (Peer Review on Nuclear Safety in Bulgaria, 2003). They found that all recommendations related to the regulatory regime in Bulgaria were adequately addressed and no further monitoring on the regulatory system of Bulgaria by the EC was necessary.

4.3. ORGANIZATION OF RADIATION PROTECTION AND RADIATION MONITORING IN THE REPUBLIC OF BULGARIA

Article 3[7] states that doses from ionizing radiation to the personnel and the population must be kept *As Low As Reasonably Achievable* (ALARA) principle. The regulation for basic standards in radiation protection (BNRP-2000) is based on international safety standards (IAEA, 1996) and EU Directive 96/29.

The gamma background in the 3 km zone around Kozloduy NPP is measured continuously with an automated system for external radiation control ("Berthold"). The system has ten monitoring stations for measuring the gamma background and the activity of ^{131}I in the ground level of the atmosphere, five water stations, and three meteorological stations. The radiation monitoring system in Kozloduy NPP is integrated with a similar system of the Ministry of Environment and Waters. Information is exchanged online.

Radiological monitoring in the country is carried out according to a program which is part of the National Automated System for Ecological Monitoring (NASEM), which provides a network of surveillance points, periodicity of sampling, and a suite of radiological indicators being monitored. Radioactive contamination of the atmosphere, soils, surface waters, ground waters, and other objects of the environment is being monitored.

Continuous surveillance of the equivalent dose rate across the territory of the Republic of Bulgaria is carried out as part of the National Automated System for continuous monitoring of the gamma background. It consists of 26 local monitoring stations, covering the whole country, and regional monitoring stations in the regional inspectorates of Varna and Vratza, and a central monitoring station in the Executive agency of environment, where the centralized database is kept. The information is gathered in real time and is transmitted to the emergency center of the NRA and the national reaction center in the State Agency for Civil Protection.

Radiological results from analyses of the main components of the environment like air, water, soil, and vegetation, as well as foods typical for the region of Kozloduy NPP, are within natural limits for the geographical latitude. Measured concentrations are many times below legal norms and are comparable to data from previous years and the period before the commissioning of the plant, 1972–1974. As in previous years, in 2005 there was no registered deviation in the radioecological parameters caused by the operation of the NPP. The radiation situation in the 100 km zone around the plant is stable and favorable. The policy for management of radioactive waste and spent nuclear fuel, information for quantities and facilities for processing and storage are presented circumstantially in the National Report on Joint Convention for safe management of radioactive waste and spent nuclear fuel (Nuclear Regulatory Agency, 2003).

The disposal of toxic waste remains a major problem for EU countries. The practice of burying the waste underground has long been found to be unsatisfactory since the wastes can leach into ground water, resulting in a significant pollution problem. Disposal of radioactive waste takes place in geological formations, for which performance assessments are carried out.

On the Kozloduy NPP site a complex for treatment, processing and storage of radioactive waste operates. The complex includes a line for supercompaction of solid radioactive waste, a line for treatment and processing of liquid radioactive waste, and interim storage for processed radioactive waste. Currently, the activities on radioactive waste in Kozloduy NPP include collection, handling, processing, and storage of liquid and solid radioactive waste on the plant site. The gaseous radioactive substances, generated during operation of the nuclear facilities on the site are released into the environment after purification, as permitted by the regulatory authority.

Financing of the radioactive waste management activities comes from the Radioactive Waste Fund. The contributions to the Radioactive Waste Fund are operational expenses of the nuclear facility operators and are included in the price of electricity. After 2001, financing of all Kozloduy NPP activities in the area of radioactive waste management is done with funds provided solely from the Bulgarian country, with the exception of the projects financed by the IAEA programs for technical cooperation. Long-term prognoses show that the resources available at the moment and estimated future payments to the fund will provide the necessary financing of the activities for radioactive waste management for the entire period of operation of the Kozloduy NPP.

The financial resources for conducting radioactive waste management are planned annually and in midterm (3-year) plans. Now the regulations for the amount of the payments, collection, spending, and control of the money in the "Decommissioning of nuclear facilities" and "Radioactive waste" funds have been actualized. There is control over the expenses – financial resources are spent only for reasonable purposes and according to the payments in the funds. At this moment the Kozloduy NPP is depositing to the two funds around 18% of the income from electricity sold, which is an exceptionally high percentage compared with other operators. According to the evaluation of foreign experts, the Bulgarian policy on this matter leads to a very high rate of accumulation of resources in the funds, and the percentage of payments is higher than all European countries.

4.4. EVALUATION OF THE SAFETY OF EXISTING NUCLEAR FACILITIES

All activities on the evaluation and justification of the safety of existing nuclear facilities were carried out in accordance with the plans presented and keeping in mind the requirements of the new Safe Use of Nuclear Energy Act[7]. In compliance with these requirements, an evaluation and licensing of all the operating units was carried out. Actualization of the existing safety analysis reports (SARs) for units 5 and 6 is being done in parallel with the implementation of measures of the modernization program and will be completed in the middle of 2006. The SAR being developed will include a completely actualized list of postulated initiating events. It is being developed in accordance with the requirements of the national regulatory basis, the applicable Russian and American contemporary standards, as well as the IAEA guidelines.

Regarding the probabilistic safety evaluation, at the moment the level 1 probabilistic safety assessment (PSA) for full power is being actualized, including analysis of internal initiating events; analysis of flooding risks; and, fire hazards analysis. Another project concerning low power PSA development is being carried out in the framework of the modernization program.

In the modernization process of units 5 and 6, each individual package of documents for changes in the design is presented to the NRA to grant permission for implementation together with the specific parts of the SAR concerning the affected equipment, systems, and components.

In accordance with the aforementioned, preparation of the safety evaluation is assured as an element of the whole design process by the designer of the modernization program measure. In the process of reviewing these documents, the operator is supported by an independent team of engineering consultants and the NRA by Technical and Scientific Support Organizations (TSOs) experts of West European regulatory bodies by a special program financed under Poland and Hungary: assistance for restructuring their economies (PHARE) – a preaccession program of the EU to assist the applicant countries of Central and Eastern Europe.

The approach for carrying out the aforementioned activities, the individual steps and the organization, including the intense dialog between the regulator and the operator, were found adequate by a peer review of the EU Atomic Question Group[9]. It was ascertained that all necessary conditions for completion of these activities within the planned deadlines are present and no monitoring on the implementation is required by the EU.

5. Conclusions

The following basic conclusions could be made as a result of this work:

1. In the Kozloduy NPP the main issues are solved in accordance with Bulgarian Legislation and international standards and the main goals are achieved to assure nuclear safety and radiation protection.
2. The radiation impact of Kozloduy NPP on the atmosphere, water, soil, plant, and animal world and protected territories, as well as the risk to the environment and the health of the population in the controlled zone are insignificant.
3. All recommendations related to the regulatory regime in Bulgaria were adequately addressed and no further monitoring of the regulatory system of Bulgaria by the EC is necessary.
4. There is no agreement on how the nuclear safety situation in the applicant countries should be assessed in connection with the EU enlargement. Therefore, operators and regulators of the applicant countries are now concerned that the new planned assessments will be a heavy burden to them.

5. Continuous interaction between researches and managers is important for addressing issues related to nuclear safety and efficiency.
6. Nuclear energy can be used to ensure energy security, without compromising environmental security.

References

Anastassov, V., 2006, Forecast of the electricity consumption. *Presentation at the Workshop "Bulgarian Energetics – which way?"* Sofia, Bulgaria, 20 March 2006.

Catalogue of Statistical Publications, 2005, *Bulgaria 2004 – Socio-Economic Development*, Sofia, 7 October 2005.

EU Directive 96/29, 13 May 1996.

Eurobarometer, 2005, European Commission. Standard Eurobarometer 62. Available at: http://ec.europa.eu/public_opinion/archives/eb/eb62/eb62_bg_nat.pdf

Gramatikov, P., 2002, Survey of the nuclear safety condition in Central/East European countries, in: *Proceedings of International 4E Symposium on Energy, Ecology, Efficiency & Economy*, Struga, FYRMacedonia, 3–4 October 2002, 1, pp. 87–96.

Gramatikov, P., 2005, Regional stability and nuclear energetics safety's management in Bulgaria, *J. Energetic. Technol.*, Serbia, October 2005, 4, v. 2, pp. 3–8.

IAEA, 1996, *International Safety Standards /BSS/, Series-115*, Vienna.

IAEA, 2001, IAEA Report, *Fundamental Principles of Physical Protection of Nuclear Materials and Nuclear Facilities*, Vienna.

IRRT, 2003, International Regulatory Review Team (IRRT) Mission to Bulgaria, July 2003.

Nuclear Regulatory Agency, 2003, National report on fulfilment of the obligations of the Republic of BULGARIA on the Joint Convention on the Safety of Spent Fuel Management and on the Safety of Radioactive Waste Management, Sofia, April 2003. Available at: http://www.bnsa.bas.bg/documents/JC_pdf/English/JC_NR_Eng.pdf

Peer Review on Nuclear Safety in Bulgaria, 2003, Sofia, 17–18 November 2003.

Safe Use of Nuclear Energy Act, 2002, Sofia.

Tarjanne, R., Luostarinen, K., 2004, Lappeenranta University of Technology (LUT), 6 April 2004, Finland.

21. ROLE OF RENEWABLE ENERGY SOURCES IN ENHANCEMENT OF ENVIRONMENTAL AND ENERGY SECURITY OF BELARUS

SEMJON P. KUNDAS, VLADIMIR V. TARASENKO,
SERGEY S. PAZNIAK, IGOR A. GISHKELUK*
International Sakharov Environmental University,
23 Dolgobrodskaya St., Minsk 220009, Belarus

Abstract: Interest in renewable energy sources (RESs) has been growing all over the world in recent years. The market for alternative energy resources is quickly growing in Western Europe and in Asia. The development of renewable energy is caused by two major factors, one of which is an environmental requirement. Its importance increases from the legislative point of view and from the conditions accepted by the Convention on Climate in December 1997. The second requirement concerns energy production capacity. According to this requirement the preference is given to production forms which can be created and developed quickly. The development of RES in many cases is closely connected with the maintenance of energy safety of the country, influencing its sovereignty and independence.

Keywords: renewable energy sources; alternative energy; Belarus

1. Introduction

Energy safety problems in Belarus are mainly characterized by the fact that the Republic must buy up to 85% of energy resources. The basic domestic direction for maintaining energy safety includes a number of fundamental ways to prevent threats, thereby reducing the probability of their occurrence and easing the consequences. On the one hand, there are energy wasteful industrial and household sectors in the country; on the other hand, there are essential energy-saving reserves (including RES use), both in the energy and in other sectors of the national economy. Nowadays, satisfaction of needs in fuel and energy

*To whom correspondence should be addressed. Gishkeluk I.A., International Sakharov Environmental University, 220009, Minsk, Belarus

R. N. Hull et al. (eds.), Strategies to Enhance Environmental Security in Transition Countries, 279–294.

resources (FERs) of Belarusian consumers, maintenance of efficient fuel–energy balance structure of the country, and the quest for additional energy sources became three of the main problems posed for the fuel and energy complex in the Republic (see "The program of increase in the use of local fuel types and alternative energy sources for 2003–2005 and up to 2010"). The involvement of RES in the economic turnover serves as an energy-saving component which is directed to realization of legal, organizational, scientific, industrial, technical, and economic measures of effective utilization of energy resources.

The Republic of Belarus does not have sufficient amount of its own FERs to maintain national economy needs. National resources of available fossil fuel are few and they are depleted practically up to 80–90%. The country imports about 84% of the consumed FERs. Obsolete capital assets in energy, industry, agriculture, and habitation are used up to 70–90% (Rusan, 2005). Therefore, as a result of low supply with its own energy sources (at a level of 15–18% from the general need) the problems of energy safety are the major components of the national and economic security for the country. The necessity for increasing energy safety is mainly caused by the need to quickly solve this problem, because, if the delivery of energy resources is restricted, the Republic can suffer a loss from gross domestic product (GDP) underproduction to the sum of $400–450 with the expectation of 1 ton of standard coal (TSC). It repeatedly exceeds the cost of FER import from any existing or new suppliers according to world prices. In case of emergency conditions in fuel systems or in the event of switching-off the heat supply systems during the winter period, the size of damage can be increased many times.

The most important branch of the economy in any developed country is power engineering. The moving forces and trends of its development reflect the processes of economical state and development of the country, geopolitical situations, and historical prospects. Realization of the measures taken by the government during 1995–2000 allowed a suspension in the decrease in production in the Republic. By the end of 1996, positive dynamics of main macroeconomic processes were achieved, year-to-year increases of GDP, and consumer and industrial goods production was provided. Investment activity has quickened. We succeeded in stabilizing the situation in the domestic consumer market as well as financial enterprises. From 2001 to 2005 GDP grew by more than 43%. In Europe, from 2000 to 2004 GDP increased on the average 1.7% a year, in the USA 2.8%, in Japan 1.9%, in Belarus this parameter reached 9.2% in 2005.

In 1990, the total FER gross consumption came to 54,965 million TSC, in 2000 it was only 34.5 million TSC, in 2005 (according to evaluation data) it was 36.8 million TSC (Figure 1). The main reasons for the sharp decrease in energy resource consumption in Belarus during 1990–2000 was the decrease in industrial production, structural changes in the energy sector of the national economy, as well as tough governmental policy on every possible economy of energy resources. At the same time, the composition of used fuel has been changing. The consumption of heavy oil fuel (black fuel, mazut) is decreasing and the consumption of imported Russian gas is increasing. The average price of Russian gas is about \$50 per 1000 m^3 now (2006) and it is constantly increasing, approaching the world level.

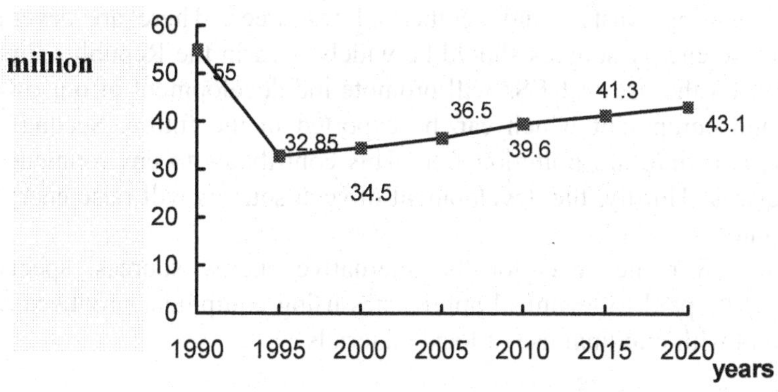

Figure 1. Time history of fuel and energy resources (FER) in Belarus (according to the data given by the Ministry of statistics of Belarus and national development energy programs).

In the long-term outlook the total FER gross consumption will grow. According to the formal forecast, by the year 2020, the size of total consumption will reach 43.1 million TSC, which is only 65–70% of the consumption level that existed in 1990. In the future, the development of the national economy of the Republic should be followed by an increase of the efficiency of energy use, related to its consumption, development, and transportation. During 2001–2005, the GDP power consumption in the Republic decreased at a rate of 4.7–5.4% per year. In 2006–2010, average annual rates of energy consumption decrease are projected to be higher – 5.1–5.9% per year, and by 2020 they will decrease to 2.2–3.0% per year.

Nowadays problems facing power engineering specialists can be divided into two parts: application of new technologies using fossil fuel and development of alternative energy sources (wind and seasonal solar energy, biomass, and wastes as local fuel).

2. Analysis and Prospects of the Use of Alternative and Renewable Energy Sources

2.1. GENERAL CHARACTERISTIC OF THE PROBLEM

According to natural, geographic, and meteorological conditions in Belarus, alternative and renewable energy sources (RESs) can be the following: firewood and wood waste products, water resources, wind-driven potential, biogas from cattle-breeding wastes, solar energy, phytomass, solid domestic wastes, plant growing wastes, and geothermal resources. There are several reasons why these energy sources should be widely used in the Republic. First of all, the work on the use of RESs will promote the development of our own technologies and equipment which can be exported in the future. Secondly, these sources, as a rule, are pollution free. This contributes to environmental security of Belarus. Thirdly, the development of such sources will raise energy safety of the state.

In order to cover the costs for the alternative energy sources, special attention should be paid to technical approaches using equipment produced in the Republic and with maximal use of local materials.

2.2. FIREWOOD AND WOOD WASTE PRODUCTS

The Republic of Belarus has huge forest resources. The total area of the forest resources on 1 January 2001 was 9,248,000 ha, forest timber inventory – 1340 million cubic meters. Annual basic increase is 32.37 million cubic meters. A systematical and stable growth of forest resources can be predicted (up to 1.8 times in 2020) if a simultaneous improvement of age and stock forest structure takes place (Tarasenko, 2005).

The centralized logging and firewood preparation in Belarus is carried out by enterprises of the Ministry of Forestry and by the Belarussian concern of wood–paper Industry. In 2003, the annual volume of firewood, sawing, and woodworking waste utilization as boiler stove fuel was 1.4 million TSC (Table 1) (Tarasenko and Poznyak, 2005). At present firewood fuel consumption for production of electric and heat energy by energy generative settings does not exceed 600,000 TSC per year.

TABLE 1. Economically expedient potential of use of fire wood and wood waste products for heat and electric energy production

Year	Firewood		Wood waste products(million TSC)	Total (million TSC)
	million cubic meter	million TSC		
2003	4.18	1.11	0.28	1.39
2004	4.51	1.20	0.29	1.49
2005	5.36	1.43	0.31	1.74
2006	6.30	1.68	0.32	2.00
2007	7.29	1.94	0.33	2.27
2008	8.08	2.15	0.35	2.50
2009	8.95	2.38	0.36	2.74
2010	9.40	2.50	0.37	2.87
2011	9.88	2.63	0.39	3.02
2012	10.15	2.70	0.40	3.10

The potential of the Republic to use wood as a fuel is estimated to be 3.5–3.7 million TSC per year, which is 2.5 times higher than in 2003. It should be pointed out that all regions of Belarus own firewood recourses. On the whole in the Republic, the annual volume of firewood, sawing, and woodworking wastes utilization was about 1.0–1.1 million TSC. A part of the firewood goes to the population via self-stocking which is estimated at 0.3–0.4 million TSC.

Limited opportunities for Belarus to use wood as a fuel can be defined from the natural annual firewood increase. It is estimated at 25 million cubic meters or 6.6 million TSC per year including that for the contaminated areas of the Gomel region – 20,000 m^3 or 5300 TSC.

In order to use wood as a fuel in these regions it is necessary to work out and to apply new technologies and equipment for gasification and parallel decontamination. According to the planned double growth of firewood storage by 2015 and taking into account volume increases of wood waste products, sawing wastes and firewood processing, in 2005 the annual volume of firewood increased up to 1.6 million TSC.

International Sakharov Environmental University (ISEU) together with the Austrian firm KÖB developed a project on installation on the territory of Educational and Research Station Volma, Dzerzhinsk region, in 2006 of two modern heat-and-power engineering stations which will use raw wood bio-material. Heat-and-power engineering station PYROT with the capacity of 250 kW (ground firewood) is shown in Figure 2 (on the right side); PYROMAT ECO with the capacity of 80 kW (firewood) is shown in Figure 2 (on the left side).

Figure 2. Overview of heat -and-power engineering stations, KÖB firm.

Along with the use of wood waste products for heating purposes, it is worthwhile to provide economically grounded involvement of wood waste products of hydrolytic factories (lignine) into the fuel balance of the Republic. In the city of Rechitsa, Gomel region, a new industrial station using lignine has been put into operation. Lignine resources are about 1 million TSC per year, and an expedient volume of use is estimated to be 50,000 TSC per year. To solve the problem we need investment support, an application of the system of fixed prices, and a normative legal base modernization specified on tax preferences for enterprises producing electric and thermal energy from firewood.

2.3. WATER POWER RESOURCES

Installed capacity of 20 hydroelectric power stations (HPSs) in Belarus was 10.9 MW on 1 January 2004. Due to water power resources about 28 million kWh of energy is produced annually. It is equivalent to the replacement of imported fuel at the rate of 7900 TSC. The potential capacity of all water channels in Belarus is 850 MW including technically accessible (520 MW) and economically expedient (250 MW) capacities (Tarasenko and Poznyak, 2005).

The main directions of the development of small hydropower engineering in Belarus are the following: construction of new HPS, reconstruction, and restoration of existing HPS. The unit capacity of each hydropower unit will be in the range of 50–5000 kW (Table 2), and the preference in that case will be given to quick mounted hydropower units of the capsular type.

TABLE 2. Real and predicted volumes of the use of water and power resources for electric energy production

Years	Input capacity (MW)	Total installed HPS capacity (MW)	Increase of replacement volume ('000 TSC per year)	Power generation (million kWh per year)	Total replacement volume ('000 TSC)
2005	0.76	11.94	0.97	34.9	9.77
2006	0.55	12.49	0.70	37.4	10.47
2007	18.37	30.86	23.50	121.3	33.97
2008	23.3	54.16	29.80	227.8	63.77
2009	20.8	74.96	26.60	322.8	90.37
2010	15.0	89.96	19.20	391.3	109.57
2011	0.29	90.25	0.40	392.8	109.97
2012	5.5	95.75	7.00	417.8	116.97

Having the capacity of hydropower units from 50 to 150 kW it is possible to use asynchronous generators as the simplest and most reliable units in operation. As a rule, all restored and newly constructed HPSs should work in parallel with power supply systems that will allow, in future, the simplification of circuit and constructive decisions.

Special attention in Belarus should be paid to the problems of cascade HPS construction on the rivers Sozh, Dnepr, and Pripyat because the possible scales of water flooding of the adjacent territories are limited by the zone contaminated with radionuclides.

2.4. WIND-DRIVEN POTENTIAL

On the territory of Belarus there are 1840 sites for the installation of wind energy stations with a theoretical energy potential of 1600 MW and annual power generation of 6.5 billion kWh (Tarasenko and Poznyak, 2005). On 1 January 2005 the total capacity of installed wind energy stations was 1.1 MW and the replacement volume was 0400 TSC (Table 3). In fact, in 2005 in Belarus there were only three wing stations («Ecodom», Naroch-2, «Areola», Minsk-1) with the total capacity of 850 kW, and one rotor wind energy station (Figure 3) with the total capacity of 250 kW situated on the territory of Educational and Research Station Volma, ISEU.

TABLE 3. Real and predicted volumes of the use of wind-driven potential for electric energy production

Years	Total installed capacity of wind energy settings (MW)	Power generation (million kWh per year)	Total replacement volume ('000 TSC)
2005*	1.2	2.15	0.60
2006	1.7	3.04	0.85
2007	2.2	3.94	1.10
2008	3.7	6.62	1.85
2009	3.7	6.62	1.85
2010	3.7	6.62	1.85
2011	5.2	9.31	2.61
2012	5.2	9.31	2.61

*Actual power for today

The development allowing the transformation of wind power into electric power by means of traditional wing wind energy stations used so far, in conditions of Belarus were economically unjustified. This fact was one of the reasons of the development of the rotor wind energy station (Figure 3). However, modern technical development allows the creation of similar wing wind energy stations with a starting wind speed from 3 m/s and with rated operation speeds of 7–8 m/s. The cost of such stations is varying from $800 to $1200 for 1 kW of the established capacity. This makes such stations more attractive for use.

The Republic of Belarus is characterized by weak continental winds with the average speed of 4–6 m/s. Therefore, choosing the sites for the wind energy stations special tests and careful studies of FER on their application are required. In order to get an objective estimation about the reserve opportunity of full wind-driven potential it is also required to complete a cycle of experimental research. The necessity of parallel work of wind energy stations with power supply system brings some complications into the general scheme and, thus, expenses for creation and operation of wind energy stations will increase considerably. At the same time while calculating the expenses also the necessity of creation and maintenance of power reserve on other types of power stations should be taken into account. According to expert predictions no more than 5% of the general potential will be developed by the year 2005, i.e., 45 million kWh which is equivalent to 12,000 TSC.

Figure 3. Overview of rotor wind energy station, ISEU.

One of the main directions of wind energy stations application will be their application for pump drive stations with low capacity (5–8 kW) and for water heating in the farming industry. These areas of application are characterized by minimal requirements for electric energy quality that allows a simplification and sharp reduction in the price of wind energy stations.

2.5. BIOGAS FROM LIVESTOCK WASTE

Tests results on biogas production from wastes of cattle-breeding complexes have proved that they are not economically competitive for only biogas production. The main reason is that it is possible to receive pollution-free and high-quality organic fertilizer without additional power expenses and as a result to reduce the power-consuming industry of mineral fertilizer production proportionally. The application of biogas allows us to improve environmental situation near the large-scale farms and cattle-breeding complexes, as well as on the areas under crops where livestock wastes are spread nowadays and in addition to receive high-quality biohumus fertilizers. Potential production of commodity biogas from cattle-breeding complexes is estimated to be 160,000 TSC per year, and by 2005 it will be no more than 15,000 TSC (Tarasenko and Poznyak, 2005).

2.6. GEOTHERMAL RESOURCES

In the Republic of Belarus geothermal resources with a density of more than 2 TSC/m^2 and with a temperature 50°C at a depth of 1.4–1.8 km, and 90–100°C at a depth of 3.8–4.2 m are found in the Gomel and Brest regions (Tarasenko and Poznyak, 2005). However, high mineralization, low productivity of available wells, the small quantity of wells and, on the whole, our poor knowledge of this resource does not allow the development of this RES for the next 10–15 years.

2.7. SOLAR ENERGY

According to meteorological data in Belarus on average there are 250 overcast, 85 rainy, and 30 clear days in a year. To satisfy electric power needs of Belarus in the volume of 45 billion kWh, 450 km^2 of heliostats are required. The price of heliostats is $450 per m^2 that is equal to $202.5 billion without the expenses for the exploitation of the synchronizers, building and construction works, cables, control systems, technical services, infrastructure, etc. The listed components will double the given sum.

Taking into account foreign experience and the experience gained from the building of a solar power station in the Crimea, specific capital investments and energy production costs are ten times higher using solar energy than using other sources. Technical progress in this area will promote the reduction of costs; however, in the case of Belarus, electric power production using solar energy will not be practical in the near future.

The main directions of solar energy utilization will be heliowater heaters and various heliostations for intensification and enhancement of drying processes as well as water heating in the farming industry.

In Belarus heliowater heaters with welded plastic collectors are worked out and are prepared for large-scale manufacture. Therefore, the expensive heavy-metal pipes are not required for solar collectors, which makes their production more economical. Having favorable conditions of economic and manufacturing facilities the widest application of heliowater heaters in southern regions of the Republic can be expected.

At the same time it is expedient to develop in Belarus the following resources:

- Self-contained power supply with capacity starting from some watts up to 3–5 kW (home equipment, lightening, power supply of residential houses, lines of communication, etc.)
- Modular photoelectric stations for rural consumers with a capacity from 0.5 to 1 kW based on the elements of modern generation

The development of such sources and stations needs a number of research efforts to create modern materials, to improve the quality of the existing materials (based on silicium), to reduce its price, and consequently the price of the finished products.

In the favorable conditions of economic and manufacturing facilities a replacement of about 25,000 TSC per year of organic fuel with solar energy can be expected by 2020.

In the ISEU the solar water heating station («Doma», Austria) with a capacity of 1.0 kW and the photoelectric station («Fotovoltaik», Austria) have been used successfully for more than 5 years. They are used for emergency lighting of the ground floor and as a training and visual appliance in the educational process (Figure 4). In the near future it is planned to install also a photoelectric station with a capacity of 1.5 kW («Stromaufwaerts» Austrian firm) in the educational–hotel building of the University in Volma.

2.8. DOMESTIC WASTE

The percentage of organic substances in domestic waste is about 40–75. Domestic waste consists of 35–40% carbon, 50–88% combustible components, and 40–70% ash. The caloric value of domestic waste is 800–2000 kcal/kg (Tarasenko and Poznyak, 2005).

In world practice, power production from domestic waste can be realized by several ways such as burning, active and passive gasification. Gasification has the greatest potential in contrast to direct burning as the latter causes environmental problems. To solve these problems, an investment twice exceeding the cost of the burning stations would be needed.

Figure 4. Overview of solar water heating and photoelectric stations, ISEU.

In Belarus about 2.4 million tons of solid domestic waste is collected annually. They are dumped or directed to two waste reprocessing plants (in Minsk and Mogilev). Annually, the following amount of solid domestic waste is collected there (in thousand tons):

- Paper – 648.6
- Food wastes – 548.6
- Glass – 117.9
- Metals – 82.5
- Textile – 70.8
- Wood – 54.2
- Leather and rubber – 47.2
- Plastic – 70.8 (Tarasenko and Poznyak, 2005)

Potential energy of solid domestic waste collected in Belarus equals 470,000 TSC. In the case of biotreatment of these wastes with the purpose of gas production, the efficiency will not exceed 20–25% which is equal to 100,000–120,000 TSC. Long-term stocks of solid domestic waste from all large cities should be taken into consideration due to the problems of its storage. It could be possible to get about 50,000 TSC from the processing of solid domestic waste into gas in the regional cities of Belarus, while in Minsk this number could be equal to 30,000 TSC. The efficiency of that direction should be estimated not only from the direct output of biogas, but also from the ecological component which is the basis in this problem. Certain characteristics of the effectiveness can be received on the basis of detailed design studies, creation, and operation of experimental–industrial area. By 2005, it can be possible to get up to 10,000 TSC.

2.9. PHYTOMASS

As a raw material for liquid and gas fuel production, a periodically RES – phytomass of fast-growing plants and trees – can be used. In the climatic conditions of the Republic a great number of plants in the amount of 10 t of dry substance which is equal to 5000 TSC are collected from 1 ha of power plantations. Using some additional agricultural methods the productivity of a hectare can be doubled. From this quality of phytomass it could be possible to get 5–7 t of liquid products equivalent to mineral oil. For raw material production the most appropriate would be the use of work-out peat deposits where no conditions for agricultural crops can be found. The area of such deposits in the Republic is about 180,000 ha which can become a stable, pollution-free source of energy raw material in a volume up to 1.3 million TSC per year.

The use of rapeseed oil as an energy resource has great potential for the Republic of Belarus. The Republic has experience in rape cultivation; there are also some rapeseed-processing plants there. Taking into account the fact that rape does not accumulate radionuclides, its cultivation on the areas contaminated after the Chernobyl accident becomes particularly important. There is some experience gained in that direction. Thus, for example, in 2005 a diesel power station with a capacity of 300 kW (electric energy) and 400 kW (thermal energy) running on rapeseed oil was installed and put into operation by the Institute of Radiology, Otto Hugo Munich University together with ISEU. This station was put into operation in the milk-processing plant in Khoiniki within the framework of a humanitarian project (Figure 5).

Figure 5. Overview of diesel power station, working on rapeseed oil in Khoiniki.

The lack of experience in the wide use of phytomass for energy production does not allow the estimation of the expenses and future prices of the fuel. Special techniques, road infrastructure, reprocessing enterprises, etc. should be developed for this purpose. However, according to the integrated calculations the price adds up to \$35 per TSC. According to the expert estimations by the year 2012, about 70,000–80,000 TSC can be received due to the use of phytomass for energy production.

2.10. PLANT GROWING WASTE

The use of plant growing waste for energy production is considered to be a fundamentally new direction in energy savings. Practical experience of plant growing waste application as the energy carrier has been acquired in Belgium and the Scandinavian countries, but not yet in Belarus. The total potential of plant growing waste is estimated to be about 1.46 million TSC per year (Table 4) (Tarasenko and Poznyak, 2005). The decisions on expedient volumes of burning plant growing waste for fuel production should be made comparing certain economic needs. By the end of the predicted period this value is estimated at a level of 40,000–50,000 TSC.

TABLE 4. Real and predicted volumes of the biogas production, domestic waste, phytomass for electric and thermal energy production

Energy resource ('000 TSC)	2007	2008	2009	2010	2011	2012
Biogas	6.6	13.2	19.8	26.4	32.9	39.5
Domestic wastes	4.9	9.9	14.8	19.8	24.7	29.6
Phytomass	12.4	24.7	37.1	49.4	61.8	74.1
Total	23.9	47.8	70.7	95.6	119.4	143.2

From Table 4 it can be seen that due to all renewable and alternative energy sources, as well as thermal secondary power resources, mineral oil, oil gas and peat, the volume of local energy carriers is estimated as 6.75 million TSC per year.

3. Conclusions

A work system on the development of the potential including educational programs for decision-making people and users of technologies should be carried out for the further development and application of alternative energy sources in Belarus. In spite of great efforts undertaken by the State Committee of Energy Efficiency and other authorities, alternative power is introduced in Belarus by a small number of pilot projects and technologies. The most successful projects are those using wood waste products for gasification with the subsequent burning and generating of thermal energy; also some small HPSs, four powerful industrial wind energy stations belonging to «Ecodom», «Areola», and ISEU; and one biogas station. Still, that is not enough. One of the low-cost ways to improve the present situation in wind-power engineering in modern economic conditions is to import already used wind energy stations with a capacity exceeding 100 kW. It is connected with the fact that in Europe wind energy stations with a capacity of 100–200 kW in the near future will be replaced by more powerful stations with a capacity from 600 kW up to 2.5 MW.

The diagram given in Figure 6 describes the prospects of RESs and local fuel type development. A percentage of total consumption of FERs is given by nongovernmental organizations and State Committee of Energy Efficiency.

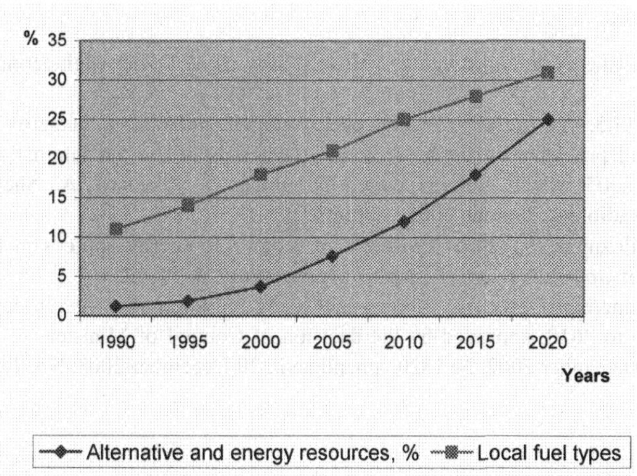

Figure 6. Prospects of the development of alternative energy sources and local fuel types up to 2020 (ratio of total consumption FER, %).

On the whole, the described tendencies of the alternative and RESs development in the Republic of Belarus meet the same tendencies in Europe. The application of new technologies of biogas production from a renewable biomass – rapeseed oil for diesel engines and ethanol for carburetor engines – shows considerable promise for Belarus. Experience of some agricultural productions in the Grodno and Gomel regions has shown ecological expediency of the application of these technologies. It is very important to complete and update these technologies to the industrial level for local conditions and to introduce them into production by 2010. During this period the drastic cost increase for liquid mineral fuel is predicted with significant reduction of its natural resources.

The application of new technologies of biomass utilization (fast-growing wood, wood waste products) as boiler-stove fuel is also a very important question during the mentioned period. In Belarus production facilities for wide introduction of biomass gasification technologies already exist; also pilot projects on development of these technologies have been carried out there with the support of UNDP and other international organizations. The development of these technologies will be especially important for cities and towns with wood-processing power and an advanced agrarian sector.

References

Rusan, V., 2005, Energy safety in villages. How to provide it? Power engineering and TEC, July 2005, No. 7.

Tarasenko, V., 2005, The use of renewable energy sources in Belarus, Sakharov Readings 2005: environmental problems of the XXI century, Materials of the 5th international conference, 20–21 May 2005, Minsk, Belarus, edited by Kundas, S., Okeanov, A., Shevchuk, V., P.1., Institute of Radiology, Gomel, 2005, – pp. 23–26.

Tarasenko, V., Poznyak, S., 2005, Prospects of renewable energy sources in the Republic of Belarus, Seibt. *Journal on modern agrarian production,* 2: 31–33.

The program of increase in the use of local fuel types and alternative energy sources for 2003–2005 and up to 2010, approved by the decision of Council of Ministers of the Republic of Belarus, 27 December 2002, № 1820, amendment, 29 December 2003, No. 1699.

22. EVALUATION OF RADIATION RISK: CYTOGENETIC AND MOLECULAR MARKERS OF LOW-DOSE RADIATION EFFECTS

S. MELNOV*, P. MAROZIK, T. DROZD
Laboratory of Molecular Markers of Environmental Factors, International Sakharov Environmental University, Minsk, Belarus

Abstract: For last 15 years, we have investigated low-dose radiation genetic effects on human populations affected by the Chernobyl accident. Cytogenetic longitudinal investigations showed that radiation markers for cleanup workers remained at an elevated level and had a trend to grow up with time. A dynamic profile of the amount of aberrations confirms that this group has symptoms of the genomic instability state formation. This situation correlated with the incidence of morbidity. The state of genomic instability correlates with accumulation in cleanup workers blood clastogenic factors, which are responsible for increased genomic instability. As a model, we used keratinocyte human line with blocked first check point of cell cycle. The same situation took place for gene mutations (investigated by T-cell receptor [TCR] test) and mitochondrial "common deletion" frequencies (*in situ* polymerase chain reaction [PCR]) in cleanup workers as for other groups in the affected population (for children, especially). The outcomes of biology dose reconstruction will be presented. Results of longitudinal investigations confirm that cytogenetic and molecular effects of irradiation can be fixed even 20 years after the Chernobyl accident.

Keywords: human populations; cytogenetic and molecular effects; markers; low dose

1. Introduction

The destabilization of the genetic machinery is closely connected with the development of many pathological states. Practically, the cause of various somatic

*To whom correspondence should be addressed. Melnov S., Laboratory of Molecular Markers of Environmental Factors, International Sakharov Environmental University, Minsk, Belarus; e-mail: sbmelnov@gmail.com

R. N. Hull et al. (eds.), Strategies to Enhance Environmental Security in Transition Countries, 295–314.

diseases lies in abnormalities of normal cell and tissue functioning, based on irregular activity of the genome. Thus, the control of genetic stability under environmental conditions of different mutagenic pressure can serve as a very important parameter for the estimation of the actual individual health state and for a perspective health risk assessment.

Today a lot of different approaches orientated on different markers (cytogenetic and molecular) determinations are developed (Baranov et al., 1995; Konoplia and Rolevich, 1996; IAEA, 1996; Melnov and Laziuk, 1997). This paper presents our results related to the investigation of cytogenetic and molecular effects for humans, affected by the Chernobyl accident.

2. Cytogenetic Analyses

The most effective approach for the assessment of genetic homeostasis is cytogenetic monitoring, which allows the tracing of the dynamics of genome destabilization, both in qualitative and quantitative respects.

It is a common knowledge that cleanup workers (liquidators) are most affected group of Chernobyl victims (Il'in, 1995; Balter, 1996). In cytogenetic studies attention was primarily focused on the most affected patients showing symptoms of acute radiation sickness. Thus, Gus'kova et al. (1987) and Piatkin et al. (1989–1991) conducted a series of cytogenetic investigations with peripheral blood lymphocytes from patients with acute radiation sickness. The doses received by these patients, which were estimated according to the frequency of dicentric chromosomes, were as follows: 10.1–13.7 Gy (7 patients), 6.1–9.5 Gy (12 patients), 4.0–5.8 Gy (16 patients), 2.0–3.2 Gy (32 patients), 1.0–1.9 Gy (19 patients), and 0.1–0.9 Gy (42 patients); 29 of the examined persons did not show any dicentrics.

For six patients who performed their tasks in the sarcophagus, Sevan'kaev et al. (1995b) demonstrated that there is good agreement between physical doses and biological doses in the dose range from 1 to 15 Gy. Pilinskaya et al. (1991, 1996) examined 51 patients with acute radiation sickness in the remote time after the CNPP accident (9 months–3 years). These authors managed to show that the frequency of cytogenetic markers depended on the acuteness of the disease.

According to the data from Lazutka (1996), collected from examination of 183 cleanup workers, only about 20% of the persons examined showed an increased frequency of dicentric chromosomes and ring chromosomes.

Thus, on the basis of the existing data we conducted a long-term study of the cytogenetic status of CNPP cleanup workers. Various techniques for routine cytogenetic analysis including the sister chromatid exchange (SCE) and T-cell receptor (TCR) tests were used in the course of this study (Melnov et al., 1986, 1987, 1999; Melnov, 1997). The results of these investigations are summarized in Table 1.

TABLE 1. Dynamic investigations of cleanup workers' cytogenetic status

Period (year)	Type of aberrations (%)				Sum of aberrations (%)	Aberrant cell frequency (%)
	Single fragments	Double fragments	Dicentrics + rings	Atypical chrom-s		
1991	1.49 ± 0.19	0.78 ± 0.10	0.28 ± 0.05	0.11 ± 0.02	2.88 ± 0.25	2.57 ± 0.24
1992	1.89 ± 0.20	1.15 ± 0.14	0.12 ± 0.03	0.09 ± 0.03	3.39 ± 0.31	3.14 ± 0.29
1993	1.89 ± 0.37	0.72 ± 0.14	0.04 ± 0.02	0.02 ± 0.01	2.70 ± 0.43	2.621 ± 0.43
1995	3.22 ± 0.20	0.96 ± 0.14	0.20 ± 0.06	0.05 ± 0.02	4.67 ± 0.52	4.28 ± 0.44
1996	2.11 ± 0.21	0.85 ± 0.15	0.11 ± 0.06	0.21 ± 0.02	3.35 ± 0.28	3.01 ± 0.23
1998–1999	1.75 ± 0.24	1.01 ± 0.14	0.16 ± 0.07	0.03 ± 0.02	3.14 ± 0.32	2.88 ± 0.280
2000	1.68 ± 0.25	1.53 ± 0.22	0.19 ± 0.08	0.04 ± 0.02	3.87 ± 0.45	3.33 ± 0.39
2001–2002	2.22 ± 0.32	2.84 ± 0.84	0.14 ± 0.07	0.09 ± 0.04	5.49 ± 0.80	5.10 ± 0.73
2004–2005	5.83 ± 0.29	2.06 ± 0.18	0.43 ± 0.08	0.00 ± 0.00	9.26 ± 0.36	8.06 ± 0.34

The data revealed diametrically different tendencies in the dynamics of the parameters of the cytogenetic status of the cleanup workers. Thus, the frequency of marker aberrations (dicentric chromosomes and ring chromosomes), as in literature (Sevan'kaev et al., 1995a) showed a tendency to constantly decrease, while the frequency of single fragments, double fragments, sum of aberrations and aberrant cells was increased.

The increase in the frequency of marker aberrations (dicentric chromosomes and ring chromosomes), noticed in the course of examination of patients in 1996, is partially connected with a more strict selection of the cohort of persons examined (only patients from the "high-dose" group were examined). Nevertheless, even in 1996 the frequency of marker aberrations did not reach the level found in 1991.

The increase in frequency of atypical chromosomes deserves special attention. This increase seems to be rather natural, since the fact of positive selection in favor of stable aberrations in comparison with unstable marker aberrations is natural and well known. Along with this no essential dynamics of the parameters of the cytogenetic status of the control group was observed (1992–1995 control, 1998–1999 control).

Basing on these data we proposed that the irradiated people got somatic cell genomic instability.

This situation looks very stable. Basing on our dynamic investigation of the control group we are able to estimate the control values for parameters of cytogenetic status for every age. We found that for every age the differences between aberration frequencies for cleanup workers' lymphocytes were higher in comparison with the control group. Even more, with time the discrepancy values had a tendency to increase (Figure 1).

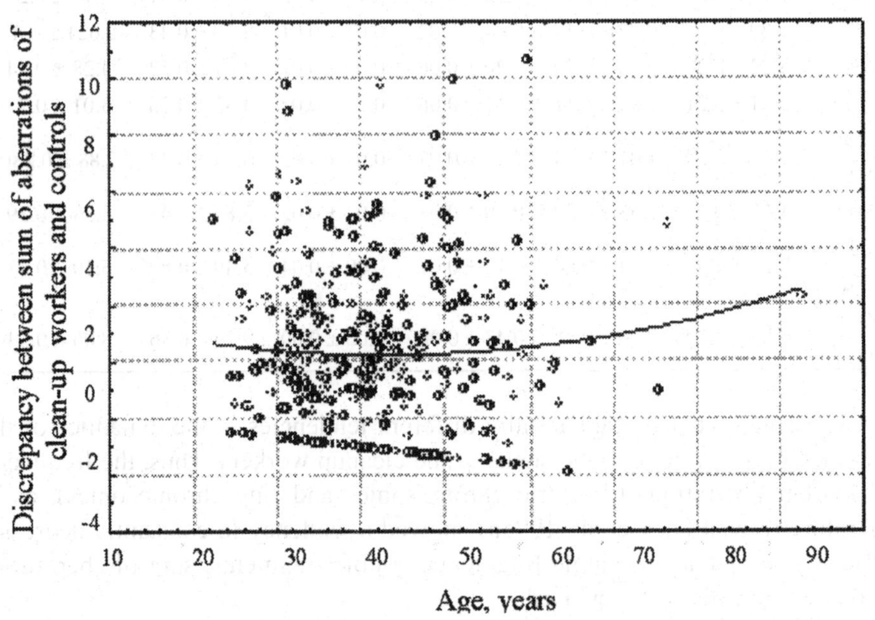

Figure 1. Age-dependent dynamics of discrepancy between sum of aberrations of cleanup workers and controls.

Data of a special interest from our point of view are data on increasing of polyploid cells frequency. They are presented in Figure 2. Reasons for this fact are not clear now.

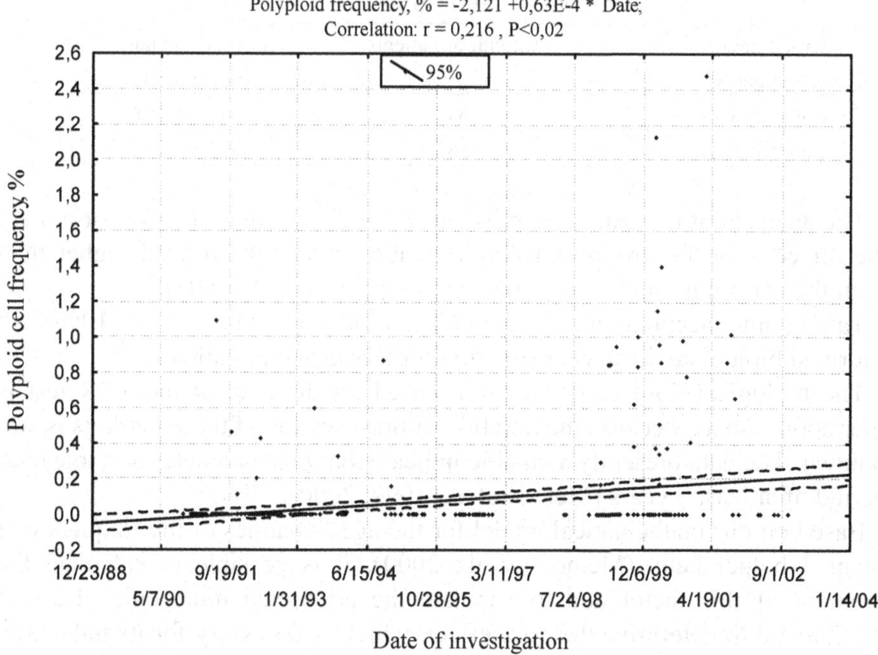

Figure 2. Increase of polyploid cell frequency with the time.

3. TCR Test Results

Chromosome aberrations are rough anomalies, which, as a rule, can have a significant effect on the viability of the affected cells. In this sense any aberration of a similar type (though in a different degree) is subject to negative selection at the population level.

Gene mutations representing a lower level of genome organization are less deleterious for the viability of the affected cells (especially, if they are located in so-called hypervariable regions) and consequently can be more informative for the assessment of an additional mutagenic effect, as well as for biological dosimetry.

We developed a mathematical model which allows the evaluation of additional radiation doses based on the frequency of mutant T-helper cells (Melnov et al., 1999). The parameter determines the frequency of mutations in two hypervariable DNA loci; it is more sensitive to radiation but not to chemical mutagenic influence (Kyoizumi et al., 1989, 1990).

As shown in Table 3, the frequency of mutant T-helper cells in the control group differs significantly from that of the group of cleanup workers.

TABLE 3. Level of mutant T-helpers in peripheral blood of cleanup workers

Group under investigation	Amount of patients	Level of mutant T-helpers cells
Cleanup workers	84	0.640 ± 0.168
Control group	38	0.353 ± 0.053

The analysis of the data shown in Table 3 indicates that the level of mutant T-helper cells in the group under investigation is about two-fold higher than that in the control group (the difference is significant at $P < 0.05$).

Taking into account our mathematical model (Melnov et al., 1999) we performed biological dose reconstruction for this group of patients.

The biological dose reconstruction, based on the data of the TCR test is rather approximate, because the duration of preservation of these markers is still unknown. The data presently available indicate that this parameter is stable over a period of at least 5 years (Melnov et al., 1999, 2000).

Based on our mathematical model for the age dynamics of the frequency of mutant T-helper cells (Melnov et al., 2000), it is possible to calculate the increment of this factor and to stipulate the additional mutagenic effect of radiation and to determine the dose effect, which is necessary for its induction. The results of these calculations are summarized in Table 4.

TABLE 4. Biologically reconstructed doses for CNPP cleanup workers 1986–1987

Dose Range (Gy)	Number of persons (%)
Background level	46.43
0.01–1.4	41.67
1.41–3.56	9.53
>3.56	2.38

The data indicate that more than 50% of the surveyed cleanup workers showed an evident increase in the frequency of mutant T-helper cells, allowing the classification of them as a potential risk group (bioindication of radiation effect) (Table 4). In 11.9% of the cases the doses were rather significant, and in 2.4% they were unreally high. Apparently, this fact can be explained as follows. The biologically reconstructed dose, as mentioned above, is stipulated not only by the effect of radiation, but is also an integral parameter that takes into account the individual radiosensitivity, the health state of the patient at the moment of irradiation, and comutagenic effects of synergetic factors.

Taking into account the special interest to evaluate the long-term cones-quences of radiation for the health state of the group of cleanup workers, this approach will allow the development of biological criteria for the assessment of genetic instability and radiosensitivity on the patient level.

It is necessary to remember that given dose assessment has only indicative character not only due to the lack of data on the stability of this parameter, but also as a consequence of the possible influence of concomitant other individual factors (such as individual radiosensitivity).

4. Mitochondrial DNA as a Marker of Irradiation

In the course of investigation of the genetic processes after irradiation, attention has been concentrated on the state of nuclear DNA. At the same time the fact of the existence of the second alternative genome (in mammalian cells – mitochondrial DNA [mtDNA]) is a well-known fact.

It is common knowledge that in vertebrate cells mtDNA is presented in multiple copies (103–104 per individual organism). The mammalian mitochondrial genome has a size of 16–18 kbp and a ringed closed structure. Intermediate product analyses of mtDNA synthesis showed the main frames of the replication cycle (Clayton, 1982, 1991).

The universal nature of mtDNA and special laws of mitochondrial genetics (more related with population genetics laws that the laws of Mendel genetics) help to understand the very high genetic heterodiversity of mitochondrial diseases, which based on the main involvement of the brain and muscular tissues got the common name "mitochondrial encephalomyopathy" (Larsson and Clayton, 1995; DiMauro et al., 1998).

It was established that in human mtDNA there are "hot spots" where mutations took place more often. The most important paradox of mitochondrial medicine (Reynier and Malthiery, 1995) consists of the inverted proportionality between number of the proteins, coded to two (nuclear and mitochondrial) genomes, and the number of diseases connected with their mutations. For the last few years in small mitochondrial genome that codes only 13 of the known approximately 1000 mitochondrial proteins (all 13 are the components of a respiratory chain), it was identified about 100 pathogenic dot mutations and reorganizations (Shadel and Clayton, 1997). And, on the contrary, the numerous nuclear genome mutations could be associated with respiratory chain defects and neural degenerative diseases as rather rare.

The accumulated information confirms the involvement of mtDNA in the wide spectrum of pathologies (Gattermann et al., 1995). There is suspicion that the level of mitochondrial mutations is one of the cancer risk factors.

With aging the frequency of mtDNA mutations is increasing (Liu et al., 1998), first of all it concerns most frequently meeting "common deletion", covering 4977 base pairs. Its frequency grows up in the cells which have undergone an ultraviolet irradiation. There is opinion that common deletion may be used as a marker of a mutagenesis level of mtDNA.

The comparative analysis of oxidizing stress effect on nuclear and mtDNA has shown that the latter is more sensitive and the arisen anomalies are saved much longer in comparison with nuclear DNA (Cai et al., 1998). The assumption was put forward that its increased sensitivity to chemical and physical mutagenic factors is connected with the absence of histones (Caron et al., 1979). This fact, and also the absence in mitochondrion of the sophisticated repair mechanism (Clayton, 1982), creates the real precondition to use the frequency of mtDNA aberrations for the purposes of biological dosimetry. Besides taking into account the fact of presence of several mtDNA copies in every mitochondrion, it is possible to admit that the aberration in one or several copies will not have a critical effect on organelle activity and cell – carrier viability as a whole and guarantees the coexistence of plenty of mutations in physiologically high-grade cells.

The first attempts to use mtDNA ("common deletion", 4977 bp) for the purposes of biodosimetry belong to Kubota et al. (1997), who showed with the help of nested PCR the increase of frequency of these deletions after irradiating.

We (Melnov et al., 1998) using "Sunrise" type primers developed a method for common deletion quantification by *in situ* PCR. The object of the investigations was human peripheral blood lymphocytes freshly purified in a density gradient with the standard technique. The cell suspension was subjected to *in vitro* irradiation in different dozes up to 2 Gy by ^{137}Cs source at the radiation power of 1 Gy/min. Then cells were transferred into the standard lymphocyte cultural medium (RPMI-1640 medium, calf serum (15%), antibiotics), supplied by (according to experiment conditions) PHA (Gibco, 15 mkg/ml). Cells were washed in warm cultural medium without PHA at +37°C, centrifugation for 5 min, 800*g* and incubated in solution of Mitotraker M-7514 (0.2 mM in cultural medium, "Molecular probes") within 30 min. Then, cells were washed in phosphate-buffered saline (PBS), subjected to hypotonic processing (PBS: distilled water = 1:1; +37°C, 8 min) and fixed three times in the cooled fixator (methanol:acetic acid = 3:1, 10 min). Intermediated centrifugation occurred for 5 min, 800*g*.

Cells were dropped out on slides, dried up, and used for PRINS PCR. In the experiments, "Sunrise" type primers (HSAS8542) were used for identification of mtDNA common deletions and standard kits «TaqMan PCR Core Reagent Core» (Perkin Elmer, H1555) were used. Due to close positions of fluorochrom and quencher in the primer, in solution it does not activate. At interaction with mtDNA primer there is a stretching out and the distance between them is increasing and fluorescence will arise. For visualization of nuclei standard DNA-specific counterstaining with DAPI was used. During the microscopic analysis the quantity of the cells with or without mutations was taken into account and expressed in percentages.

Method development was carried out on lymphocyte culture of the patient with Piers syndrome (higher frequency of common deletion).

Optimization of PRINS PCR regime (20–25 cycles) has allowed the identification, with a high degree of reliability, of mutations of mtDNA with the help of the specified methodical approach – practically all cells contained mitochondria with mutant DNA, and the color varied significantly depending upon the number of mutant mtDNA copies from yellow to orange, up to intensively red.

At the analysis of the peripheral blood lymphocytes, irradiated by ^{137}Cs source in various doses (up to 2 Gy) we fixed elevated levels of mutations (Table 5), statistically significantly distinguished from the control.

TABLE 5. Dynamics of the mtDNA common deletion frequency in peripheral blood lymphocytes after irradiation in various doses (time of incubation – 0 h)

Dose (Gy)	Amount of investigations	Cells in analyses	Frequency of mutant cells (%)
0	3	487	1.23 ± 0.50
0.5	3	500	1.40 ± 0.53
1.0	3	477	1.89 ± 0.62
2.0	3	501	2.60 ± 0.71

The analysis of the collected data specifies presence of statistically significant differences only between control samples and the samples irradiated at a dose of 2 Gy (according to Mann–Whitney criterion – $P < 0.05$). Change of the mutant cells frequency depends on a dose and may be described in the frameworks of a regression model $\beta = 0.74021$, $P < 0.01$ (Figure 3).

Unfortunately, the quantitative estimation of the mtDNA mutant frequency per one cell now is not possible, and in thanks to such situation we prefer to use as a basic criterion the frequency of mutant cells without dependence on the frequency of mutant organelles (or the damaged copies of mtDNA).

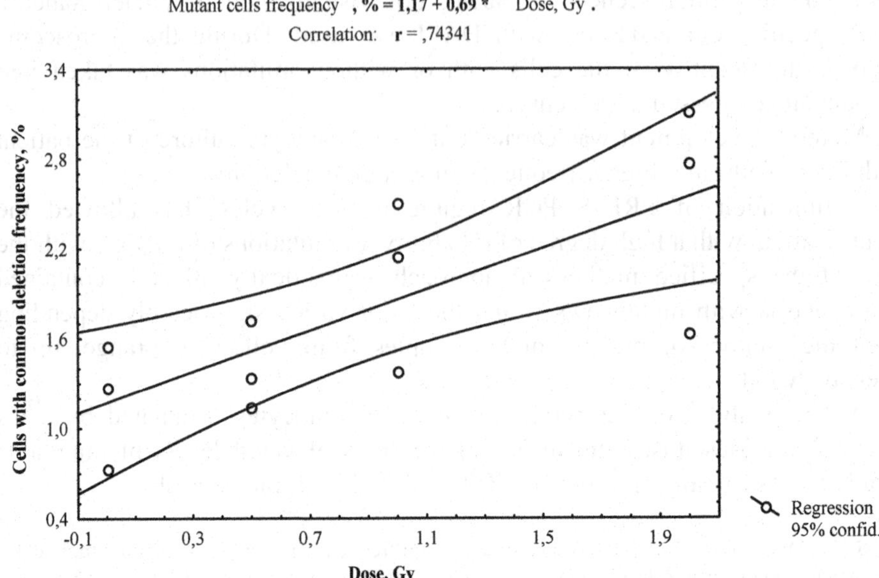

Figure 3. Dynamics of the mtDNA common deletion induction after additional irradiation.

Taking into account a possible role of the time factor and proliferation potential in realization of radiation effects, we carry out the analysis of frequency of an induction of deletions of an mtDNA for the same dose range in 96 h (time factor) after an irradiation without or at presence PHA (proliferation). Results are summarized in Tables 6 and 7.

TABLE 6. Dynamics of the mtDNA common deletion frequency in peripheral blood lymphocytes after an irradiation in various dozes (time of incubation – 96 h, without PHA)

Dose (Gy)	Amount of investigations	Cells in analysis	Frequency of mutant cells (%)
0	3	754	0.93 ± 0.35
0.5	2	990	3.43 ± 0.58
1.0	3	960	15.73 ± 1.18
2.0	3	1082	16.73 ± 1.14

The analysis of the data presented in Table 6 testifies that frequency of cells with deleted mtDNA is really increased in the dose range 0–1 Gy (Mann–Whitney criterion, for all cases – $P < 0,05$). In the dose range 1–2 Gy, the output of mutant cells exits on a plateau.

TABLE 7. Dynamics of the mtDNA common deletion frequency in peripheral blood lymphocytes after an irradiation in various doses (time of incubation – 96 h, with PHA)

Dose (Gy)	Amount of investigations	Cells in analysis	Frequency of mutant cells (%)
0	3	938	1.60 ± 0.41
0.5	3	1158	13.04 ± 0.99
1.0	3	1102	22.60 ± 1.26
2.0	3	1205	27.14 ± 1.28

At the same time, in case of stimulation, cellular proliferation frequency of mutant cells is much higher, and the difference between points 1 and 2 Gy is statistically significantly different (Mann–Whitney criterion, for all cases – $P < 0.05$).

It is clear that in both cases the dose–effect dependence may be interpreted in the framework of linear regression – in the case nonstimulated lymphocytes $\beta = 0.890$ ($P < 0.001$); after PHA stimulation – $\beta = 0.930$ ($P < 0.00005$) (Figure 4).

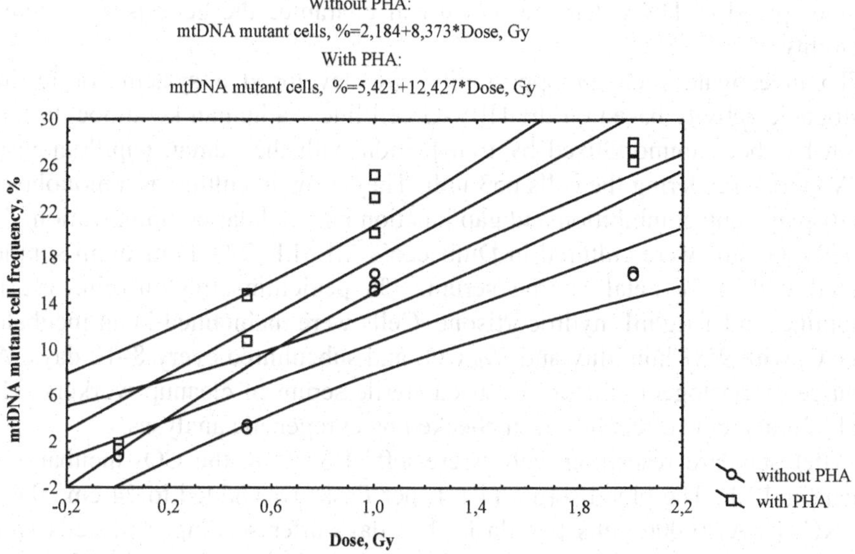

Figure 4. Dynamics of mtDNA common deletion induction after additional irradiation (incubation time, 96 h) – without the PHA stimulation.

Special attention is deserved for the fact that the frequency of mutant cells submits to the same law both in PHA-stimulated and in PHA-unstimulated cell populations that enables the assumption of relative independence of the specified parameter from proliferation activity of lymphocytes, and the marked quantitative difference can be attributed to the account of duplication of cells – carriers of a mutated mtDNA.

At the same time, the comparison of the regression equations, devoted to the dose–effect dependencies in these cases testifies that between them there are real statistically significant discrepancies ($P < 0{,}01$). The above-mentioned data allow making the following conclusions:

- The analysis of the nonnuclear DNA condition is a reliable parameter for biological dosimetry
- Relative independence of the frequency of mutant cells from cell proliferation activity of the lymphocytes allows the assumption of the possibility of long-time stability of this parameter.

5. Clastogenic Effect of Low-Dose Radiation

The recently investigated bystander effect and work dealing with clastogenic factors are evidence of the existence of special transmitters which are mediating the spread of DNA damage which can guarantee the genesis of genomic instability.

To investigate such an opportunity, we developed a system for testing clastogenic activity based on the HPV-G cell line – a human keratinocyte line, which has been immortalized by transfection with the human papillomavirus (HPV) virus, rendering the cells p53 null. They grow in culture as a monolayer, and display contact inhibition and gap junction intracellular communication.

HPV-G cells were cultured in Dulbecco's MEM:F12 (1:1) medium supplemented with 10% Fetal bovine serum, 1% penicillin–streptomycin, 1% L-glutamine, and 1 μg/mL hydrocortisone. Cells were maintained in an incubator at 37°C, with 95% humidity and 5% CO_2 and subcultured every 8–10 days. As a source of clastogenic factors we used sterile serum of cleanup workers. The level of mutagenic effect has been checked by cytogenetic analyses.

After standard reseeding, cells were left at 37°C in the CO_2 incubator to attach for 12 h. The blood serum (1 mL per flask) was added to 24 cm^2 flasks (NUNC, USA) (6000 cells per flask) 1–2 days after seeding, and cells were cultivated in the incubator for 1–2 h. Then, cytochalasin B was added (7 μg/mL concentration) and the cells were incubated for 24 h.

After this cell culture medium was removed, the cells were washed with PBS and fixed with chilled Karnua solution (1 part glacial acetic acid and 3 parts methanol, 10–15 mL three times for 10–20 min). Coverslips were dried and stained by 10% Giemsa solution. Using mounting medium (Sigma), coverslips were attached to the microscope slides. The micronuclei count was carried out under inverted microscope (× 400). Micronuclei were counted only in binucleated cells (1000 binucleated cells per flask). All the data are calculated as the micronuclei number recorded per 1000 binucleated cells (micronuclei were analyzed only in such cells). Cytogenetic analysis has been fulfilled as before. Results are summarized in Table 8.

TABLE 8. Cytogenetic status of lymphocytes and micronuclei frequency in HPV-G cells after treatment with serum samples (average data)

Parameters		Groups of populations		*P*
		Control	Main group	
Results of cytogenetic analysis				
People analyzed		61	32	–
Average age (years)		37.66 ± 1.5	41.74 ± 0.98	–
Cytogenetic status (%)	N of metaphases	11,179	8739	–
	Single fragments	1.67 ± 0.12	5.78 ± 1.97	<0.05
	Double fragments	0.81 ± 0.08	1.89 ± 0.1	<0.01
	Dicentrics + rings	0.11 ± 0.03	0.31 ± 0.22	>0.05
	Atypical chromosomes	0.02 ± 0.01	0.02 ± 0.01	>0.05
	Polyploidic cells	0.08 ± 0.03	0.22 ± 0.19	>0.05
	Aberrant cells	2.48 ± 0.14	7.59 ± 2.37	<0.05
	Sum of aberrations	2.63 ± 0.15	8.59 ± 2.94	<0.05
Results of micronuclei test				
Frequency of micronuclei in cells (‰)	Number of binucleated cells	5554	14,045	–
	Number of cells with 1 MN	69.50 ± 12.21	189.66 ± 20.74	<0.01
	Number of cells with 2 MN	5.00 ± 3.08	24.51 ± 8.05	<0.05
	Number of cells with 3 MN	0.26 ± 0.21	4.56 ± 3.14	>0.05
	Number of cells with MN	74.76 ± 12.69	218.69 ± 21.90	<0.01
	Total frequency of MN	80.29 ± 13.14	251.18 ± 22.96	<0.01

The results of polynomial statistical analysis of individual data are presented in Figures 5 and 6.

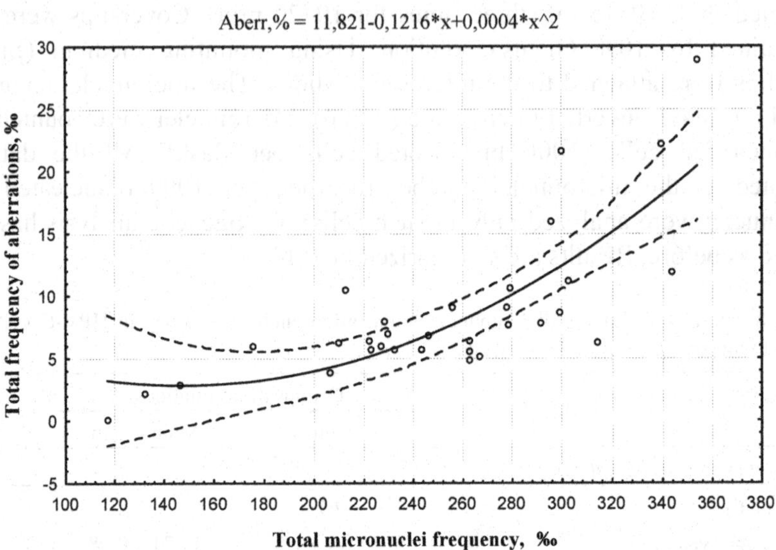

Figure 5. Dependence of total frequency of micronuclei from total frequency of aberrations (polynomial fitting).

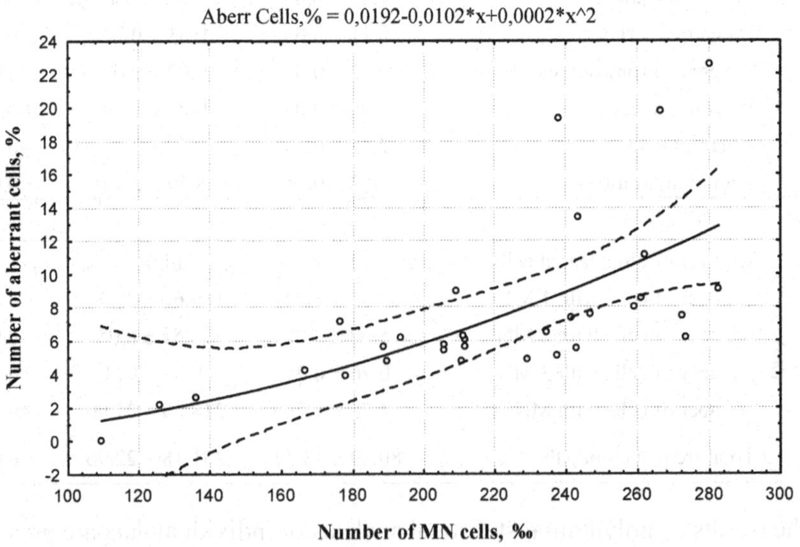

Figure 6. Dependence of total number of cells with micronuclei from total number of aberrant cells (polynomial fitting).

Figure 7 and Table 9 present the cluster analysis showing linkages between parameters "micronuclei frequency, ‰" and number of single, double fragments, dicentric and ring chromosomes, atypic chromosomes, polyploidic cells, aberrant cells frequency, and aberrations frequency.

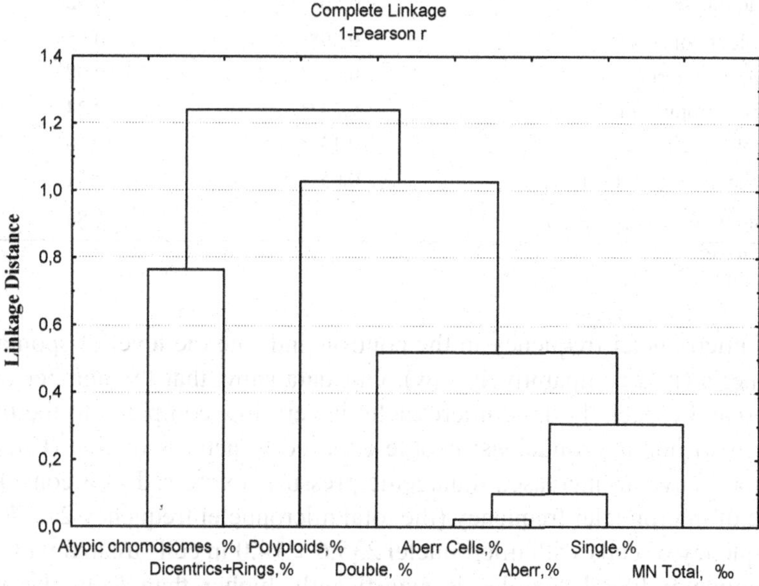

Figure 7. Joint tree (cluster analysis) comparison of cytogenetic status and bystander serum effect (MN induction).

TABLE 9. Linkage distance between cytogenetic status parameters and bystander-inducing effect of blood serum from patients affected after the Chernobyl accident

	Parameter	1	2	3	4	5	6	7
1	Total MN frequency (‰)							
2	Single fragments (%)	0.31						
3	Double fragments (%)	0.42	0.52					
4	Dicentrics and rings (%)	0.83	0.81	0.80				
5	Atypical chromosomes (%)	1.24	1.21	1.09	0.76			
6	Polyploidic cells (%)	0.90	1.03	0.96	0.79	1.10		
7	Aberrant cells (%)	0.28	0.10	0.28	0.53	1.08	0.94	
8	Total number of aberrations (%)	0.24	0.06	0.33	0.63	1.12	0.94	0.02

TABLE 10. Correlation matrix (by Spearman) for cytogenetic parameters compared to induced effect

Parameters of comparison (%)	Number of cells with MN (‰)	Total MN number (‰)
Single fragments	0.50*	0.62*
Double fragments	0.49*	0.55*
Dicentric and ring	0.07**	0.08**
Atypic chromosomes	–0.24**	–0.24**
Polyploid cells	0.13**	0.10**
Number of aberrant cells	0.65*	0.71*
Total number of aberrations	0.76*	0.84*

*$P < 0.005$
**$P > 0.1$

The micronuclei frequency in the controls indicate the level of spontaneous mutagenesis (it is comparatively low). Our data show that the number of cells with two and especially three micronuclei is very low compared to the number of cells with one micronucleus; people exposed to acute radiation (Chernobyl liquidators) have an increased mutagenic pressure, expressed as a considerable increase of micronuclei frequency (the total micronuclei frequency 273.7 ± 22.4 and frequency of cells with micronuclei 235.6 ± 14.0 in cells treated with serum from liquidator blood samples is significantly higher than from the control group (in both cases $P < 0.01$)). At the same time an increase in the number of cells with more than one micronucleus was observed. Thus, the data clearly indicate that acute intensive radiation exposure significantly influences the level of bystander factor accumulation.

These results correlate well with the data from the Hiroshima population (Goh and Sumner, 1968). The data on the bystander effect of the blood serum from liquidators confirms an earlier hypothesis on the possible effects of acute exposures to low doses of radiation.

As can be seen from Table 8, people affected after the Chernobyl accident have significantly higher levels of chromosome aberrations compared to controls ($P < 0.05$) as a result of a high level of nonspecific aberrations (single and double fragments, $P < 0.05$ and $P < 0.01$ compared to controls, respectively).

At the same time frequency of specific marker aberrations (dicentric and ring chromosomes) is almost three times higher compared to controls, but not statistically significant. This data is confirmed by data from Sevan'kaev et al. (2005), showing that with time a significant part of these nonstable marker aberrations is lost, decreasing the effectiveness of biological evaluation of radiation doses.

Summary analysis of micronuclei frequency in HPV-G cells after treatment with blood serum samples indicates that activity of serum samples from people affected after the Chernobyl accident was almost three times higher than in controls (251.18 ± 22.96 and 80.29 ± 13.14, respectively; $P < 0.01$).

Thus, preliminary comparison analysis of obtained data shows that blood serum samples from people, affected after the Chernobyl accident, even after a prolonged period after the Chernobyl accident (19 years to analysis date) may be influenced as a result of prolonged effects of radiation.

In order to understand interactions between parameters (total number of cells with aberrations and micronuclei, and total frequency of aberrations and micronuclei) the dependence between them using polynomial analysis has been studied.

From data presented in Figures 5 and 6 there is evident correlation dependence observed between parameters. The strongest dependence is observed between total micronuclei frequency and total frequency of aberrations. Also direct dependence, but not as evident, is between parameters of total number of cells with micronuclei and total number of aberrant cells. Thus, data analysis suggests that the level of damaging bystander factors in blood serum samples from people affected be the Chernobyl accident (micronuclei frequency) closely correlates with the level of genome damage (aberrations frequency).

Using cluster analysis, linkages between parameters "micronuclei frequentcy, ‰" and number of single, double fragments, dicentric and ring chromosomes, atypic chromosomes, polyploidic cells, aberrant cells frequency, and aberrations frequency have been studied.

As can be seen from Table 9 and Figure 7, the closest linkage is observed between parameters "total number of aberrations, %" and "total number of aberrant cells, %" (linkage distance $d = 0.02$). Parameter "total micronuclei frequency, ‰" has the closest linkage with parameters "number of aberrant cells, %" ($d = 0.28$) and "aberrations frequency, %" ($d = 0.24$).

Using Spearman correlation analysis, we built a correlation matrix for cytogenetic parameters compared with bystander (micronuclei) parameters. From Table 10 one can conclude that the strongest dependence is observed between total micronuclei frequency and total frequency of aberrations (Spearman coefficient of correlation $r = 0.84$; $P < 0.005$). Also direct dependence, but not as evident, is observed between parameters "total number of cells with micronuclei" and "total number of aberrant cells" ($r = 0.65$; $P < 0.005$). The levels of single and double fragment have statistically significant correlation with the level of micronuclei.

Thus, presented data allow us to conclude that, after radiation influence in serum samples from victims of the Chernobyl accident, there are factors appearing and continuing for almost 20 years, inducing the bystander effect.

These factors are able to cause genomic instability. The level of bystander factors in serum significantly correlates with frequency of chromosome aberrations in human lymphocytes.

As a consequence, our results allow us to suggest that the total level of genomic instability may correspond from on the one side the level of radiation influence, and from the other – individual radiosensibility of examined people. The level of factors–inducers of bystander effect in serum has direct correlation with genomic instability. Thus, even in a remote period after acute and chronic low-dose radiation influence in human blood *in vivo* there is accumulation of specific factors, transferring through serum and able to induce an increased level of chromosome aberrations in intact somatic cells, providing genomic instability in tissues of irradiated organism.

References

Balter, M., 1996, Poor dose data hamper study of cleanup workers, *Science* 272: 360.

Baranov, A.E., Gus'kova, A.K., Nadejina, N.M., Nugis, V.Yu., 1995, Chernobyl experience: biological indicators of exposure to ionizing radiation, *Stem Cells* 13(Suppl. 1): 69–77.

Cai, J., Yang, J., Jones, D.P., 1998, Mitochondrial control of apoptosis: the role of cytochrome c. *Biochem. Biophys. Res. Commun.* 1366: 139–149.

Caron, F., Jaco, C., Rouviere-Yaniv, J., 1979, Characterisation of a histone-like protein extracted from yeast mitochondria, *Proc. Natl. Acad. Sci. USA* 76: 4265–4269.

Clayton, D.A., 1982, Replication of animal mitochondrial DNA, *Cell* 28: 693–705.

Clayton, D.A., 1991, Nuclear gadgets in mitochondrial DNA replication and transcription, *Trends Biochem. Sci.* 16: 107–111.

DiMauro, S., Bonilla, E., Davidson, M., Harino, M., Shon, E.A., 1998, Mitochondria in neuromuscular disorders, *Biochim. Biophys. Acta* 1366: 199–210.

Gattermann, N., Berneburg, M., Heinisch, J., Aul, C., Schneider, W., 1995, Detection of the ageing-associated 5-kb common deletion of mitochondrial DDNA in blood and bone marrow of hematologically normal adults. Absence of the deletion in clonal bone marrow disorders, *Leukemia* 9: 1704–1710.

Goh, K., Sumner, H., 1968, Breaks in normal human chromosomes: are they induced by a transferable substance in the plasma of persons exposed to total-body irradiation? *Radiat. Res.* 35(1): 171–81.

Gus'kova, A.K., Baranov, A.E., Barabanova, A.V., Gruzdev, G.P., Piatkin, E.K., 1987, Acute radiation effects in victims of the accident at the Chernobyl nuclear power station, *Med. Radiol.* (Mosk) 32: 3–18.

IAEA, 1996, International Conference: One Decade after Chernobyl. Summing up the Consequences of the Accident, Vienna, Austria, 8–12 April 1996, *IAEA Bull.* 38: 3.

Il'in, L.A., 1995, *Chernobyl: Myths and Reality*, Megapolis, Moscow.

Konoplia, E.F., Rolevich, I.V. (eds.), 1996, Ecological, medical and biological, social and economic consequences of the accident at CNPS in Belarus. Ministry on Emergency Situations and Protection of Population against the Consequences of the Accident at

Chernobyl NPS of Republic of Belarus, Radiobiology Institute of the Academy of Sciences of Belarus, Minsk.
Kubota, N., Hayashi, J.-I., Inada, T., Iwamura, Y., 1997, Induction of a particular deletion in mitochondrial DNA by X rays depends on the inherent radiosensitivity of the cells, *Radiat. Res.* 148(1): 395–398.
Kyoizumi, S., Akiyama, M., Hirai, Y., 1989, Technical Report Series RERF, TR 22-89.
Kyoizumi, S., Umeki, S., Akiyama, M., Hirai, Y., Kusunoki, Y., Nakamura, N., Endoh, K., Konishi, J., Sasaki, M., Mori, T., Cologne, J.B., 1990, Technical Report Series RERF, TR 10-90.
Larsson, N.-G., Clayton, D.A., 1995, Molecular genetic aspects of human mitochondrial disorder, *Annu. Rev. Genet.* 29(151): 78.
Lazutka, J.R., 1996, Chromosome aberrations and rogue cells in lymphocytes of Chernobyl clean-up workers, *Mutat. Res.* 350: 315–329.
Liu, V.W.S., Zhang, C., Nagley, P., 1998, Mutations in mitochondrial DNA accumulate differentially in three different human tissues during ageing, *Nucleic Acids Res.* 26(5): 1268–1275.
Melnov, S., Zullo, S.J., Hamel, C.J.C., Prasanna, P.G.S., Pogozelski, W.K., Fischel-Ghodsian, N., Merril, C.R., Blakely, W.F., 1998, Cytological detection of the 4977-bp "common" mitochondrial DNA deletion using an in situ PCR assay. Mitohondria: interaction of two genomes, Mitohondria interest group minisymposium mitohondria, pp. 25–26.
Melnov, S.B., 1997, Dosimetry control. Materials of International Symposium on Urgent Problems of Dosimetry, Minsk, Belarus, pp. 79–90.
Melnov, S.B., Laziuk, G.I., 1997, *Cytogenetic Aspects of Biodosimetry Methodological Recommendations*, Minsk, pp. 1–28.
Melnov, S.B., Koryt'ko, S.S., Rybalchenko, O.A., Shimanets, T.V., Dudarenko, O.I., Paplevka, A.A., Blizniuk, A.I., 2000, Imunogenetic status of liquidators, 1986–1987 years, Annual "Ecological Anthropology", Minsk, pp. 274–279.
Melnov, S.B., Minenko, V.F., Demidchik, E.P., 1999, Analysis of the frequency of mutant T-helpers as a parameter for biodosimetry, Reports of the NAS of Belarus 43: 63–66.
Piatkin, E.K., Chirkov, A.A., Solovijov, V.J., Nugis, V.J., 1990, Direct application of cytogenetic data for creation of prognostic curves of dynamics of the number of neutrophiles after acute relatively even gamma-irradiation (on the material of the affected during the accident at Chernobyl NPS), *Med. Radiol.* 2: 29–32.
Piatkin, E.K., Nugis, V.J., Chirkov, A.A., 1989, Estimation of acquired dose according to the results of cytogenetic investigations of lymphocyte cultures in the affected by the accident at Chernobyl NPS, *Med. Radiol.* 6: 52–57.
Piatkin, E.K., Nugis, V.J., Chirkov, A.A., 1991, Analysis of chromosomal aberrations and prediction of medullar syndrome gravity in acute radiation affection of man, *Haematol. Transfusiol.* 10: 21–26.
Pilinskaya, M.A., Shemetun, A.M., Dybski, S.S., Redko, D.V., 1996, Genetic effects in somatic cells of persons working under the conditions of chronic exposure of various intensity after the accident at Chernobyl NPS, *Cytol. Genet.* 30: 17–25.
Pilinskaya, M.A., Shemetun, A.M., Red'ko, D.V., Shepelev, S.E., 1991, Dynamics of cytogenetic effect in repeatedly examined persons, who participated in liquidation of the accident at Chernobyl NPS, in various periods after the exposure, *Cytol. Genet.* 25: 3–9.
Reynier, P., Malthiery, Y., 1995, Accumulation of deletions in mtDNA during tissue aging: analysis by long PCR, *Biochem. Biophys. Res. Commun.* 217(1): 59–67.
Sevan'kaev, A.V., Lloyd, D.C., Braselmann, H., 1995a, A survey of chromosomal aberrations in lymphocytes of Chernobyl liquidators, *Radiat. Prot. Dosim.* 58: 85–91.

Sevan'kaev, A.V., Lloyd, D.C, Edwards, A.A., Moiseenko, V.V., 1995b, High exposures to radiation received by workers inside the Chernobyl sarcophagus, *Radiat. Prot. Dosim.* 59: 85–91.

Sevan'kaev, A.V., Lloyd, D.C., Edwards, A.A., et al., 2005, A cytogenetic follow-up of some highly irradiated victims of the Chernobyl accident, *Radiat. Prot. Dosim.* 113(2): 152–161. Epub 30 November 2004.

Shadel, G.S., Clayton, D.A., 1997, Mitochondrial DNA maintenance in vertebrates, *Annu. Rev. Biochem.* 66: 409–435.

23. BIOINDICATION APPROACH FOR AN ASSESSMENT OF TECHNOGENIC IMPACT ON THE ENVIRONMENT

STANISLAV A. GERAS'KIN*, VLADIMIR G. DIKAREV, ALLA A. OUDALOVA, DENIS V. VASILIEV, NINA S. DIKAREVA
Russian Institute of Agricultural Radiology and Agroecology, Kievskoe shosse, 109 km, 249020, Obninsk, Russia
TATIANA I. EVSEEVA
Institute of Biology, Komi Scientific Center, Ural Division RAS, Kommunisticheskaya 28, 167982, Syktyvkar, Russia

Abstract: The results of both *Allium*-test application for the environment quality assessment and long-term field experiments that have been carried out on different species of wild and agricultural plants are discussed. Plant populations growing in areas with relatively low levels of pollution are characterized by the increased level of both cytogenetic disturbances and genetic diversity. The chronic low-dose exposure appears to be an ecological factor creating pre-conditions for possible changes in the genetic structure of a population. A long-term existence of some factors (either of natural origin or man-made) in the plants environment activates genetic mechanisms, changing a population's resistance to exposure. However, in different radioecological situations, genetic adaptation of plant populations to extreme edaphic conditions could be achieved at different rates. The findings presented clearly indicate that an adequate environment quality assessment cannot rely only on information about pollutant concentrations. This conclusion emphasizes the need to update some current principles of ecological standardization, which are still in use.

Keywords: radioactive and chemical contamination; bioindication; plant populations; environment quality assessment

*To whom correspondence should be addressed. Stanislav A. Geras'kin, Russian Institute of Agricultural Radiology and Agroecology, Kievskoe shosse, 109 km, 249020, Obninsk, Russia; e-mail: stgeraskin@gmail.com

R. N. Hull et al. (eds.), Strategies to Enhance Environmental Security in Transition Countries, 315–328.

1. Introduction

Contamination of the environment has become a worldwide problem. Therefore, a clear understanding of all the dangers posed by environmental pollutants to both human health and ecologic systems are needed. With this in mind, considerable efforts have been undertaken to develop effective methods for assessing the quality of the environment.

Generally, two approaches are used. The more classical one is to take samples of air, water, and soil and analyze them in a laboratory using routine chemical–physical techniques. An evaluation of true exposure characteristics is complicated, however, since most quantification techniques are capable of recognizing just a specific compound or its metabolites. Consequently, this approach gives only a part of the knowledge necessary to evaluate the harmful potential of pollutants.

The other approach is to score the biological effects in animals or plants that could be exposed at contaminated sites. An obvious advantage of bioindicators is that these show the results of the pollutants' action on living nature. The use of such a bioindication approach could remove much of the uncertainty associated with current ecological risk assessments and provide meaningful indicators of biological damage. In contrast to the specific nature of assessments on exposure, studies of biological effects integrate the impacts of all the harmful agents, including synergistic and antagonistic effects. The biomarkers may also illuminate previously unsuspected chemical or natural stressors in the study area or reveal that damage has been caused by a pollutant that has since degraded and is no longer detectable by residue analysis. Therefore, this approach is particularly useful for assessing unknown contaminants, complex mixtures, or hazardous wastes.

Certainly, it will never be possible to replace direct physical–chemical measurements of pollutant concentrations entirely by the determination of effects on bioindicators. It is obvious that a correct estimation of environment pollution risk needs to be derived from biological tests and pollutant chemical control in ecosystem compartments. Both need to be carried out simultaneously. The simultaneous use of chemical and biological control methods allows an identification of the relationships between the pollutant concentrations and the biological effects that they cause. In turn, such relationships may allow identification of the contribution of each specific pollutant to the overall biological effect observed. An increased understanding of fate and exposure pathways of harmful substances in various test systems is also needed for a better prediction of what has happened in the field. The knowledge generated makes it possible to limit an effect of unfavorable factors on biota and to predict the further ecological alterations in regions submitted to intensive industrial impact.

To assess the quality of the environment, we mostly used plants as test objects, for several reasons. Plants are higher eukaryotes and essential components of any ecosystem. Moreover, plants seem especially well suited for an environmental assessment since they have fast growth rates and provide a large number of offspring. Owing to their settled nature, plants are constantly exposed to the pollution agents and, therefore, can characterize the local environment in the best way. Plants do not have a predetermined germline (Walbot, 1985); the germ cells are produced during plant development from somatic cells, and, consequently, mutations occurring during somatic development can be inherited. Furthermore, in many cases, plant bioassays are the simplest and most cost-effective among test systems for an environmental assessment.

Interaction of contaminants with biota first takes place at the cellular level making cellular responses not only the first manifestation of harmful effects, but also suitable tools for an early and reliable detection of exposure. Cellular changes would initially be less obvious than the direct visible effects of pollutants, but in the long run they could be more significant. Because the targets of protection in ecological risk assessment most often are populations, communities, and ecosystems, the biomarker response must be tightly linked to responses at these higher levels. It is becoming increasingly clear (Theodorakis et al., 1997) that cellular alterations may in the long run influence biological parameters important for populations such as health and reproduction. These types of effects are of special concern because they can manifest themselves long after the source of contamination has been eliminated. Therefore, it is genetic test systems that should be used for an early and reliable demonstration of the alterations resulting from human industrial activity.

2. Results

In most bioindication studies, standard indicator plant species such as *Tradescantia*, *Allium cepa*, or *Vicia faba* were used. Among the test systems suitable for toxicity monitoring, the *Allium*-test is well known and commonly used in many laboratories. Results from the *Allium*-test have shown a good agreement (Fiskesjo, 1985) with results from other test systems, eukaryotic as well as prokaryotic. As a genotoxicity test, the *Allium*-based assay of chromosome aberration in anaphase–telophase is for many reasons especially useful for the rapid screening of pollutants posing environmental hazards. In addition, a good toxicity indicator is given by the mitotic index. In Table 1 our results of *Allium*-test application for the environment quality assessment are briefly summarized.

TABLE 1. *Allium*-test application for the environment quality assessment

Site	Contamination	Results
Radium production industry storage cell territory, Komi Republic, Russia	Heavy natural radionuclides and chemical pollution	All water samples caused a significant increase in the chromosome aberration frequency. Genotoxic effect was a result of chemical toxicity mainly (Evseeva et al., 2003)
Underground nuclear explosion site, Perm region, Russia	Radionuclides	^{90}Sr significantly contributes to the induction of cytogenetic disturbances (Evseeva et al., 2005)
Radioactive waste storage facility, Obninsk, Russia	Radionuclides and chemicals	All water samples caused a significant increase in the chromosome aberration frequency. Genotoxic effect was a result of chemical toxicity mainly (Oudalova et al., 2006)
Upper Silesian Coal Basin, Poland	Heavy natural radionuclides and chemical pollution	All water and sediment samples caused a significant increase in the chromosome aberration frequency. Ra, Ba, Sr, and Cu contribute significantly to the induction of cytogenetic disturbances
Semipalatinsk Test Site, Kazakhstan	Radionuclides	Data analysis in progress

As an illustration of the *Allium*-test application, the frequency of aberrant cells in root meristem of bulbs, grown on sediment sampled from the post-mining areas of the Upper Silesia, is presented in Figure 1. Cytogenetic damage to meristem cells of *A. cepa* is evident in all noncontrol variants (1–4) which are significantly higher than the control value of variant 5. So, the clear genotoxic effect of sampled sediment is shown. Also, an important contribution of such severe types of cell damage as chromosome bridges and laggings is demonstrated.

Concentrations of 12 heavy metals and 4 radionuclides were measured in the sediments. All samples contained extremely high concentrations of Ba, 5–50.5 times over the maximum permissible level. In variants 3 and 4, concentrations of Ba, Cu, Mn, and Zn were above the permissible limits.

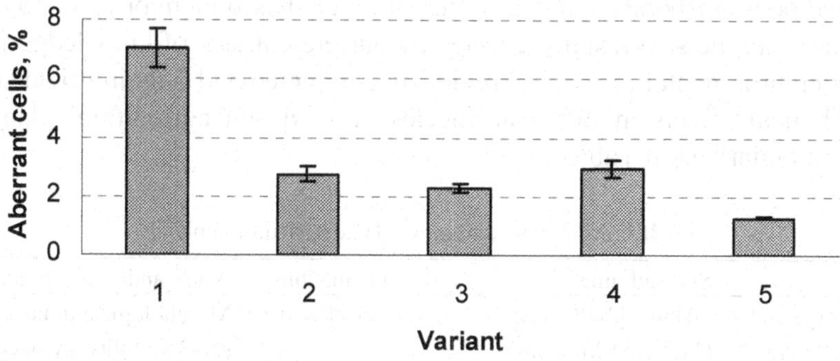

Figure 1. Aberrant cells frequency in root meristem cells of *Allium cepa*, grown in sediment sampled from the post-mining areas in the Upper Silesia, Poland. Sampling variants: 1: underground galleries, 2–4: settling ponds, 5: control.

Specific activities of radium nuclides in the samples from the underground gallery (variant 1) and the bank of the Rontok pond (variant 2) were over the corresponding values of the minimum significant specific activity. Key pollutants governing biological effect observed were revealed, through an adaptation of mathematical and statistical techniques of multivariate analysis, including correlation analysis, step-by-step inclusion or exclusion of essential predictors, and multivariable linear regression. From preliminary assessments, the best model to describe an aberrant cell's occurrence in root meristem cells of *A. cepa* germinated on the sediment sampled is:

$Y (\%) = (3.32 \pm 0.28) - (0.104 \pm 0.009) \cdot [Cu] + (2.10 \pm 0.18) \cdot 10^{-2} \cdot [Ba] + (4.94 \pm 0.46) \cdot 10^{-4} \cdot [Ra_{\Sigma}]$ where [Cu] and [Ba] are concentrations of chemical elements in soil, in mg/kg; $[Ra_{\Sigma}]$ is total specific activity of radionuclides analyzed in soil, in Bq/kg; the determination coefficient $R^2 = 60.2\%$; Fisher statistics $F = 49.5$.

The other possible way to assess the quality of the environment is the use of plants directly growing in contaminated sites. This approach is particularly useful for assessing long-term ecotoxical effects induced by chronic low-dose-rate and multipollutant exposure at contaminated sites. Up to now we have known little about responses of plant and animal populations to environmental pollutants in their natural environments. Although radionuclides and heavy metals cause primary damage at the molecular level, there are emergent effects at the level of populations, nonpredictable solely from the knowledge of elementary mechanisms of the pollutants' influence. The usefulness of data gathered both in laboratory-based studies and field-based monitoring observations may, therefore, be significantly affected by our present lack of knowledge in this area of environmental research. Previously completed and ongoing field studies on biological effects in different species of wild and agricultural plants are briefly summarized in Table 2.

TABLE 2. Field studies on wild and agricultural plants

Species	Site and time	Contamination	Assay and/or end points
Winter rye and wheat, spring barley and oats	10 km ChNPP zone (11.7–454 MBq/m^2), 1986–1989	Radionuclides	Morphological indices of seeds viability, cytogenetic disturbances in intercalar and seedling root meristems (Geras'kin et al., 2003a)
Scots pine, couch grass	30 km ChNPP zone (250–2690 μR/h), 1995	Radionuclides	Cytogenetic disturbances in seedling root meristem (Geras'kin et al., 2003b)
Scots pine	Radioactive waste storage facility, Leningrad Region, Russia, 1997–2002	Mixture	Cytogenetic disturbances in needles intercalar and seedling root meristems (Geras'kin et al., 2005)
Wild vetch	Radium production industry storage cell, Komi Republic, Russia (73–3300 μR/h), 2003	Heavy natural radionuclides and chemical pollution	Embryonic lethals, cytogenetic disturbances in seedling root meristem (Evseeva et al., 2007)
Scots pine	Sites in the Bryansk Region radioactively contaminated in the Chernobyl accident (451–2344 kBq/m^2), 2003–2006	Radionuclides	Cytogenetic disturbances in seedling root meristem, enzymatic loci polymorphism analyses
Scots pine	10 km ChNPP zone (1100 μR/h), 2004	Radionuclides	Morphological modifications in pine needles, cytogenetic disturbances in seedling root meristem
Koeleria gracilis Pers., *Agropyron pectiniforme* Roem. et Schult.	Semipalatinsk Test Site, Kazakhstan (74–3160 μR/h), 2005–2006	Radionuclides	Cytogenetic disturbances in seedling root meristem

With each passing year since the Chernobyl accident of 1986, more questions arise about the potential for organisms to adapt to radiation exposure. It is often thought to be attributed to somatic and germline mutation rates in various organisms. Research into the mechanisms of plant adaptation to environmental stresses and increased radiation levels still lags far behind many other areas of plant molecular biology. Adaptation is a complex process (Kovalchuk et al., 2004) by which populations of organisms respond to long-term environmental stresses by permanent genetic change. Studies of multiple generations exposed to radiation were rarely undertaken due to the difficulties of creating a suitable model to study the effects of chronic exposure. While limited in number, these radioecological studies have attracted a great deal of interest in the question of how organisms adapt to ionizing radiation. In 1987–1989, an experimental study on the cytogenetic variability in three successive generations of winter rye and wheat, grown at four plots with different levels of radioactive contamination, was carried out within the 10 km ChNPP zone (Geras'kin et al., 2003a). In the autumn of 1989, aberrant cell frequencies in intercalary meristem of winter rye and wheat of the second and third generations significantly exceeded these parameters for the first generations (Figure 2). In 1989, plants of all three generations were maintained in the identical conditions and accumulated the same doses, which is why the most probable explanation of the registered phenomenon relates to a genome destabilization in plants grown from radiation-affected seeds. Such detailed analysis of several generations of plants exposed to radiation provides some insight into possible mechanisms of plant adaptation to chronic radiation exposure. It relates to higher-order ecologic effects, as well as to contaminant-induced selection of resistant phenotypes. From these viewpoints, the results observed in this study and indicating a threshold character of the genetic instability induction may be a sign of an adaptation processes beginning. That is, the chronic low-dose irradiation appears to be an ecological factor creating preconditions for possible changes in the genetic structure of a population.

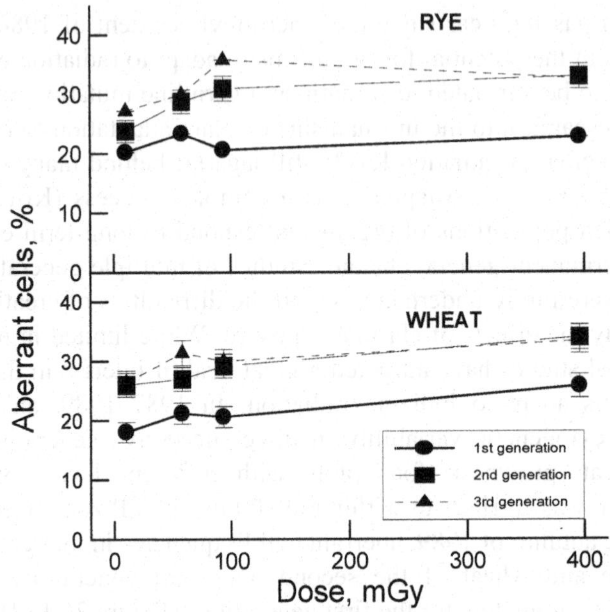

Figure 2. Aberrant cells frequency in three successive generations of winter rye and wheat, grown on contaminated plots within the 10 km ChNPP zone (Geras'kin et al., 2003a).

The adaptation processes in impacted wild plant populations were also investigated within the framework of other studies. The results of these experiments indicate (e.g., see Table 3) that an increased level of cytogenetic disturbance is a typical phenomenon for plant populations growing in areas with relatively low levels of pollution.

TABLE 3. Aberrant cell frequency in seedling root meristem of Scots pine growing in the Bryansk Region of Russia, radioactively contaminated as a result of the Chernobyl accident

Test site	^{137}Cs contamination density (kBq/m^2)	Dose rate (mGy/year)[a]	Aberrant cells (mean ± SE) (%)	
			2003	2004
Reference	–	0.14	0.90 ± 0.09	0.88 ± 0.09
VIUA	451	7.40	1.47 ± 0.15[b]	1.59 ± 0.14[b]
Starye Bobovichy	946	15.3	1.32 ± 0.12[b]	1.37 ± 0.14[b]
Zaborie 1	1730	28.3	1.69 ± 0.17[b]	1.67 ± 0.17[b]
Zaborie 2	2340	37.8	1.63 ± 0.15[b]	1.68 ± 0.17[b]

Seeds were collected in 2003 and 2004

[a]Absorbed doses are estimated for γ- and β-radiation

[b]Difference from the reference population is significant, $P < 5\%$

In the period 1949–1963, 116 air and ground-surface explosions for nuclear and hydrogen bomb testing was carried out in the Semipalatinsk Test Site. A study of *Koeleria gracilis* Pers. populations, a typical wild crop for Kazakhstan, showed that not only did the total frequency of cytogenetic disturbances increase significantly with the dose rate measured in sampling points (Figure 3), but that the relative contributions of such severe disturbances as double (chromosome) bridges and tripolar mitoses increased as well.

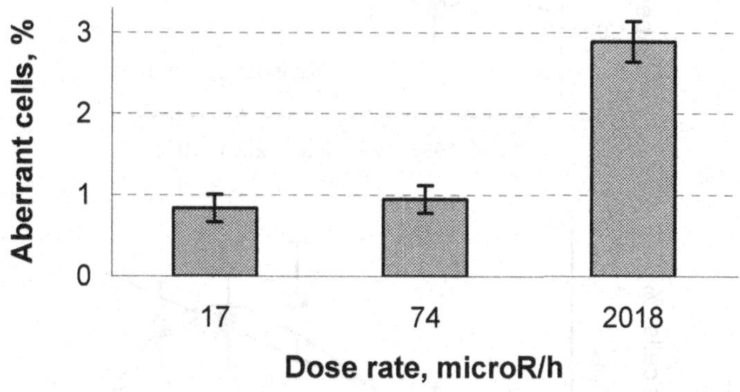

Figure 3. Frequency of aberrant cells in coleoptiles of *Koeleria* seedlings.

Forest trees have gained much attention in recent years as nonclassical model eukaryotes for population, evolutionary, and ecological studies (Gonzalez-Martinez et al., 2006). Because of low domestication, large open-pollinated native populations, and high levels of both genetic and phenotypic variation, they are ideal organisms to unveil the genetic basis of population adaptive divergence in nature. In the field study (Geras'kin et al., 2005), Scots pine populations were used for an assessment of the genotoxicity originating from an operation of a radioactive waste storage facility. Specifically, frequency and spectrum of cytogenetic disturbances in reproductive (seeds) and vegetative (needles) tissues sampled from Scots pine populations were studied to examine whether Scots pine trees have experienced environmental stress in areas with relatively low levels of pollution. The temporal changes of the cytogenetic disturbances in seedling root meristem from 1997 to 2002 are shown in Figure 4. There are essential differences between these relationships for the reference and impacted Scots pine populations. Statistical analysis revealed (Geras'kin et al., 2005) that cytogenetic parameters at the reference site (Bolshaya Izhora) tend to follow cyclic fluctuations in time, whereas in technogenically affected populations ("Radon", LWPE, and Sosnovy Bor) this relationship could not be revealed with confidence. Thus, man-made impact in this region is strong enough to destroy natural regularities.

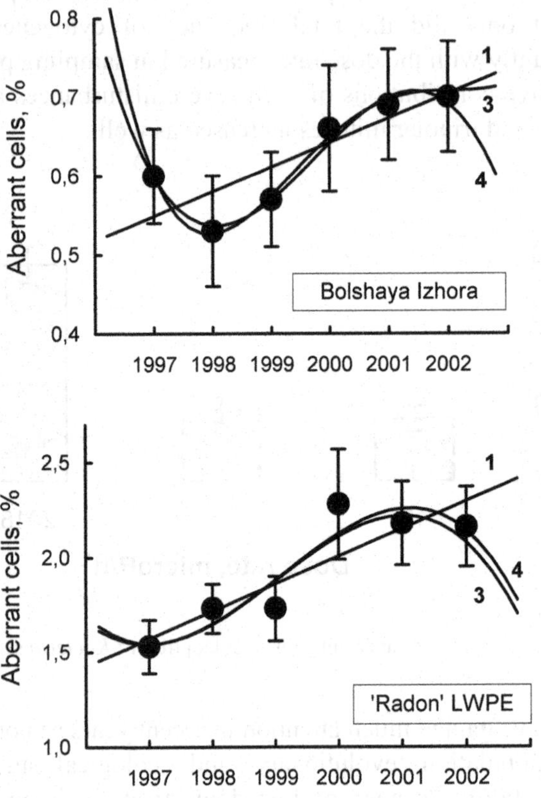

Figure 4. Aberrant cells percentages in seedling root meristem of Scots pine trees in dependence on year and their approximation by the best models. 1: linear model, 3 and 4: polynomial models of the third and fourth degrees, correspondingly.

Pollution is well-documented selective force (Breitholtz et al., 2006), which has been found to induce metal tolerance in plants, pesticide resistance in insects, and pollution tolerance in a variety of aquatic organisms. To study possible adaptation processes in impacted plant populations, a portion of the seeds collected were subjected to an acute γ-ray exposure (Geras'kin et al., 2005). The seeds from the Scots pine populations experiencing man-made impacts showed (Figure 5) a higher resistance than the reference ones. Al. though the picture of adaptation is far from complete, there is convincing proof

(Shevchenko et al., 1992) that the divergence of populations in terms of radioresistance is connected with a selection for changes in the effectiveness of the repair systems.

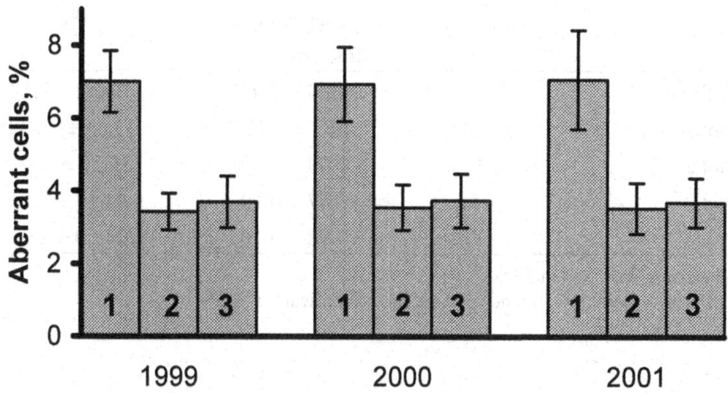

Figure 5. Aberrant cell frequency in root meristem of Scots pine seedlings grown from seeds sampled in the Leningrad Region in 1999–2001 and exposed to an acute γ-ray dose of 15 Gy. 1: Bolshaya Izhora (reference), 2: Sosnovy Bor town, 3: "Radon" LWPE.

It is well known (Macnair, 1993; Rajakaruna, 2004) that genetic adaptation in plant populations to extreme environmental conditions can take place quite rapidly, even within a few generations. But this rule does not apply to all the cases. A study of the wild vetch (*Vicia cracca* L.) population for more than 40 years inhabiting a site with an enhanced level of natural radioactivity has shown that transformations of genetic structure of the population are still ongoing. In this site, both low doses of external exposure and incorporated heavy natural radionuclides have significant effects on the variability in plant populations and their potential for an adaptation (Evseeva et al., 2007). External exposure appears to be, as a selection factor, increasing the frequency of embryonic lethal mutations (Table 4). In addition, in contrast with (Geras'kin et al., 2005), the acute γ-irradiation has demonstrated that seeds from the experimental population show rather high radiosensitivity. In a special study, the inherited character of this phenomenon was established (Alexakhin et al., 1990). These findings remain to be investigated, but at this point some speculation could be made. A specific radioecological situation within the uranium–radium anomaly, where a decisive role is played by α-emitters, is probably the cause of this amazing phenomenon. However, the mechanisms involved in this plant response are still unclear.

TABLE 4. Embryonic lethal and chromosome aberration frequency in seedling root meristem of wild vetch inhabiting the site with enhanced level of natural radioactivity (Komi Republic) (Evseeva et al., 2007)

	Dose rate (μR/h)	Chromosome aberrations frequency (%)	Embryonic lethals (%)
Reference	9	0.80 ± 0.07	24.62 ± 2.12
Plot 1	3300	1.33 ± 0.12[a]	33.63 ± 23.32[a]
Plot 2	2400	1.00 ± 0.09	35.38 ± 2.06[a]
Plot 3	758	0.87 ± 0.10	25.17 ± 2.24
Plot 4	73	0.98 ± 0.10	24.60 ± 2.08

Seeds were collected in 2003
[a]Difference from the reference population is significant, $P < 5\%$

3. Conclusions

Findings presented here clearly indicate that an adequate environment quality assessment cannot rely only on information about pollutant concentrations. This conclusion emphasizes the need to update some current principles of ecological standardization, which are still in use today. When assessing potential hazards from radionuclide and chemical pollution for ecosystems, a harmonized approach based on ecotoxicology principles should be applied. At the first stage, it is advisable to use biological testing that result in an integral assessment of effects from all substances presented in the environment. If bioassays give positive responses, a more detailed survey should be undertaken including physical–chemical analysis, geochemical barriers, and study of migration parameters of the contaminants for a certain landscape. An effective linking of bioindication screening assays to well-established environmental pollution monitoring is a way of improving and upgrading an existing system of public and environment protection in order to meet requirements of consistency between current scientific knowledge and decision-making processes.

A better understanding of the genetic aspects of population, community, and the whole ecosystem responses to toxic agent's exposure is vital to future environmental management programs. The results described here provide evidence that plant populations growing in areas with relatively low levels of pollution are characterized by the increased level of both cytogenetic disturbances and genetic diversity. Man-made pollution may influence an evolution of exposed populations through a contaminant-induced selection process. The long-term existence of some factors (either of natural origin or man-made) in the plants' environment activates genetic mechanisms, changing a population's resistance to exposure. However, in different radioecological situations, genetic adaptation to extreme edaphic conditions in plant populations could be achieved at different rates. Such evolutionary effects are of special concern because they are able to negatively affect population dynamics and local extinction rates. These processes have a genetic basis; therefore, understanding changes at the genetic level should help in identifying more complex changes at higher levels. Recent studies have shown (Whitham et al., 2006) that heritable traits in a single species have predictable effects on community structure and ecosystem processes. Therefore, we can begin to apply the principles of population and quantitative genetics to place the study of complex communities and ecosystems. Finally, in spite of the wealth of information collected so far, much more still remains to be explained in order to fully understand the basis of plant populations' adaptation to a harmful environment.

Acknowledgments

Works presented were supported by the EU-Project WaterNorm (contract no. MTKD-CT-2004-003163), Ministry of Science and Innovation of Russian Federation (contract no. 02.445.11.7463), and ISTC projects no. 3003 and K-1328.

References

Alexakhin, R.M., Arkhipov, N.P., Barhudarov, R.M., Vasilenko, I.Y., Drichko, V.F., Ivanov, U.A., Maslov, V.I., Maslova, K.I., Nikiforov, V.S., Polykarpov, G.G., Popova, O.N., Sirotkin, A.N., Taskaev, A.I., Testov, B.V., Titaeva, N.A., Fevraleva, L.T., 1990, *Heavy Natural Radionuclids in Biosphere. Migration and Biological Effects on Population and Biocenosis*, Nauka Publishers, Moscow (in Russian).

Breitholtz, M., Ruden, C., Hansson, S.O., Bengtsson, B.E., 2006, Ten challenges for improved ecotoxicological testing in environmental risk assessment, *Ecotoxicol. Environ. Safety* 63: 324–335.

Evseeva, T.I., Geras'kin, S.A., Shuktomova, I.I., 2003, Genotoxicity and toxicity assay of water sampled from a radium production industry storage cell territory by means of *Allium*-test, *J. Environ. Radioact.* 68: 235–248.

Evseeva, T.I., Geras'kin, S.A., Shuktomova, I.I., Taskaev, A.I., 2005, Genotoxicity and toxicity assay of water sampled from the underground nuclear explosion site in the north of the Perm region (Russia), *J. Environ. Radioact.* 80: 59–74.

Evseeva, T.I., Majstrenko, T.A., Geras'kin, S.A., Belych, E.S., 2007, Genetic variance in wild vetch cenopopulation inhabiting site with enhanced level of natural radioactivity, *Radiats. Biol. Radioecol.* 47: 54–62. (in Russian)

Fiskesjo, G., 1985, The *Allium* test as a standard in environmental monitoring, *Hereditas* 102: 99–112.

Geras'kin, S.A., Dikarev, V.G., Zyablitskaya, Ye.Ya., Oudalova, A.A., Spirin, Y.V., Alexakhin, R.M., 2003a, Genetic consequences of radioactive contamination by the Chernobyl fallout to agricultural crops, *J. Environ. Radioact.* 66: 155–169.

Geras'kin, S.A., Zimina, L.M., Dikarev, V.G., Dikareva, N.S., Zimin, V.L., Vasiliyev, D.V., Oudalova, A.A., Blinova, L.D., Alexakhin, R.M., 2003b, Bioindication of the anthropogenic effects on micropopulations of Pinus sylvestris L. in the vicinity of a plant for the storage and processing of radioactive waste and in the Chernobyl NPP zone, *J. Environ. Radioact.* 66: 171–180.

Geras'kin, S.A., Kim, J.K., Oudalova, A.A., Vasiliyev, D.V., Dikareva, N.S., Zimin, V.L., Dikarev, V.G., 2005, Bio-monitoring the genotoxicity of populations of Scots pine in the vicinity of a radioactive waste storage facility, *Mutat. Res.* 583: 55–66.

Gonzalez-Martinez, S.C., Krutovsky, K.V., Neale, D.B., 2006, Forest-tree population genomics and adaptive evolution, *New Phytologist* 170: 227–238.

Kovalchuk, I., Abramov, V., Pogribny, I., Kovalchuk, O., 2004, Molecular aspects of plant adaptation to life in the Chernobyl zone, *Plant Physiol.* 135: 357–363.

Macnair, M., 1993, The genetics of metal tolerance in vascular plants, *New Phytologist* 124: 541–559.

Oudalova, A.A., Geras'kin, S.A., Dikarev, V.G., Dikareva, N.S., Kruglov, S.V., Vaizer, V.I., Kozmin, G.V., Tchernonog, E.V., 2006, Method preparation on integrated assessment of ecological situation in the vicinity of the Russian nuclear facilities, in: *Proceedings of the Regional Competition of the Natural-Science Projects*, Issue 9, Kalugian scientific centre, Kaluga, pp. 221–239 (in Russian).

Rajakaruna, N., 2004, The edaphic factor in the origin of plant species, *Int. Geol. Rev.* 46: 471–478.

Shevchenko, V.A., Pechkurenkov, V.L., Abramov, V.I., 1992, *Radiation Genetics of Natural Populations: Genetic Consequences of the Kyshtym Accident*. Nauka Publishers, Moscow, 221 p. (in Russian).

Theodorakis, C.W., Blaylock, B.G., Shugart, L.R., 1997, Genetic ecotoxicology I: DNA integrity and reproduction in mosquitofish exposed in situ to radionuclides, *Ecotoxicology* 6: 205–218.

Walbot, V., 1985, On the life strategies of plants and animals, *Trends Genet.* 1: 165–170.

Whitham, T.G., Bailey, J.K., Schweitzer, J.A., Shuster, S.M., Bangert, R.K., LeRoy, C.J., Lansdorf, E.V., Allan, G.J., DiFazio, S.P., Potts, B.M., Fischer, D.G., Gehring, C.A., Lindroth, R.L., Marks, J.C., Hart, S.C., Wimp, G.M., Wooley, S.C., 2006, A framework for community and ecosystem genetics: from genes to ecosystems, *Nature Rev. Genet.* 7: 510–523.

24. MANAGEMENT OF AGRICULTURAL BUILDINGS TO PROTECT ANIMAL HEALTH AND ENSURE BUILDING MAINTENANCE

IOANA TANASESCU*
University of Agricultural Sciences and Veterinary Medicine, Cluj-Napoca, Faculty of Animal Breeding and Biotechnology, Civil Engineering and Technical Design Department, Romania

Abstract: Agricultural integration within the integration of our country in the European Union (EU) is an important aspect not only in political, but also in economical, environmental, and educational terms. The agricultural built stock no longer respond neither to the users requirements, nor to the new standards and regulations regarding the animal welfare, environment protection, energy saving and quality of the technological processes and products (Official Journal of the EU/L001, 2003; Law Nr. 10, 1995). Basic knowledge about the interdependence between technology and plants and animals, microbiology, and physical phenomena must be put together in the functional design and layout of the facilities, which also means requirements for: visual and esthetic quality, environmental sustainability, demand for animal welfare and health, food quality and safety, economical competitiveness, and other requirements.

Keywords: fungi, agriculture, livestock health

1. Introduction

Since the beginning of the 1990s, agriculture in Central and Eastern European Countries has changed significantly. Despite vast natural resources, in terms of area, agriculture has not been able to use this potential to its full extent. In spite of huge efforts, restructuring agriculture and the food industries is far from being complete.

*To whom correspondence should be addressed. IOANA TANASESCU, University of Agricultural Sciences and Veterinary Medicine, Cluj-Napoca, Faculty of Animal Breeding and Biotechnology, Civil Engineering and Technical Design Department, Romania; e-mail: ioanatanasescu@usamvcluj.ro

R. N. Hull et al. (eds.), Strategies to Enhance Environmental Security in Transition Countries, 329–338.

Accession of the Central and Eastern European Countries will increase the European Union (EU) agricultural area by 45% and the population by 28% (Official Journal of the EU/C085, 2003). The agricultural sector in Romania underwent significant change since transition started in 1989. General agricultural production has fluctuated and livestock production fell substantially.

A recent study made in Transylvania, related to the existing construction solutions in dairy farming, emphasizes a large diversity of buildings and interior arrangements. They were differentiated in function of the designed technological solutions used by the farmers, the level of the feeding and disposal mechanization, solutions that in the majority of the cases do not respond to the interdependence of the functional factors.

These constructions represent the "built stock" of former collective and state farms. Most of them were privatized during the last 16 years, but in many cases, the buildings have suffered severe degradation.

The inside climate is highly influenced by construction elements that have direct contact with the outside environment. The total surface area of these elements represents the envelope, which determines the heat and mass transfers between the inside and outside environment. The inside climate affects the properties of the building envelope via its effect on condensation of the air creating special climate conditions. In long run, the presence of high concentrations of aggressive gases (like ammonia) might be more important in combination with moist air via their effect on the degradation of the building materials which finally leads to the deterioration of their insulating effect. Furthermore, the presence of unwanted climate conditions and wet surfaces are ideal conditions for all kind of respiratory diseases to break out. Pathogens and parasites, spread in this way, find an ideal breeding place inside the wet building materials where they can enhance the degradation process. Therefore, in this study, the state of the envelopes of several cow barns was determined by mineralogical analysis, biological analysis, and by investigations of the relation between the inside and outside climate under winter conditions.

The quality of agricultural buildings are affected by the balance between the investment price, technological facilities, and production costs, together with the personnel requirements, and livestock needs as users of these constructions.

2. Results and Discussion

The quality of agricultural buildings, including their envelope elements, constitutes the basis of a new approach to evaluating the value of buildings affected by aggressive environmental conditions. This evaluation can be combined with resource conservation, economic development, and pollution reduction, as the main instruments in assuring users' requirements.

The "aggressive environment" problem was often treated as an old, complex but accidental issue, of short duration. However, it has become more and more a current problem of old agricultural buildings, with increasing technical and economic importance and implications.

For the animal houses this is a result of the technological processes and poor maintenance. The aggression is manifested not only upon the animals and plants, but also on the building stock placed in the same area; by the appearance of some significant degrading phenomena of the built environment. The biotic system has specific environmental needs and its production and well-being will be influenced by the environment. At the same time, the biotic system will strongly influence conditions within the air space of the construction as shown in Figure 1.

From the analysis of all the requirements referring to comfort, animal welfare, security, and sustainability, results their interdependence and their continuous evolving transformation.

The "aggressive environment" existing inside animal shelters as a result of the biological and technological processes submit in some cases the construction elements to strains with negative effect upon the animal's health and implicit upon the production and personnel.

Within the building there are a number of moisture sources such as moisture from the animals and their physiological mechanisms, evaporating forage and manure moisture, open water tank, cleaning solutions, and other technological processes. The building volume exchanges air to and from outside environment.

There are now many numerical models to describe heat and moisture transfer within the building envelope. Generally, these models are developed for buildings used by people and are centered on the performances of the various layers of the building envelope.

On the other hand, for barns housing dairy cow and other species of farm animals, several other physical processes are involved, in direct relation with the metabolic process, animal age and weight, housing, and ventilation (Albright, 1990).

The automatic surveying system used for the climate control experiment consists of several transformers attached to measuring and transmitting stations and a central survey station. The signal transmission is made with radio channels. It can assure the configuration of some alarm and signal barriers in case of exceeding the admissible limit values for moisture.

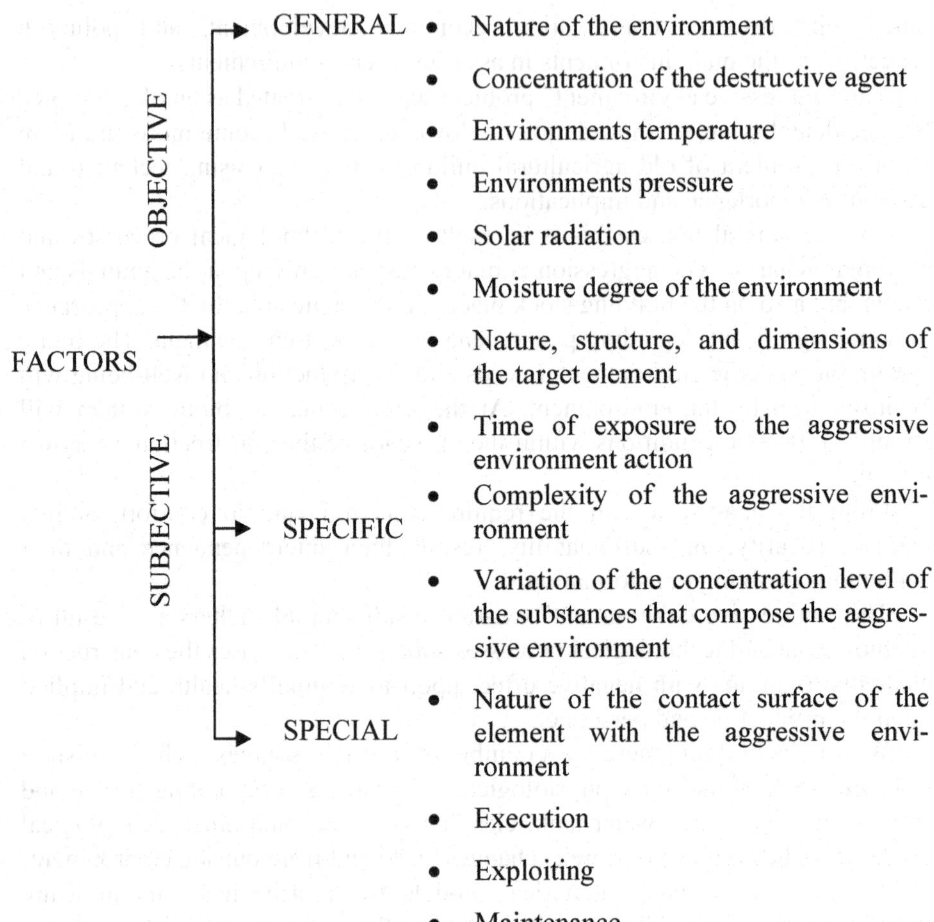

Figure 1. Factors that influence inside conditions of agricultural buildings.

The measuring and transmitting stations and transformers were placed: inside the building on the exterior wall at the animal level in lay down position (~80 cm high), with sensors for moisture and inside air temperature measurement; and outside the building at the same level, with sensors for the determination of air temperature, humidity, and wind velocity. The soil temperature and global radiation were not used in this experiment, but were automatically registered by the station.

The measurements were carried out between 7 and 13 January 2006 in the dairy cow barn of the experimental farm of the University of Agricultural Sciences and Veterinary Medicine. Significant variations of temperature and humidity were registered both inside and outside the building. In Figure 2, as examples, the values of temperature and humidity from 12 January are illustrated.

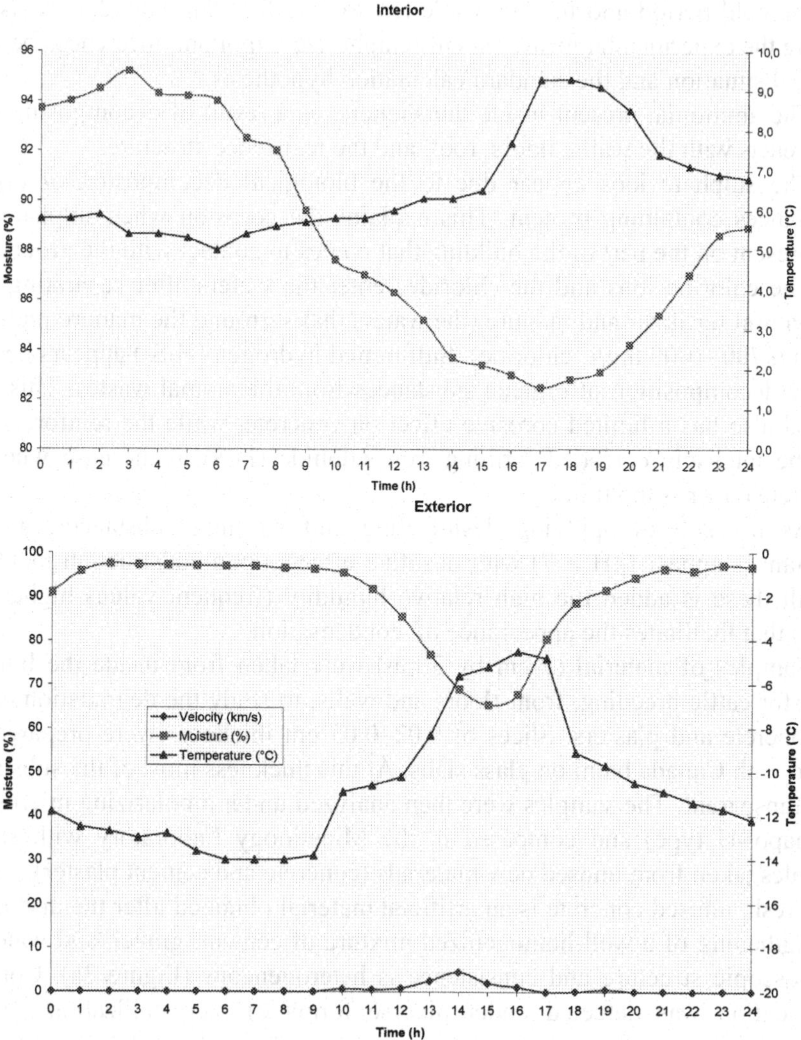

Figure 2. Example for inside and outside air temperature and relative humidity, measured on 12 Jan. 2006.

During the cold season, the values for relative humidity are very high. This implies that the condensation was unavoidable most of the time. The increasing air temperature during the day together with the effect of solar radiation on the envelopes surface determines the reduction of atmospheric humidity. The fog phenomenon and the condensation observed inside the cow barn during the measuring period can be explained in light of the specific indoor conditions. As the standard calculation values of relative humidity of the exterior air are 85% for the cold period and for the whole territory of Romania, the data registered during the experimental measurements emphasize important differences between the real situation and the standard calculation hypothesis.

The ammonia, present inside the shelters, as a result of decomposing manure, reacts with the walls, floors, roof, and the resistence structure.

The sulphate ions appear due to the biological decomposion of organic substances containing protein. This explains the corosion when sulphate ions are present on the part of the building that comes in contact with the ground.

The chlorine ions and the chlorides enter the waters after separating from the natural fertilizer and manure (the waters that surround the manure platforms contain 700–1000 mg/L chlorine). Sulfuretted hydrogen (H_2S_3) appears because of the decomposition of organic substances from the animal wastes. This compound also has a limited corosive effect on concrete, while the reinforcements can be intensely corroded forming iron sulphide (FeS) in the case when the concrete cover is too thin.

As a result of applying desinfecting and washing substances, alkaline substances appear ($pH > 7$) with negative effects upon construction materials. To all these is added the high relative humidity (frequent values higher than 85%) that facilitates the appearance of condensation.

Samples of material (2 cm by 3 cm) were taken from inside the building used for cattle breeding, from floors and walls, to study the degradation causes of concrete and plasters. Slices of 0.02–0.03 cm thickness were prepared and glued with Canada balm on glass slabs. At this thickness most of the substances are transparent. The samples were then analyzed under a polarizing microscope (Yenapol-U type) and compared in the Mineralogy Laboratory with similar samples taken from unused new materials (concrete and cement plaster).

Fresh, unused concrete is an artificial material obtained after the drying and strengthening of a well-homogenized mixture of cement, gravel, and water. Its microscopic structure and consistence is heterogeneous (Figure 3a). Concrete degradation was detected under diverse forms of recrystallizations, as the aggressive agents from the exterior environment acted on the concrete rock (Figure 3b):

- The aggresive ions reacted with the cement replacing the positive ions of calcium (Ca^{2+}), new reaction products without binding properties appear. This type of corrosion is caused by acid solution, salts, and strong alkali.
- On large concrete surfaces the corrosion by expansion due to the formation of some salts that crystallize with a great amount of water, destroys the stiffened cement because of the volume increase. The concrete is distroyed by the accumulation of expansive salts.

- The aggregates used to prepare the concrete were not affected by the aggresive agents.

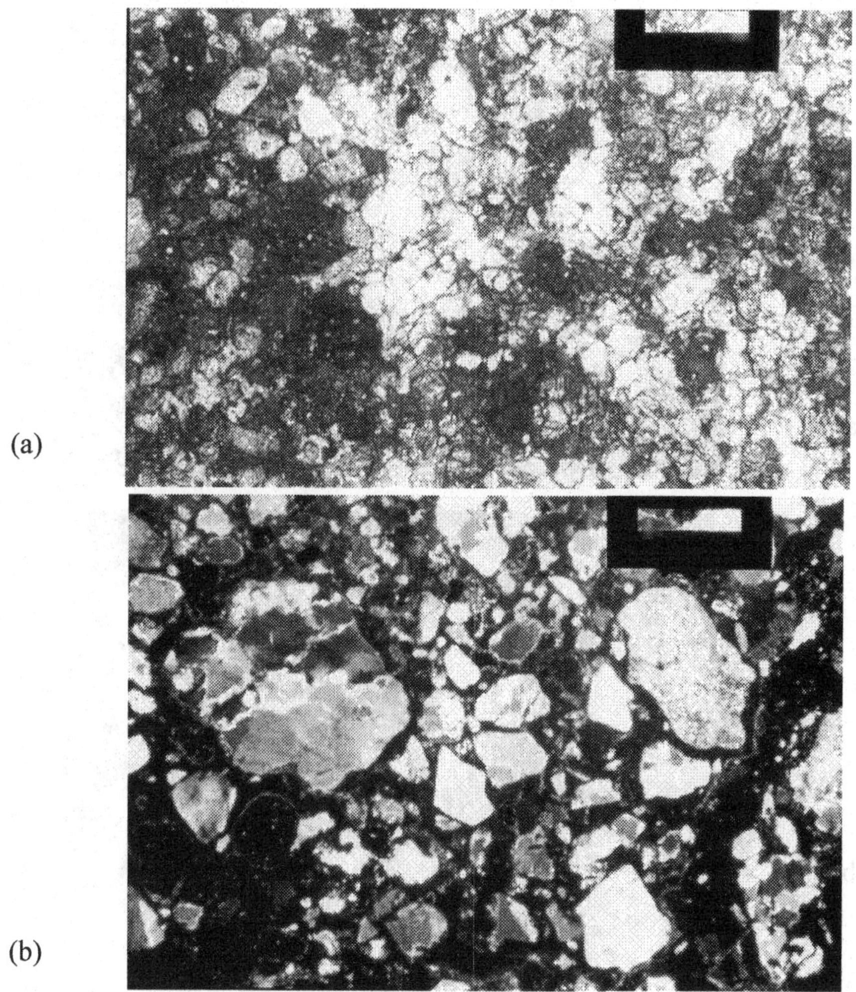

(a)

(b)

Figure 3. Microscopic image of concrete samples in polarized light, showing the degradation stage of the floor: (a) unused concrete material and (b) sample from the concrete floor.

Newly made mortar is an artificial, solid, and homogenous mixture of fine sand, cement, and water. Under the polarizing microscope, a sample slice (Figure 4a) revealed a relatively homogenous microgranular material, which consists of quartz granules of small size but variable shape and a matrix with amorphous aspect. The samples from the interior plaster of the exterior and the inside walls (Figure 4b), show that the corrosive process is indicated on plaster but not on bricks. The plaster with 3.5 cm thickness made with fine aggregates, favored the water infiltration through capillary action.

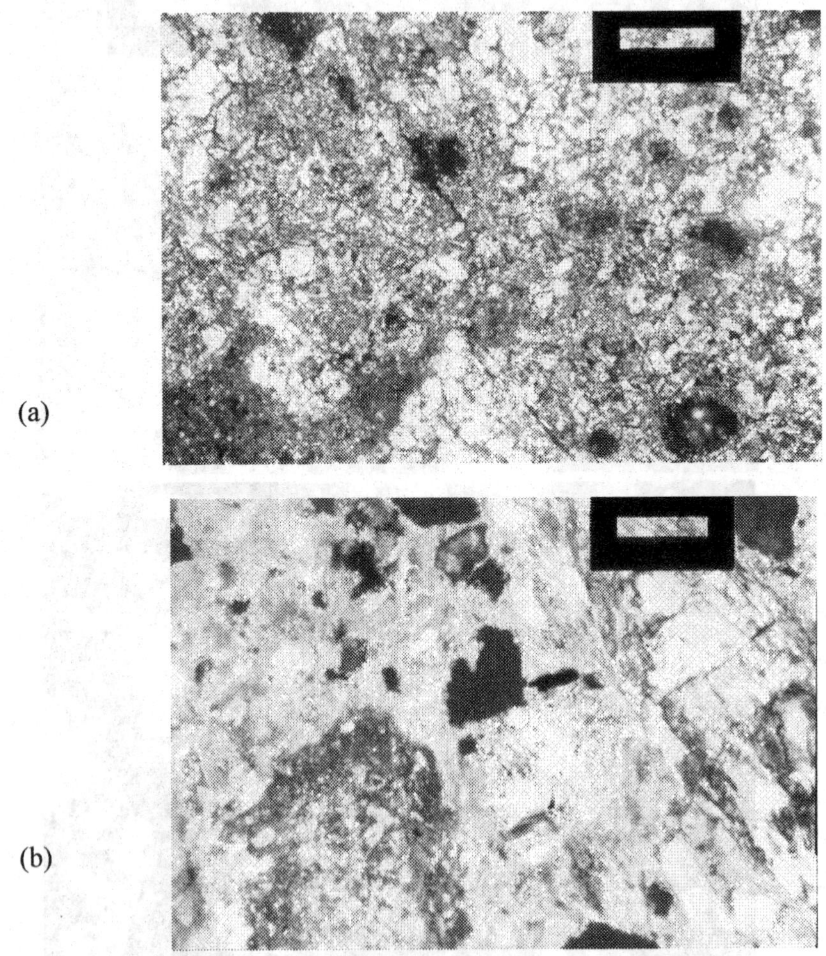

Figure 4. Microscopic image in polarized light showing the degradation stage of the wall plaster: (a) unused mortar material and (b) sample from the wall plaster.

The plaster detachment from the bricks surfaces is caused by the low temperatures during the cold season, when around doors and window frames were measured negative temparatures. The frost–defrost phenomena associated with the aggresive action of the environment leads to their degradation.

Inside the construction, the microbiocenosis, favored by the environment are extremely complex and heterogenous.

Using prepared cultures in the phytopathology laboratory from the material sampled (concrete and plaster) from walls and floor, eight species of saprophyte fungus and one species of nematode were identified.

Alternaria tenuis and *Cladosporium herbarum* – from Dematiaceae Family, are characterized by a white-gray mycelium that form brown conidiophores, simple, or branched. Macroscopically, the mycelium formations and the fruit forms appear as a velvet shape that proliferates and maintain a high rate of humidity into the affected zone.

Aspergillus niger, *Botrytis*, *Penicillium*, and *Stemphilium* – from Mucedinaceae Family, with a white-gray mycelium that develops characteristic spores, extremely numerous, that release a colorless powder mycelium, or colored, with effects in time on respiratory system of animals and personnel.

Giberella – from Tuberculariaceae Family has a simple or branched and heterogeneous unicellular spores, or multicellular white-pink colored and different sizes.

Rhizopus nigricans – from Mycelia Sterilia Order is characterized by abundant white-gray mycelium filaments that create a massive wool appearance on the infected surface.

Ditylenchus – from Tylenchidae Family that register tiny cylinder worms with oblong bodies of 1.0–2.5 mm, thinned at the ends, transparent white-yellow colored, covered with a ringed cuticle under which there is a stratum of longitudinal muscular cells and at the surface, projections, and thorns. Generally, these nematodes arrive inside the specific microbiocenosis of the animal houses within the forage and straws.

The humid environment existing in the tested areas is favorable to the development of microorganisms that cause the degradation of construction materials.

3. Conclusions

The design of animal shelters that can assure the optimum necessary conditions is essential for developing a normal, physiological life inside the buildings. The internal climate conditions as: temperature, relative moisture, and ventilation are determinant factors for obtaining high animal productions. The correct sizing of the closing peripheral elements, from the building physics point of view, associated with proper ventilation maintain the environment between normal limits at all parameters.

The humidity regime of the construction elements is closely tied with their thermal regime. Due to the increasing of the humidity rate, the thermal conductivity increases, the thermal resistance decreases, reducing the animal welfare, and raising the exploitation costs.

Although the animal breeding farms do not represent an immediate risk in the environment domain, their special conditions of exploitation, technologies, waste disposal, and microclimate conditions may contribute to accumulations of polluting factors that need to be long-term monitored.

A better integration of both agricultural buildings and quality of life in environmental policies is still an open issue. This requires consistent reporting and appropriate indicators to support evaluation and adjusting public environmental security and health action plans for reducing and preventing these risks.

References

Albright, L.D., 1990, *Environment Control for Animals and Plants*, American Society of Agricultural Engineers, 453 p., ISBN 0-929355-08-3.

Law Nr. 10, 1995, Quality in Constructions.

Official Journal of the EU/L001, 2003, Directive EC91/2002 of the European Parliament and of the Council, on Energy Performance in Buildings.

Official Journal of the EU/C085, 2003, European Economical and Social Committee.

PART IV

ISSUES RELATED TO METALS IN THE ENVIRONMENT

25. WHICH HUMAN AND TERRESTRIAL ECOLOGICAL RECEPTORS ARE MOST AT RISK FROM SMELTER EMISSIONS?

RUTH N. HULL*
Cantox Environmental Inc. 1900, Minnesota Court, Suite 130, Mississauga, Ontario L5N 3C9, Canada

STEVEN R. HILTS
Teck Cominco Metals Ltd., Trail, British Columbia, Canada

Abstract: Base-metal smelters, which operated more than 100 years ago and may continue to operate, are found in many countries. These smelters have released sulfr dioxide (SO_2) and metal particulates (e.g., arsenic, cadmium, copper, nickel, lead, and zinc) into the local and regional environment. Human health risk assessments (HHRAs) and ecological risk assessments (ERAs) are required by contaminated site regulatory programs in many countries. Regional- or landscape-scale risk assessments are nearing completion for areas surrounding several of these smelters in Canada, the USA, and Europe. Valued ecosystem components (VECs) that were assessed in the terrestrial components of the ERAs are similar to those assessed in any terrestrial ERA, including plant communities, soil invertebrate communities, birds, and mammals. Results from completed ERAs clearly illustrate which VECs are most at risk. Although the emissions histories, ecosystems, metals, and VECs may be different, the conclusions are similar; plant communities are most at risk around smelters, and impacts to plant communities can have an impact on wildlife due to changes in habitat. In transition countries, such as Romania, the focus remains on human health impacts. Results from an HHRA conducted for a lead smelter in Romania are consistent with those for a lead smelter in Canada and results from smelters from other countries. Children are most at risk from ingestion of deposited dust from airborne emissions from lead smelters, but also from soil ingestion and dust inhalation. This can be monitored via a program to measure blood lead levels in young children. Various programs can be implemented to decrease exposure of children to metals around smelters. In addition, improvements in human health, plant communities (and wildlife habitats) can be made through the decrease in SO_2 and metal emissions.

*To whom correspondence should be addressed. Ruth N. Hull, Cantox Environmental Inc., 1900 Minnesota Court, Suite 130, Mississauga, Ontario L5N 3C9 Canada; e-mail: rhull@cantoxenvironmental.com

R. N. Hull et al. (eds.), Strategies to Enhance Environmental Security in Transition Countries, 341–348.

Keywords: risk assessment; smelter; ecological risk; human health risk

1. Introduction

Many of the base-metal smelters (e.g., copper, nickel, lead, and zinc) around the world, which are currently operating or recently closed, operated for over 100 years. Environmental practices and regulations have changed dramatically over this time. However, only recently have these sites been subject to formal environmental assessment. The purpose of these assessments is to identify the human health and ecological risks and impacts, and focus remediation activities to address these impacts.

A large number of different groups of people and terrestrial ecological organisms may be exposed to contaminants released from smelters, including adults and children living nearby, or in specific communities; trees and other plants, including fruits, vegetables, and lichens; soil organisms and insects; and, birds and mammals.

Assessing human health and ecological risks around a smelter is complicated by the large geographic area which often has been affected; the different exposures for people and wildlife living in different areas (relative to the prevailing winds, habitats, types of food grown or eaten, etc.); and, the fact that exposure can be to contaminants in air, water, soil, dust, and food.

Environmental assessments for smelters in countries such as the USA and Canada are multiyear, multimillion dollar studies. A review of these studies has allowed the identification of priority health and ecological issues, the data that contributed most to remediation decision making, and potential mitigation strategies. This information can contribute to focused evaluations around smelters in transition countries, where resources are limited, to ensure the most important issues are addressed in the most cost-effective manner.

2. Priority Human Health Issues around Smelters

2.1. PEOPLE AT GREATEST RISK AND DOMINANT EXPOSURE PATHWAYS

Human health risk assessments (HHRAs) estimate risks to adults and children, males and females, and sensitive subgroups of the population. However, children often are the most at-risk group, due to their higher exposure and greater sensitivity to contaminants. Children generally eat and breathe more than adults, on a body weight normalized basis; children may incidentally or intentionally consume more soil than adults; and, children are still growing and developing, making them more sensitive to contaminants.

Risk assessments assess exposure from multiple media and via several exposure pathways (Figure 1). Typically, exposures from ingestion of drinking water, soil, dust, and food are assessed. Dermal and inhalation exposures also are included in the assessment.

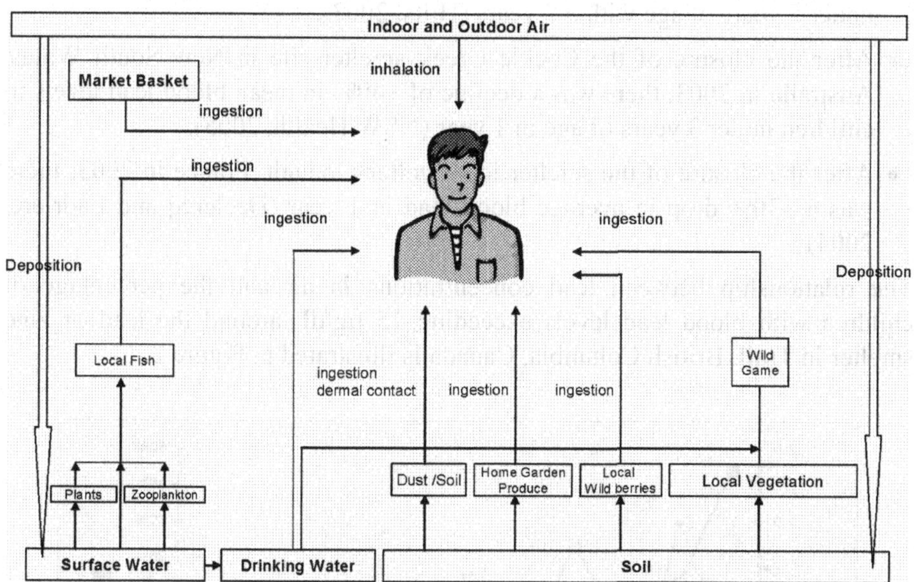

Figure 1. Contaminated media and exposure pathways by which humans may be exposed to contaminants released from a smelter.

While a smelter is operating, significant exposure may occur via ongoing emissions (ingestion of recently deposited dust, inhalation of lead in air, etc.). After a smelter is closed, ingestion of soil and remnant dust is the main contributor to risk at most metal-contaminated sites (US EPA, 2004). The estimated relative contribution of various exposure pathways and media to overall exposure may differ depending on the metal, whether local drinking water is contaminated, and whether concentrations in the food supply (termed "market basket" exposures) are taken into account in the exposure assessment.

It may be important to distinguish between the contributions of soil and dust to overall human exposure. However, this is difficult to do in human monitoring programs. The obvious tendency is to focus most attention on soil; remediation programs in the past generally have involved the large-scale removal of soil from a community. However, soil is coarser and more weathered than dust. Dust is finer and generally the metals in dust are more bioavailable than those in soil. Therefore, people may receive greater exposure via ingestion of metals in dust than soil. Around several smelters, blood lead levels decreased when emissions decreased (i.e., lower lead concentrations in air and lower dust loadings) even though no soil removal or remediation was implemented:

- In Trail, British Columbia, an 80% reduction in airborne emissions in 1997 resulted in a reduction of ~50% in average blood lead levels in children under 5 years of age within 2 years (Hilts, 2003).
- After the closure of the Cockle Creek smelter site in New South Wales, Australia in 2003, there was a decline of ~40% in mean blood lead levels in children under 3 years of age in 1 year (NSW Health, 2005).
- After the closure of the smelter in Noyelles-Godault, France in 2003, there was a ~30% drop in average blood lead in 1 year (Declercq and Ladriere, 2004).

The relationship between lead concentrations in air and the percentage of children with blood lead levels exceeding 15 μg/dL around the lead or zinc smelter in Trail, British Columbia, Canada is illustrated in Figure 2.

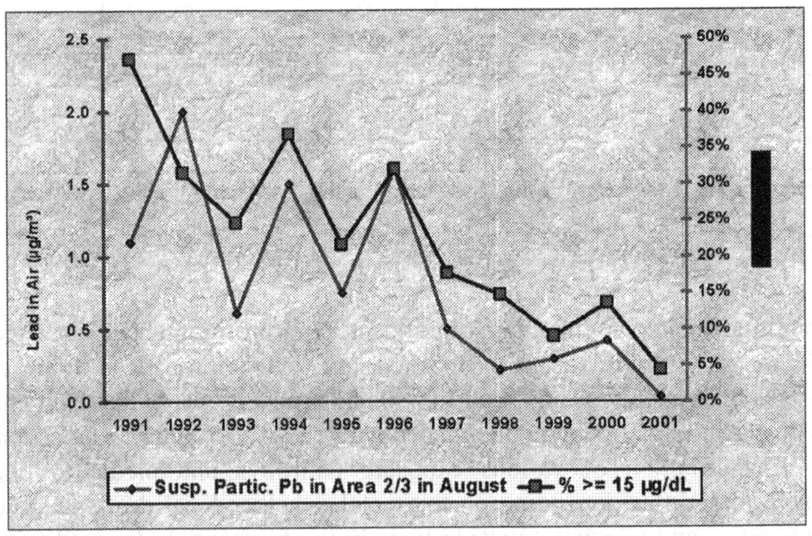

Figure 2. Percentage of children with elevated blood lead levels relative to lead concentrations in air.

In summary, experience around several smelter sites shows that children are most at risk, and that exposure to dust (where smelters are still operating) and soil are the main contributors to risk. However, it also may be important to consider other exposure pathways and media (e.g., ingestion of drinking water and food) when assessing risks to people.

2.2. KEY DATA TO ASSESS HEALTH RISKS

The critical aspect to assessing health risks is to characterize exposure as accurately as possible. Biological monitoring tools are available for some metals. Urinary arsenic levels in adults and children may be monitored where elevated arsenic concentrations are a concern. Blood lead monitoring of children may be conducted around lead smelters. These monitoring techniques are standard, relatively simple, and provide better indications of exposure than other modeling methods.

An understanding of the bioaccessible fraction of metals in soil or dust also contributes to a more accurate exposure assessment. The bioaccessible fraction is the mass fraction of a substance that is converted to a soluble form, and is therefore potentially available for uptake, under conditions of the external part of the membrane of interest. If one is evaluating bioaccessibility via the oral route, it is the fraction of a substance that becomes solubilized within the gastrointestinal tract (i.e., stomach and small intestine). The analytical methods to determine the bioaccessible fraction *in vitro* are still being developed and validated. Efforts to validate *in vitro* methods against animal models have to date focused on arsenic and lead (Ruby et al., 1996, 1999; Rodriguez and Basta, 1999; Kelley et al., 2002).

It also is important to understand other sources of exposure to metals to which people may be exposed, in addition to smelter-related emissions. For example, metal concentrations in a drinking water source (whether natural or anthropogenic) can significantly influence health risk predictions.

In summary, a meaningful assessment of health risks should consider methods to more accurately assess exposures, from both smelter- and nonsmelter-related sources.

3. Priority Terrestrial Ecological Issues around Smelters

3.1. TERRESTRIAL ECOLOGICAL RECEPTORS AT GREATEST RISK

Terrestrial ecological risk assessments (ERAs) assess risks to plants, invertebrates, birds, and mammals. Most metals do not biomagnify; therefore, birds and mammals at lower trophic levels are more at risk than top carnivores. Lower trophic level organisms that often are assessed include small mammals (e.g., mice, shrews, voles) and songbirds. These organisms frequently also have small home ranges, and thus may be exposed to elevated concentrations of metals in soil near the smelter.

Birds and mammals are assessed via modeling of exposure to metals in soil, water, and food. Exposure from water rarely contributes significantly to total risk. Rather, exposures from incidental ingestion of soil and ingestion of food are the main contributors to risk. Incidental ingestion of soil is a particular concern for wildlife that forage on the ground and that consume significant quantities of earthworms. Wildlife that consumes earthworms and other soil invertebrates are at greater risk than herbivores (Beyer et al., 1985). Many metals, such as lead, accumulate in roots of plants, but the lead is not translocated to aboveground plant material (Pahlsson, 1989). Therefore, metal concentrations in plants generally are lower than in soil invertebrates, particularly earthworms.

Plants are the most obviously adversely affected group, due to their exposure and sensitivity to sulfur dioxide (SO_2). Smelter technologies from 20–100+ years ago released high levels of SO_2, which killed plants (trees, shrubs, herbs, etc.) (Figure 3). SO_2 also resulted in acid rain, which may decrease soil pH, resulting in increased bioavailability of metals. Declines in soil microbial community activity (e.g., litter decomposition) and subsequent impairment of soil development also contribute to impacts to plants.

Figure 3. Adversely affected plant community around a smelter.

Loss of plants can contribute to other ecological impacts. On hillsides, loss of plants can lead to soil erosion. This loss of soil then inhibits revegetation after emissions reduction. Changes in plant communities also result in changes to wildlife habitat suitability. Loss of certain plant species, or changes in plant community structure can cause a decrease in abundance of some species, and increases in others (e.g., birds that prefer open habitats). Therefore, direct impacts to plants can have indirect impacts on wildlife.

3.2. KEY DATA TO ASSESS ECOLOGICAL RISKS

In contrast to HHRA, the critical aspect to assessing ecological risks is not related to exposures, but rather to characterize existing impacts as accurately as possible. The primary reason for this is that toxicological data in the literature (particularly for plants and invertebrates) do not allow an adequate evaluation of potential risks based on metal concentrations in soil, water, or tissues. Rather, standard methods exist to evaluate impacts on plant communities directly, rather than via modeling approaches.

Field surveys may be conducted of the plant communities, to assess parameters such as plant community structure, diversity, and presence of sensitive species. In addition, parameters which influence plant communities may be monitored, such as soil development, litter decomposition, condition of downed woody debris, and soil parameters (e.g., nutrient and organic matter levels). All of these parameters together give an indication of the naturalness of the plant community, and the potential of the plant community to be self-sustaining.

4. Mitigation Strategies

When unacceptable human health or ecological risks are predicted or observed, then various mitigation strategies must be considered and implemented to decrease these risks or ameliorate these impacts.

The best mitigation strategy is to decrease the source of contaminants (i.e., decrease SO_2 and metal emissions from the smelter). This is becoming less of an issue, as regulations require smelters to meet stricter emissions targets.

There are other approaches which can be used to reduce exposures of people in the community to metals. These include: (1) decreasing the transport of metals from off the smelter site and into the community (e.g., via vehicle washing), (2) street washing, and (3) various dust control measures. In addition, education plays a key role in reducing exposure, particularly for children (e.g., stressing the importance of hand washing).

Monitoring for results of these exposure–reduction strategies is critical, so that success of the program can be assessed. This allows particular focus on cases where targets (e.g., blood lead levels) are not being met.

Regreening (restoration of plant communities) often requires soil amendments (e.g., to raise pH, add nutrients, and organic matter), and may require erosion control measures. Restoration of plant communities should be done with human use and wildlife management (i.e., habitat suitability) in mind.

5. Conclusions

Impacts from smelters have been dramatic in the past, especially related to children's health, and plant communities. However, improvements in health and the environment are beginning to be seen around these sites. Source reduction is the primary strategy. Then, many approaches can be used to decrease exposure of people, particularly children, to metals. Restoration of plant communities can assist with human exposure reduction as well as improving aesthetics and wildlife habitat.

References

Beyer, W.N., Pattee, O.H., Sileo, L., Hoffman, D.J., Mulhern, B.M., 1985, Metal contamination in wildlife living near two zinc smelters, *Environ. Pollut.* (Series A) 38: 63–86.

Declercq, C., Ladriere, L., 2004, Programme de Depistage du Saturnisme Infantile Dans 9 Communes du Nord et du Pas-de-Calais, Public Health agencies presentation of blood lead survey results.

Hilts, S.R., 2003, Effect of smelter emission reductions on children's blood lead levels, *Sci. Tot. Environ.* 303: 51–58.

Kelley, M.E., Brauning, S.E., Schoof, R.A., Ruby, M.V., 2002, Assessing oral bioavailability of metals in soil, Battelle Press, Columbus Ohio, 124 p.

New South Wales Health, 2005, North Lake MacQuarrie Blood Lead Monitoring Services, Summary of Results July 2004 to July 2005.

Pahlsson, A.B., 1989, Toxicity of heavy metals (Zn, Cu, Cd, Pb) to vascular plants, *Water Air Soil Pollut.* 47: 287–319.

Rodriguez, R.R., Basta, N.T., 1999, An in vitro gastrointestinal method to estimate bioavailable arsenic in contaminated soils and solid media, *Environ. Sci. Technol.* 33(4): 642–649.

Ruby, M.V., Davis, A., Schoof, R., Eberle, S., Sellstone, C.M., 1996, Estimation of lead and arsenic bioavailability using a physiologically based extraction test, *Environ. Sci. Technol.* 30(2): 422–430.

Ruby, M.V., Schoof, R., Brattin, W., Goldade, M., Post, G., Harnois, M., Mosby, D.E., Casteel, W., Berti, W., Carpenter, M., Edwards, D., Cragin, D., Chappell, W., 1999, Advances in evaluating the oral bioavailability of inorganics in soil for use in human health risk assessment, *Environ. Sci. Technol.* 33(21): 3697–3705.

US EPA, 2004, Issue Paper on Metal Exposure Assessment, US Environmental Protection Agency Risk Assessment Forum, August 2004. http://cfpub.epa.gov/ncea/raf/recordisplay.cfm?deid=59052

26. INTEGRATION OF METAL BIOAVAILABILITY IN RISK ASSESSMENT POLICY DECISION MAKING

EUGEN S. GURZAU*, ANCA ELENA GURZAU, IULIA NEAMTIU, ALEXANDRU COMAN
Environmental Health Center, Cetatii 23 A, 400166 Cluj Napoca, Romania

Abstract: Metals are used in construction, automobiles, aerospace, electronics, and other manufacturing industries as well as in paints, pigments, and catalysts. Many metals accumulate in selective organs and tissues and therefore have potential for chronic toxicity. Models may be used to predict environmental distribution of metals for which available measurements must be considered. Prevention of chronic toxicity remains a challenge for children exposed to metals, particularly lead. New concepts like social marketing should be used to decrease exposure to metals. Risk assessment is being employed in legislation as a mechanism to establish responsibility to impose liability on those creating risks to humans and the environment, to formulate and implement appropriate measures to eliminate or control such risks, to decide what products should be marketed, and to assess compliance with regulations. It is difficult to say where risk assessment ends and risk control begins. Decision makers using risk assessment must adopt criteria for establishing how much risk is acceptable, taking into account the willingness of the public to tolerate the risks in return for benefits.

Keywords: heavy metals; risk assessment/communication/management; policy decision making; environmental rehabilitation

1. Introduction

We all know that the human environment consists of very basic elements: (1) the air we breathe, (2) the water we drink, (3) the food we eat, (4) the climate

*To whom correspondence should be addressed. Eugen S. Gurzau, Environmental Health Center, Cetatii 23 A, 400166 Cluj Napoca, Romania; e-mail: egurzau@ehc.ro

R. N. Hull et al. (eds.), Strategies to Enhance Environmental Security in Transition Countries, 349–368.

surrounding our bodies, and (5) the space available for movement. In addition, we exist in a social and spiritual environment, which is of great importance for our mental and physical health.

The diverse and valuable physical characteristics of metals have resulted in their extensive use in industry. Metals are used in construction, automobiles, aerospace, electronics, and other manufacturing industries, as well as in paints, pigments, and catalysts (Department of environmental health, 1991a, b; Tenorio and Espinosa, 2001). Because of such widespread potential exposure, familiarity with the toxicity of metals is critical for the health and safety of the workers and of surrounding communities (Klein et al., 1994). Metals exert biologic effects often through the formation of stable complexes with sulfhydryl groups and other ligands (LaDou, 1990). These binding properties also form the basis for chelation therapy in the treatment of metal toxicity (LaDou, 1990). Toxicity varies with the physical form of the metal, which is the major consideration of this paper (LaDou, 1990; Stroebel, 1992).

Metals are stable, which explains their wide use in structural materials and their pervasiveness in the environment (Morlot, 1994; Pyatt, 2001). Many metals accumulate in selective organs (Gonzalez-Reimers et al., 1999) and tissues and therefore have the potential for chronic toxicity (Bae et al., 2001). Metals are rarely used in their pure form, usually being present in alloys. They may be bound to organic materials which alters their physical characteristics and toxicity (Hamel et al., 1999).

There are different sources of exposure for different types of metals, such as: in Mexico, traffic for lead (Romieu et al., 1992) and industry for heavy metals (hot spots) (Gurzau et al., 1995); in the USA, lead in paint (Squinazi, 1994), lead in soil in Colorado and Butte, Montana (Department of environmental health, 1991b); in Chile, China, Inner Mongolia (Aposhian et al., 2000), and Romania, natural arsenic in underground water and drinking water (Gurzau et al., 2001); in Egypt, lead in freshwater macroinvertebrates (Canivet et al., 2001); in Hungary, lead in paprika; and in developing countries, occupational exposure and "hazardous pay" (Gurzau et al., 1995) (Table 1).

TABLE 1. Lead emissions of two smelters from Baia Mare, Romania

Year	1992	1993	1994	1995
Tons/year Phoenix	80	60	15	7
Tons/year Romplumb	500	300	160	40

Metals play a very important role in environmental health in Eastern and Central Europe (Fabianova et al., 2000) as a result of both their widespread distribution at high levels in the environment and of the associated health outcomes.

2. Risk Assessment of Heavy Metals

The process of risk assessment has wide application. It is being increasingly employed in legislation as a mechanism to establish responsibility to impose liability on those creating risks to humans and the environment, to formulate and implement appropriate measures to eliminate or control such risks, to decide what products should be marketed, and to assess compliance with regulations (Health and Safety Executive, 1996). Generally speaking, assessing risk seems to be simple, in principle. It involves identifying hazards or examining a particular situation for adverse health effects, and, of course, assessing the likelihood that the effect will actually be experienced by a specified population including the kind of consequences to be observed (Staessen et al., 1992). The overall objective is to determine how to manage the risk or to compare the risk with other risks.

This paper will give examples of risk assessment in some countries in Central and Eastern Europe, but mostly examples on risk control (the prioritization of risks and the introduction of measures that might be put in place to reduce, or prevent, the harm from occurring).

Risk assessment as part of the sequence of a process for policy decision making, consists of (Health and Safety Executive, 1996):

- Hazard identification which can be performed through epidemiological studies, animal studies, *in vitro* tests, and biochemical activities and structure activity relationships.
- Exposure assessment which is the determination or estimation (qualitative or quantitative) of the magnitude, frequency, duration and route of exposure and nature and size of the exposed population.
- Dose–response relationship.
- Risk characterization.

For metals which accumulate in the body or the environment past exposure is very important. Exposure assessment has been defined by Committee E-47, Biological effects and environmental fate of the American Society for Testing Materials, as the contact with a chemical or physical agent. In the case of metals, the magnitude of the exposure is determined by measuring or estimating the amount of a metal available at the exchange boundaries, e.g., lungs, gut, and skin, during some specified time (Bae et al., 2001; Lyngbye et al., 1990).

Exposure pathways address how a metal moves from the source to the exposed population or subject (Lundberg et al., 1999). It may result in an estimation of the geographic and temporal distribution of concentration of the metal in the various contaminated environmental media (Feng and Barratt, 1999). Models may be used to predict environmental distribution of metals based upon available measurements (Table 2). In Baia Mare, Romania, the distribution of lead in soil was estimated based on some measurements of lead level in soil (Gurzau et al., 1995) (Figure 1).

TABLE 2. Lead concentration in air, Baia Mare, Romania

Year	1992	1993	1994	1995	1996
ug/m^3	7	7.8	4.1	1.3	2

m^3 = cubic meter

A metal once released in the environment may be transported (e.g., through the water or atmosphere), may undergo chemical transformation (Ground water issue, 1995) (e.g., Cr VI–Cr III in soil (Sheehan et al., 1991)) and generally may accumulate in one or more media (soil, dust) (Gurzau et al., 1995). Metals also can accumulate by being taken up by plants. Persistence of metals varies in different media being very low in air, and high in soil and dust.

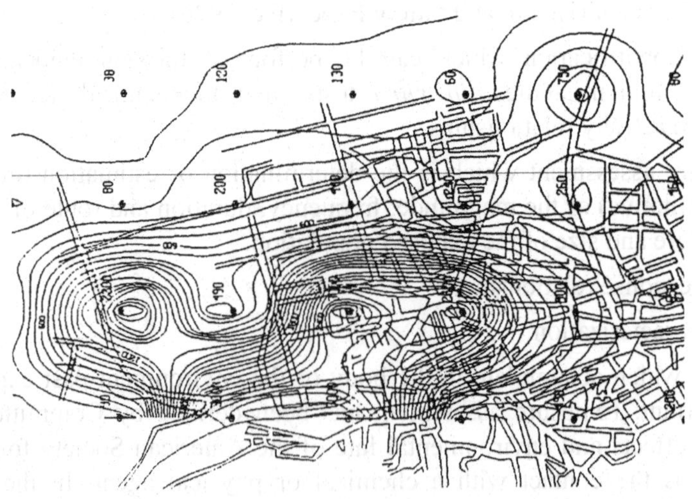

Figure 1. A plot of isopleths of estimated lead concentration in soil based on measurements, Baia Mare area, Romania.

Measurements are used to identify releases and, in the pathways and fate assessment, to quantitatively estimate both releases and environmental concentrations (Butler and Howe, 1999; Cabridenc et al., 1994; Morlot, 1994). Measurements are direct sources of information for exposure analysis (Plomb et environnement, 1994) (Tables 2 and 3). There are many methods to measure metals such as atomic absorption spectroscopy (AAS) with graphite furnace (GF), inductively coupled plasma (ICP) with mass spectrometry, anodic stripping voltametry (ASV), KX ray fluorescence (perfect for soil – very low cost per analysis, fast results) (Butler and Howe, 1999; Cabridenc et al., 1994). For some media, measurement methods are time consuming, e.g., for measuring metals in different particulate sizes in air (Gurzau et al., 1995). In general, the concentration of some metals in natural drinking water, soil, and dust is little changed over time (Gurzau et al., 1995). Arsenic concentrations were measured in drinking water samples from deep underground wells in some villages in Arad county, Romania (Lyngbye et al., 1990) (Table 3).

TABLE 3. The importance of quality control/assessment for measuring metal concentration in different media

Locality	Chisineu Cris	Zerind	Sepreus	Ciumeghiu
Our analysis				
As^{V}(μg/L ± SD)	2.8[a]	23.5 ± 0.1[b]	29.1 ± 0.7	87 ± 2[c]
As^{III} (μg/L±SD)	nd	5.0 ± 0.3[b]	55 ± 2	74 ± 1[c]
Sum	2.8	29	84	161
Government data As (μg/L)	6	42	99	171

[a]Drinking water in this area is obtained from a central water facility dug in 1976 that is 100 m deep
[b]This value is for water from the old well of this village
[c]By our analysis, the new well contained As^{V} (46.8 + –0.7 ug/L) and As^{III} could not be detected

Concentrations of metals should be estimated for all environmental media that might contribute to significant exposure. Sometimes an estimate of annual average concentrations is sufficient (Aposhian et al., 2000) (e.g., natural As in drinking water). It is important to quantitatively estimate the amount of the metal from each exposure pathway (e.g., natural As from drinking water to assess for other pathways to adjust for confounders such as food, occupational exposure) (Aposhian et al., 2000, 2001) (Table 3).

When exposures involve more than one exposure route, the relative amounts of a metal absorbed is usually route dependent (e.g. inorganic lead, respiratory, and GI tract; organic lead, skin) (Loghman-Adham, 1997). Measurement of lead concentration in air in Baia Mare region, Romania, pointed out high values (Gurzau et al., 1995) (Table 2).

Quality assurance or control for metals measurement in different media should be a very important component in risk assessment of metals. A study conducted by Aposhian showed important differences between two different techniques for measuring arsenic concentration in drinking water from artesian wells in Arad county, Romania (Aposhian et al., 2000, 2001) (Table 4).

TABLE 4. Urinary arsenic species (%)

Water (μg As/L)	Inorg As (%)	MMAV (%)	MMA^{III} (%)	DMA (%)	Summation (%)
85.5	7	14	11	76	108
161.2	7	13	7	76	103
Literature	12–20	14	–18	63–70	89–108

MMAV: monomethylarsonic acid; MMA^{III}: monomethylarsnous acid; DMA: dimethyl arsenic acid

Epidemiological studies on heavy metals (Sargent et al., 1997) worldwide are used for risk assessment and also as important information databases for risk perception or communication as part of policy decision making. In spite of the amount of data, there is a lack of unitary methodologies conducted on comparable databases. There are many debates on whether epidemiological data should or should not be used in policy decision-making processes (Corvalan et al., 1999; Hertz-Picciotto and Brunekreef, 2001). In general, the traditional concerns of environmental epidemiology have been the contamination of air, water, and food. In some Eastern and Central Europe countries epidemiological studies as part of hazard identification within the risk assessment process have focused on heavy metals. One of the priorities in these countries is lead which is a result of industry and an additive in gasoline.

In Romania many large populations are still being exposed at very high levels of environmental lead because of the continued use of leaded gasoline, the presence of old smelters still running, and mostly the lack of protection from industrial processes, wastes and even consumer products (Gurzau et al., 1995). Sometimes the policy relevance of epidemiologic data can be improved through changes in the way results are presented (Hertz-Picciotto and Brunekreef, 2001; Hertz-Picciotto, 1999). As an example, the data from Zlatna case study, mentioned above, may be considered too sparse to accurately determine dose–response relationships or even to identify susceptible population subgroups, therefore in a published paper the results might be criticized. However, that type of information is highly relevant to policy decision-making processes.

A major priority in an epidemiological study is the choice of the population selected for the study which will be identified as a result of the source and fate studies. Populations, group of individuals with potentially high exposure should be identified. Also populations of higher susceptibility should be identified. Children age 0–6 (Andrews et al., 1994; Lasky et al., 2001; NRC, 1993), pregnant women (Gurzau et al., 1995; Klein et al., 1994) and women in lactation (Gurzau et al., 1995) are the most susceptible groups to lead poisoning. In Copsa Mica, Romania higher blood lead levels were found in young children and pregnant women (CDC, 1991) (Table 5).

TABLE 5. Blood lead level in children aged 7–11 Copsa Mica, Romania

PbL	<10	10–15	15–30	30–50	50–70	70–100	>100
%	0	0	14.2	57.2	21.4	7.2	0

PbL: blood lead level; %: % of the investigated children

Biological measurements of human body fluids and tissues for metals can be used to estimate current or past exposure to metals (Table 5). We can use biomarkers of exposure which represent a xenobiotic or its metabolite, or the product of an interaction between a xenobiotic agent and some target molecule or cell that is measured within a comportment of an organism. The preferred biomarkers of exposure for metals are generally the metals or substance-specific metabolites in readily obtained body fluid or excreta. Biomarkers of exposure are very specific for metals, as these are indestructible as compared to other substances (e.g., aromatic hydrocarbon compounds) for which measurement of metabolites is often needed (Pott et al., 2001). Metals are specific biomarkers of exposure, but sometimes are very limited for assessing acute versus chronic exposure (Isnard, 1994). Other times metabolites (like MMA^{III} – methylated inorganic arsenic) are more important to assess exposure and toxicity (Aposhian et al., 2000). Using research to identify more sensitive and specific biomarkers for metal exposure should be a very high priority (Needleman et al., 1990). MMA^{III} in human urine of subjects exposed to inorganic arsenic was measured for the first time in a study conducted by Aposhian (Aposhian et al., 2001) (Table 7). As it was previously mentioned in Hungary, Slovakia, and Romania (Gurzau and Gurzau, 2001) arsenic levels in groundwater sources tend to be higher than in surface water sources (i.e., lakes and rivers) of drinking water (Wang et al., 2001). In most drinking water sources, the inorganic form of arsenic tends to be more predominant than organic forms. Inorganic arsenic in drinking water can exert toxic effects after acute (short-term) or chronic (long-term) exposure. Studies link inorganic arsenic ingestion to a number of health effects, including cancer (Abernathy et al., 1999): skin, bladder (Cuzick et al., 1992), lung, kidney (Surdu et al., 1997), nasal passages, liver and prostate cancer (Pott et al., 2001); and noncancerous effects (Fabianova et al., 2000; Hopenhayn-Rich et al., 2000; Lewis et al., 1999; Vahter and Concha, 2001): cardiovascular (Abernathy et al.,

1999; Surdu et al., 1997; Wu et al., 1989), pulmonary, immunological, neurological (Gerr et al., 2000), and endocrine (e.g., diabetes) effects. Arsenic species in human urine in Arad County, Romania, were measured in order to assess arsenic toxicity (Aposhian et al., 2000; Gurzau et al., 2001) (Table 6).

It is possible to use models to predict a biomarker of exposure based on environmental distribution (media concentration). Such a model is the United States Environmental Protection Agency (US EPA) Uptake/Biokinetic Model for Lead elaborated by US EPA. Research into the health effects of low-level exposure to metals (Alexander et al., 1993; Brenner, 1996; Needleman et al., 1990) continues to suggest that physiological alternatives may occur at levels that previously had been considered safe (Andrews et al., 1994; Schwartz, 1991).

TABLE 6. MMA^{III} in human urine of subjects from Arad, Romania, exposed to inorganic arsenic from drinking water

Group	Water(μg As/L)	Number in group	Subjects having MMA^{III} in urine	MMA^{III} in urine(μg/11 h ± SE)	MMA^{III}(μg/L ± SE)
A	2.8	14	0	nd	nd
B	28.5	14	1[a]	12.0	4.8
C	85.5	14	4[b]	4.51 ± (7–14)	5.7 ± (nd–12)
D	161.2	16	5[c]	5.12 ± (3–10)	6.9 ± (nd–16)

[a]1 male
[b]1 male and 3 females
[c]4 males and 1 female

Neurological and neuropsychological effects (Gerr et al., 2000; Jacqmin-Gadda et al., 1996; Pococks et al., 1994), reproductive toxicity, teratogenicity (Trombini et al. 2001) (including behavioral alterations (Balbus-Kornfeld et al., 1995; McMichael et al., 1988; Munoz et al., 1993; Schwartz et al., 1994)), and carcinogenicity (Hayes, 1997; Infante-Rivard et al., 2001) remain at the forefront of current research. Medical surveillance and biological monitoring programs are important for identifying the safe level of exposure (Table 7).

3. Integration of Risk Assessment in Policy Decision Making

In general, risk assessment can be scientific and objective which is not the case for risk control because risk control integrates the results of an assessment of risks with other critical inputs (US EPA, 1986). These include economic analysis, which is very important in countries with transitional economies such as the countries in Central and Eastern Europe; perception of risks, which is specific for each region; the need for and availability of alternative technologies; concerns about equity; consistency across a range of risks; and so on. Therefore, it could be difficult to say where an assessment of risk ends and risk control begins. In other words, a risk assessment is in practice a mixture of science and policy (Nurminen et al., 1999).

TABLE 7. Interpretation of blood lead test results and follow-up activities; risk – class of child based on blood lead concentration in Zlatna Case Study, Romania

Risk	Blood lead level (ug/dL)	Comment/messages/action
Low	<20	At blood lead levels (BPb) up to 15 adverse health effects are possible; parents education to avoid lead exposure of their children; parents education for children's diet, control, personal hygiene measures, etc.; if exposure based on questionnaire evaluation is low, BPb retest after 1 year
Medium low	20–34	Adverse health effects are possible; parents education to avoid lead exposure of their children; evaluation of lead concentration in child's environment and eliminate the lead sources if possible; parents education for children's diet, control, personal hygiene measures, etc.; BPb retest after 1 year
Medium high	35–39	Possibility of adverse health effects development is very high; parents education to avoid lead exposure of their children; evaluation of lead concentration in child's environment and eliminate the lead sources if possible; activities and behavior changes to decrease lead exposure; working with children's parents; a proper calcium diet (600–800 mg/day), vitamins and iron medical checkup, medical evaluation medical management
High	>60	Possibility of adverse health effects development is very high; medical emergency, medical management must begin immediately; urgent BPb retest; parents education to avoid lead exposure of their children, to cut the pathways of exposure; environmental management; eliminate the source activities and behavior changes to decrease lead exposure; working with children's parents; a proper calcium diet (600–800 mg/day), vitamins and iron

Risk perception or communication plays a very active role in metals risk assessment policy decision making (Teret, 2001). In general, risk communication should be a democratic process. It is based on the idea that risk is real, that it is easily measured, particularly when direct biomarkers of exposure are available such as with metals; that scientists can agree on risk; and that it is necessary to translate scientific knowledge into education for the public. In technical terms, this conventional approach to risk communication is based upon positivist conceptions of the nature of risk (Weiss, 2001). It assumes that risks are unambiguously real and measurable, and that a community of scientific risk assessors can agree on their seriousness (Pastides, 1999). Risk communication in some areas in the region is used to legitimate actions to mitigate hazards (mostly lead and arsenic related) (Tchounwou et al., 1999). It also involves free discussion and debate. Risk communication is part of a larger process of understanding and managing hazards. Risk assessment, risk communication, and risk management represent a part of a political activity during which risk is socially constructed (Payne-Sturges and Breugelmans, 2001).

Working together, scientists, workers, public health officials, political representatives, and ordinary citizens can define, assess, and manage risk to heavy metals in some "hot spot" areas in Central and Eastern Europe. It is not the role of scientists involved in the risk assessment of heavy metals to tell the people, especially those responsible for action, about what should be safe enough, or how they could make it safer. Risk assessors can describe the distribution of heavy metals within the environment, body burden and expected health-associated outcomes. The latter will be mostly based on epidemiological studies (Epidemiology and Society, 1999) and also some international experience. There is information available for some metals (e.g., lead, cadmium, chromium, and arsenic) (Aposhian et al., 2000, 2001; Gurzau et al., 1995), but less for other metals which also reach high concentrations in the environment. Therefore, those involved in risk management of metals in the region have to address the question of what is to be done and who is responsible for acting or preventing action. One of the cornerstones is to have institutions and procedures for agreeing on risk policies (Weiss, 2001) and also to select individuals and organizations with the responsibility for implementing these policies. In view of the heavy metals health-related problems in Central and Eastern Europe (e.g., Romania) it is also important to develop the conceptual and practical insights into heavy metal body burdens and effects, and also to share experience to define priorities, and to provide methodological guidance on the quantitative assessment of the heavy metals burden of disease at national and also regional levels (Pruss et al., 2001). Increasingly, countries wish to use information about disease burdens for priority setting and resource allocation in health, including environmental health (Pruss et al., 2001).

Many scientists assume that when you state risk probabilities the risk communication job is over. However, knowledge does not necessarily change behavior. In the same respect setting new standards, e.g., as a request to became part of the European Union (EU); does not necessarily change organization or industry behavior. We should take into account the existing well-documented gap between attitudes and behavior. Informing the public of facts about risky actions does not necessarily make them act more prudently. It is difficult to avoid drinking water with (Brown and Fan, 1994) elevated arsenic concentration when there are no other alternatives to provide the comunity with safe drinking water or facilities to provide safe bottled water. People know their behavior is harmful to their health, using that kind of water because it's "spring water", but they do not act on that knowledge for a variety of reasons. Strategies resulting in changing behavior, not just imparting information, could help the people in the region to avoid the exposure to natural arsenic from drinking water, to cut the pathway, and to decrease the associated risks. Risk communication in this case involves promoting behavior changes. We need to develop program strategies through creative interpretation of research findings, which are accessible after the risk assessment is carried out. Therefore, once again the importance of risk assessment in policy decision making is obvious.

Programs to change individual behavior, such as Zlatna case study presented in this paper, should reflect social and psychological research on how to change actual behavior.

On the other hand, because social marketing is an approach and not a solution, there is no program template for others to copy (which means we can transfer some of the experience from Zlatna case study to other hot spots in the region, but the equation being different the program could differ). We need social forces to influence people and enterprises in order to make them act on their knowledge of risk.

Economic and political interests of local authorities and leaders are not always similar to those of the public to whom risk communication is directed (Health and Safety Executive, 1996). For this reason, in our region, risk communication is sometimes, when it is not well conducted, interpreted as factory or government propaganda (Gurzau et al., 1995). Communication must be trusted by the public. Trust, especially in Central and Eastern Europe, is the most important criterion for democratic risk communication. Countries in this region, like Romania, are leaving systems where many old structures that communicated risk to the public were arms of propaganda. The region needs to have institutions that provide reliable data on risk assessment to heavy metals and to carefully translate it into ordinary or common language (EHP, 1995).

In some countries, like Romania prevention of chronic toxicity remains a challenge for children exposed to metals, particularly lead. New concepts like social marketing may help (Griffits, 1997). Social marketing is a systematic approach to solving problems, in this case public health problems, related to service utilization, product development, and acceptance and behavior adoption. It is the application of fundamental marketing principles to social program design and management. Social marketing has as its objective changes in behavior, not just imparting information. Social marketing concentrates on the half of the marketing equation that is often ignored – creation of demand. Social marketing requires creativity in message design. Social marketers conceptualize their task in four main phases:

- Strategy development
- Strategy formulation
- Strategy implementation
- Strategy assessment or evaluation

A good example of this new concept is the Zlatna case study, funded by US Agency for International Developments (USAID) in Romania. The Environmental Health Project (EHP) was asked in June 1994 to conduct a 1-year technical assistance project in Zlatna, Romania, as part of the USAID efforts to improve environmental conditions (EHP, 1995). The main source of atmospheric pollution and also the principal employer in this town of 8000 inhabitants is the 250-year-old Ampelum SA copper smelter and refinery. The town's mayor has made community health one of his main priorities and has requested designation as a calamity zone, with no response from the national government. At the request of the Romanian Government, USAID selected Zlatna and Ampelum SA copper smelter as a demonstration site with the goal of providing the smelter and community groups with both technical and equipment assistance to achieve short-term improvements to reduce the environmental and health problems caused by the smelter (Gurzau et al., 1995). The following concerns were identified in an initial report produced by the Work Environment Center:

- Reducing the exposure of young children to lead
- Air monitoring and control
- Occupational health and safety protection and training

The initial objectives of the project remained consistent throughout the 3-year project:

- Address immediate health issues in the short-term and initiate long-term efforts to reduce smelter emissions.
- Assist in developing a reliable ambient air quality database and monitoring program for use in public health education and evaluating plant improvements.

- Assist in reducing the conflict between maintaining employment and protecting human health and the environment during a period of social and economic transition.

The project provided assistance for equipment (half of project budget), for technical assistance, interinstitutional teacher development, and support for Romanian institutions and the Zlatna Community (Gurzau et al., 1995). Three working groups were developed to address environmental health issues and included individuals from the following stakeholder groups: the smelter, the Zlatna hospital, the Alba County Environmental Protection Agency, the Alba County Sanitary Police, the Center for Medical Research in Cluj, and the Zlatna kindergarten. A participatory process was implemented throughout the entire program, providing for an exchange of ideas, agreements regarding implementation of the action plans, and sharing of expectations. It was clear that the collaborative problem-solving process provided momentum and energy for action. The project consisted of the following steps, implemented between June 1994 and June 1995:

- Meetings with stakeholders to evaluate proposed activities and identify equipment and resource needs.
- Working groups to reevaluate health issues of concern, goals, activities, and equipment needs.
- Delivery of equipment and associated training, including blood collection and analysis and air quality and workplace-monitoring equipment and respirators.
- Support for a locally designed family and community-wide health education and counseling program with accompanying education materials to reduce childhood lead exposure.
- Identification and assessment of potential opportunities for emission reduction at the smelter.
- Training in air quality monitoring.
- Occupational health and safety training.
- Initiation of a community effort to create safe play areas.
- Testing of children for blood lead levels (revealing high levels of blood lead poisoning).
- Evaluation of the project.

Preliminary data on child lead levels (170 soil samples and 170 children) demonstrated blood lead levels of 35–60 ug/dl and soil levels of 50–4000 ppm. All children have stayed in the community (Gurzau et al., 1995). Depending on the child's blood lead level the following actions were prescribed: (1) education and retesting; (2) avoidance of exposure and alleviation of sources; (3) calcium supplements; and (4) medical intervention. Child exposure and medical management was a new program in the town and cooperative efforts needed to be established between health care workers and teachers to develop methods to avoid exposure, including interventions and education about risky behaviors (Gurzau et al., 1995).

Several factors have emerged that could contribute to program sustain ability including: (1) clear demonstration by working group participants of an increased knowledge and understanding of environmental health concerns; (2) establishment of a new institutional capacity to conduct social marketing activities; increased levels of participation and desire for technical assistance; (3) desire by participants for follow-up to consolidate gains and disseminate lessons learned to other Romanian communities; and (4) the introduction of an interagency or public approach to addressing environmental health issues.

The next step in the project consists of training sessions on environmental health data analysis and risk communication; support for the workgroups in implementing their action plans and assistance in disseminating lessons learned; and risk assessment, communication and social marketing.

The initial evaluation was conducted by a small technical team. At that time, agencies and communities at the local, county, and regional levels have had little or no experience working together to solve environmental health problems. Three lessons were learned regarding implementation: (1) there was a need for interinstitutional working groups to provide a model for collaborative problem solving; (2) there was a need to empower individuals in their abilities to change environmental conditions; and (3) there was a need for frequent written communication with local officials during the entire period of the project, as well as host–country communication at all levels.

A tremendous decrease of about 47% of blood lead levels in very young children accrued after 3 years working with local authorities, scientists, mass media, local community, changing attitudes or behavior and creating a coalition, several action plans were implemented.

Decreasing emissions occurred through improving technology (Lahmeyer International, 1996; Seux and Dab, 1994) with reliable filters (Figure 2). Other measures such as site remediation and rehabilitation are needed. Some of these measures require significant resources such as soil removal, physical separation or chemical extraction, SAREX chemical fixation process, xanthate treatment, cyclone furnace, solidification, and surfactant gmethods (Member agencies of the federal remediation technologies roundtable, 1993).

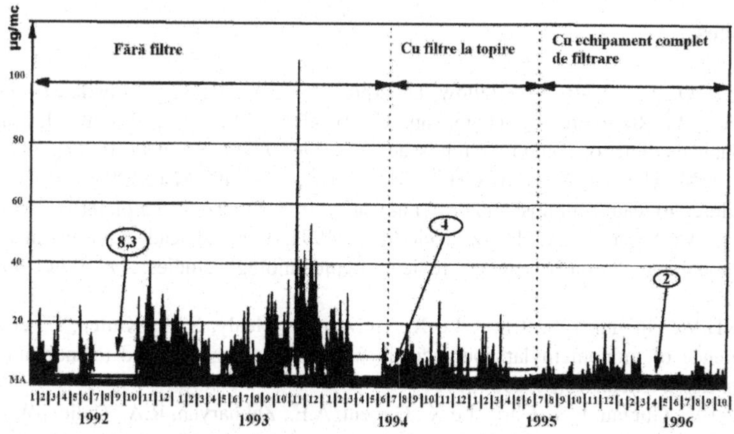

Figure 2. Filters installation and lead concentration in air of Baia Mare area, Romania.

Another example is the Air Management Information System (AMIS) (Schwela, 1999) which is an information-exchange system in the scheme of the Global Air Quality Partnership providing information on all issues of air quality management between its participants: municipalities, national environmental protection agencies, international organizations, World Bank and international development banks, and nongovernmental organizations. Public health studies of air pollution-induced health effects are an important ingredient for decisions with respect to the management of air quality. Air is playing an important role in the transfer of heavy metals to soil and vegetables particularly from the nonferrous metallurgical industry. Such information systems facilitate the use of high-quality data in the decision-making process.

Finally, decision makers using risk assessment must adopt criteria for establishing appropriate management responses in accordance with the risks the community is prepared to accept, taking into account the willingness of the public to tolerate the risks in return for benefits (Advances in lead research: implications for environmental health, 1990).

There is a need to use unitary methodologies in order to get comparable databases and formulate, develop, and implement the best intervention strategies in the region. "Think globally and act locally" should be an important priority in solving heavy metals environmental problems in Central and Eastern Europe. Therefore, the need for reliable information about funding opportunities could play an important role in these countries. A good example of how such information could serve as a source for funding public and private investments opportunities in environmental protection and remediation is the directory of financing sources for environmental investments in Romania, prepared and published under the auspices of the USAID (EAPS, 2000)

References

Abernathy, C.O., Liu, Y.-P., Longfellow, D., Aposhian, H.V., Beck, B., Fowler, B., Goyer, R., Menzer, R., Rossman, T., Thompson, C., Waalkes, M., 1999, Arsenic: health effects, mechanisms of actions, and research issues, *Environ. Health Perspect.* 107(7): 593–597.

Alexander, L.M., Heaven, A., Delves, H.T., Moreton, J., Trenouth, M.J., 1993, Relative exposure of children to lead from dust and drinking water, *Arch. Environ. Health* 48(6): 392–400.

Andrews, K.W., Savitz, D.A., Hertz-Picciotto, I., 1994, Prenatal lead exposure in relation to gestational age and birth weight: a review of epidemiologic studies, *Am. J. Ind. Med.* 26(1): 13–32.

Aposhian, H.V., Gurzau, E.S., Chris, Le X., Gurzau, A., Healy, S.M., Xiufen, L., et al., 2000, Occurence of monomethylarsnous acid in urine of humans exposed to inorganic arsenic, *Chem. Res. Toxicol.* 13(8): 693–697.

Aposhian, H.V., Gurzau, E.S., Chris, Le X., Gurzau, A.E., Zakharyan, R.A., Cullen, W.R., Healy, S.M., Gonzalez-Ramirez, D., Morgan, D.L., Sampayo-Reyes, A., Wildfang, E., Radabaugh, T.R., Petrick, J.S., Mash, E.A. Jr., Aavula, R.B., Aposhian, M.M., 2001, The discovery, importance and significance of monomethylarsonous acid (MMA^{III}) in Urine of humans exposed to inorganic arsenic, in: *Arsenic Exposure and Health Effects*, Chappell, W.R., Abernathy, C.O., Calderon, R.L. (eds.), Elsevier Science B.V., Amsterdam, pp. 305–314.

ATSDR, 1992, Chromium, US Department of Health and Human Services, Atlanta, GA.

Bae, D.S., Gennings, C., Carter, W.H. Jr., Yang, R.S., Campain, J.A., 2001, Toxicological interactions among arsenic, cadmium, chromium and lead in human keratinocytes, *Toxicol. Sci.* 63(1): 132–142.

Balbus-Kornfeld, J.M., Stewart, W., Bolla, K.I., Schwartz, B.S., 1995, Cumulative exposure to inorganic lead and neurobehavioural test performance in adults: an epidemiological review, *Occup. Environ. Med.* 52(1): 2–12.

Brenner, H., 1996, Correcting for exposure misclassification using an alloyed gold standard, *Epidemiology* 7(4): 406–410.

Brown, J.P., Fan, A.M., 1994, Arsenic: risk assessment for California drinking water standards, *J. Hazard Mater.* 39(2): 149–159.

Butler, O.T., Howe, A.M., 1999, Development of an international standard for the determination of metals and metalloids in workplace air using ICP-AES: evaluation of sample dissolution procedures through an interlaboratory trial, *J. Environ. Monit.* 1(1): 23–32.

Cabridenc, R., Carbonnier, F., Cordonnier, J., Legenti, L., Rizet, M., 1994, Le plomb dans l'environnement, Techniques sciences methodes, *genie urbain genie rural* 89(2): 64–69.

Canivet, V., Chambon, P., Gilbert, J., 2001, Toxicity and bioaccumulation of arsenic and chromium in epigean and hypogean freshwater macroinvertebrates, *Arch. Environ. Contam. Toxicol.* 40(3): 345–354.

CDC, 1991, Preventing lead poisoning in young children, US Department of Health and Human Services, Atlanta, GA.

Corvalan, C.F., Kjellstrom, T., Smith, K.R., 1999, Health, environment and sustainable development: identifying links and indicators to promote action, *Epidemiology* 10(5): 656–660.

Cuzick, J., Sasieni, P., Evans, S., 1992, Ingested arsenic, keratoses and bladder cancer, *Am. J. Epidemiol.* 136(4): 417–421.

Department of environmental health, 1991a, Final work plan for sampling of analysis of lead occurrence in and immediately adjacent to residential areas, University of Cincinnati, Cincinnati, OH.

Department of environmental health, 1991b, The Bute Silver Bow Environmental health lead study, Final report, University of Cincinnati, Cincinnati, OH.

EAPS, 2000, Directory of financing resources for environmental investments in Romania, USAID.

EHP, 1995, Activity report No. 13, Summary of activities in Zlatna, Romania, 1994–1995, USAID.

Fabianova, E., Hettychova, L., Koppova, K., Hruba, F., Marko, M., Maroni, M., Grech, G., Bencko, V., 2000, Health risk assessment for inhalation exposure to arsenic, *Cent. Eur. J. Public Health* 8(1): 28–32.

Feng, Y., Barratt, R., 1999, Distributions of lead and cadmium in dust in the vicinity of a sewage sludge incinerator, *J. Environ. Monit.* 1(2): 169–176.

Gerr, F., Letz, R., Ryan, P.B., Green, R.C., 2000, Neurological effects of environmental exposure to arsenic in dust and soil among humans, *Neurotoxicology* 21(4): 475–487.

Gonzalez-Reimers, E., Arnay-De-La-Rosa, M., Velasco-Vazquez, J., Galindo-Martin, L., Delagado-Ureta, E., Santolaria-Fernandez, F., 1999, Bone lead in the prehistoric population of Gran Canaria, *Am. J. Human Biol.* 11(3): 405–410.

Gurzau, E.S., Gurzau, A.E., 2001, Arsenic in drinking water from groundwater in Transylvania, Romania: an overview, in: *Arsenic Exposure and Health Effects*, Chappell, W.R., Abernathy, C.O., Calderon, R.L. (eds.), Elsevier Science B.V., Amsterdam, pp. 181–184.

Gurzau, E.S., Niciu, E., Surdu, S., Bodor, E., Costin, I., Maier, A., 1995, Environmental health assessment of irritants and heavy metals in Transylvania, Romania, *CE J. OEM* 1(1): 70–75.

Gurzau, E.S., Ponoran, C., Ponoran, S., Micka, M.A., Billig, P., Silberschmidt, M. Zlatna, 1995, Case study, *Proceedings of the 6th Annual Symposium on Environmental and Occupational Health during Social Transition in Central and Eastern Europe*, 5–10 June 1995, Eforie Nord, Romania, JSI 1996.

Hamel, S.C., Ellickson, K.M., Lioy, P.J., 1999, The estimation of the bioaccessibility of heavy metals in soils using artificial biofluids by two novel methods: mass-balance and soil recapture, *Sci. Total Environ.* 243–244: 273–283.

Hayes, R.B., 1997, The carcinogenicity of metals in humans, *Cancer Causes Control* 8(3): 371–385.

Hertz-Picciotto, I., Brunekreef, B., 2001, Environmental Epidemiology: where we've been and where we're going, *Epidemiology* 12: 479–481.

Hertz-Picciotto, I., 1999, What you should have learned about epidemiologic data analysis, *Epidemiology* 6: 778–783.

Hopenhayn-Rich, C., Browning, S.R., Hertz-Picciotto, I., Ferreccio, C., Peral, L., Gibb, H., 2000, Chronic arsenic exposure and risk of infant mortality in two areas of Chile, *Environ. Health Perspect.* 108(7): 667–673.

Infante-Rivard. C., Olson, E., Jacques, L., Ayotte, P., 2001, Drinking water contaminants and childhood leukemia, *Epidemiology* 12: 13–19.

Isnard, H., 1994, Toxicite du plomb et saturnisme infantile, Techniques sciences methodes, *genie urbain genie rural* 89(2): 70–72.

Jacqmin-Gadda, H., Commenges, D., Letenneur, L., Dartigues, J.F., 1996, Silica and aluminium in drinking water and cognitive impairment in the elderly, *Epidemiology* 7(3): 281–285.

Klein, M., Kaminsky, P., Barbe, F., Duc, M., 1994, Saturnisme au cours de la grossesse, *Presse Med.* 23(12): 576–580.

LaDou, J., 1990, *Occupational Medicine*, Appleton & Lange, Norwalk, CA.

Lahmeyer International, 1996, Demonstration pilot project on air self monitoring in Baia Mare, Romania, EC Phare.

Lasky, R.E., Laughlin, N.K., Luck, M.L., 2001, The effects of elevated blood lead levels and succimer chelation therapy on physical growth in developing rhesus monkeys, *Environ. Res.* 87(1): 21–30.

Lewis, D.R., Southwick, J.W., Ouellet-Hellstrom, R., Rench, J., Calderon, R.L., 1999, Drinking water arsenic in Utah: a cohort mortality study, *Environ. Health Perspect.* 107(5): 359–365.

Loghman-Adham, M., 1997, Renal effects of environmental and occupational lead exposure, *Environ. Health Perspect.* 105(9): 928–938.

Lundberg, M., Hallqvist, J., Diderichsen, F., 1999, Exposure-dependent misclassification of exposure in interaction analyses, *Epidemiology* 10(5): 545–549.

Lyngbye, T., Hansen, O.N., Grandjean, P., 1990, Predictors of tooth-lead level with special reference to traffic. A study of lead-exposure in children, *Int. Arch. Occup. Environ. Health* 62: 417–422.

McMichael, A.J., Baghurst, P.A., Wigg, N.R., Vimpani, G.V., Robertson, E.F., Roberts, R.J., 1988, Port Pirie cohort study: environmental exposure to lead and children's abilities at the age of four years, *N Engl. J. Med.* 319(8): 468–475.

Member agencies of the federal remediation technologies roundtable, 1993, Synopsis of federal demonstrations of innovative site remediation technologies, US EPA, Department of defense, Department of energy, Department of interior.

Morlot, M., 1994, Aspects analytiques du plomb dans l'environnement, Techniques sciences methodes, *genie urbain genie rural* 89(2): 79–87.

Munoz, H., Romieu, I., Palazuelos, E., Mancilla-Sanchez, T., Meneses-Gonzalez, F., Hernandez-Avila, M., 1993, Blood lead level and neurobehavioral development among children living in Mexico City, *Arch. Environ. Health* 48(3): 132–139.

Needleman, H.L., Schell, A., Bellinger, D., Leviton, A., Allred, E.N., 1990, The long-term effects of exposure to low doses of lead in childhood. An 11-year follow-up report, *N Engl. J. Med.* 322(2): 83–88.

NRC, 1993, Measuring lead exposure in infants, children and other sensitive populations, National academy press, Washington, DC.

Nurminen, M., Nurminen, T., Corvalan, C.F., 1999, Methodologic Issues in epidemilogic risk assessment, *Epidemiology* 10(5): 585–593.

Pastides, H. (ed.), 1999, Methods for health impact assessment in environmental and occupational health. Consultation – World Health Organization and International Labor Office, Geneva, Switzerland, 9–11 July 1997, *Epidemiology* 10: 569–660.

Payne-Sturges, D.C., Breugelmans, J.G., 2001, Local lead data are needed for local decision making, *Am. J. Public Health* 91(9): 1396–1397.

Pococks, S,J., Smith, M., Baghurst, P., 1994, Environmental lead and children's intelligence: a systematic review of the epidemiological evidence, *Br. Med. J.* 309(6963): 1189–1197.

Pott, W.A., Benjamin, S.A., Yang, R.S., 2001, Pharmacokinetics, metabolism, and carcinogenicity of arsenic, *Rev. Environ. Contam. Toxicol.* 169: 165–214.

Pruss, A., Corvalan, C.F., Pastides, H., de Hollander, A.E.M., 2001, Methodological considerations in estimating the burden of disease from environmental risk factors at national and global level, *Int. J. Environ. Occup. Health* 7: 58–67.

Pyatt, F.B., 2001, Copper and Lead bioaccumulation by Acacia retinoides and Eucalyptus torquata in sites contaminated as a consequence of extensive ancient mining activities in Cyprus, *Ecotoxicol. Environ. Saf.* 50(1): 60–64.

Romieu, I., Palazuelos, E., Meneses, F., Hernandez-Avila, M., 1992, Vehicular traffic as a determinant of blood-lead levels in children: a pilot study in Mexico City. *Arch. Environ. Health* 47(4): 246–249.

Sargent, J.D., Bailey, A., Simon, P., Blake, M., Dalton, M.A., 1997, Census tract analysis of lead exposure in Rhode Island children, *Environ. Res.* 74(2): 159–168.

Schwartz, 1991, Lead, blood pressure, and cardiovascular disease in men and women, *Environ. Health Perspect.* 91: 71–75.

Schwartz, J., 1994, Low-level lead exposure and children's IQ: a meta-analysis and search for a threshold, *Environ. Res.* 65(1): 42–55.

Schwela, D.H., 1999, Public Health and the Air Management Information System (AMIS), *Epidemiology* 10(5): 647–655.

Seux, R., Dab, W., 1994, Evaluation de l'exposition au plomb et strategie de prevention, Techniques sciences methodes, *genie urbain genie rural* 89(2): 73–78.

Sheehan, P.J., Meyer, D.M., Sauer, M.M., Paustenbach, D.J., 1991, Assessment of the human health risks posed by exposure to chromium-contaminated soils, *J. Toxicol. Environ. Health* 32(2): 161–201.

Squinazi, F., 1994, Le plomb dans les vieilles peintures. Du saturnisme professionel au saturnisme infantile, Techniques sciences methodes, *genie urbain genie rural* 89(2): 88–93.

Staessen, J.A., Lauwerys, R.R., Buchet, J.P., Bulpitt, C.J., Rondia, D., Vanrenterghem, Y., Amery, A., 1992, Cadmibel Study Group. Impairment of renal function with increasing blood lead concentrations in the general population, *N. Engl. J. Med.* 327(3): 151–156.

Stroebel, R., 1992, Etat comparatif de resultats de mesure issus de divers reseaux de surveillance de la pollution atmospherique en France, en Europe et en Amerique du Nord – polluants majeurs (SO_2, NO_2, CO, O_3, Pb et particules), *Pollution Atmosherique* (133): 5–23.

Surdu, S., Rudnai, P., Gurzau, A., Dora, C., Fletcher, T., Leonardi, G., 1997, Natural arsenic in drinking water and adverse health effects in Romania, *Proceedings of the International Symposium on Environmental Epidemiology in Central and Eastern Europe*, 1997, Smolenice, Slovak Republic.

Tchounwou, P.B., Wilson, B., Ishaque, A., 1999, Important considerations in the development of public health advisories for arsenic-containing compounds in drinking water, *Rev. Environ. Health* 14(4): 211–229.

Tenorio, J.A., Espinosa, D.C., 2001, Treatment of chromium plating process effluents with ion excresins, *Waste Manag.* 21(7): 637–642.

Teret, S., 2001, Policy and science: should epidemiologists comment on the policy implications of their research? *Epidemiology* 12: 374–375.

Trombini, T.V., Pedroso, C.G., Ponce, D., Almeida, A.A., Godinho, A.F., 2001, Developmental lead exposure in rats: is a behavioral sequel extended at F2 generation? *Pharmacol. Biochem. Behav.* 68(4): 743–751.

US EPA, 1986, The risk assessment guidelines of 1986, US EPA Office of health and environmental assessment.

Vahter, M., Concha, G., 2001, Role of metabolism in arsenic toxicity, *Pharmacol. Toxicol.* 89(1): 1–5.

Wang, H.C., Wang, H.P., Peng, C.Y., Liu, S.H., Yang, Y.W., 2001, Speciation of As in the blackfoot disease endemic area, *J. Synchrotron Radiat.* 8(Pt2): 961–962.

Weiss, N.S., 2001, Policy emanating from epidemiologic data: what is the proper forum? *Epidemiology* 12: 373–37433.

Wilcox, A.J., 2001, Our policy on policy, *Epidemiology* 12(4): 371–372.

Wu, M.M., Kuo, T.L., Hwang, Y.H., Chen, C.J., 1989, Dose–response relation between arsenic concentration in well water and mortality from cancers and vascular diseases, *Am. J. Epidemiol.* 130(6): 1123–1132.

Advances in lead research: implications for environmental health, 1990, *Environ. Health Perspect.* 89: 240.

Epidemiology and Society, 1999, What you should have learned about epidemiologic data analysis, *Epidemiology* 10(6): 778–783.

Epidemiology and Society, 2001, Update on World Health Organization's initiative to assess environmental burden of disease, *Epidemiology* 12(2): 277–289.

Ground water issue, 1995, Natural attenuation of hexavalent chromium in ground water and soils, Document no.Epa/540/S-94/505.

Health and Safety Executive, 1996, Use of Risk Assessment within Government Departments, Report prepared by the Interdepartmental Liaison Group on Risk Assessment, HSE, London, 3–4, p. 13.

Plomb et environnement, 1994, Techniques sciences methodes, *genie urbain genie rural* 89(2): 63–93.

27. ATMOSPHERIC HEAVY METAL DEPOSITION IN ROMANIA AND NEIGHBORING COUNTRIES: COMPARATIVE EVALUATION ON THE BASIS OF ALL EUROPEAN MOSS MONITORING

OLEG BLUM*
M. M. Gryshko National Botanical Garden, National Academy of Sciences of Ukraine, Timiryazevs'ka 1, 01014 Kyiv, Ukraine

Abstract: Anthropogenic emissions of heavy metals represent potentially hazardous agents for individual elements of the environment and for mankind itself. At elevated levels, in soil and ground water, heavy metals can inhibit plant growth, consequently reduce crop yield, and via the food chain cause harm to human health. Therefore, long-term monitoring programs of large-scale atmospheric deposition of heavy metals are crucial for observation of contamination levels by especially hazardous elements in several countries. Mosses provide an effective and cheap method for monitoring trends in heavy metal pollution of the environment. The concept and technique of analyzing deposition of heavy metals (As, Cd, Cr, Cu, Fe, Hg, Ni, Pb, V, and Zn) in mosses was introduced in the Nordic countries. Afterwards, based on this biomonitoring technique, the All-European project *European Survey of Atmospheric Heavy Metal Deposition* was launched. Upon the results of this research, repeated at 5-year intervals, the comparison of atmospheric heavy metal deposition in Romania with neighboring countries such as Ukraine, Poland, Slovakia, Hungary, participating in the above-mentioned moss surveys is conducted. The most important potential sources of heavy metal emissions into the atmosphere in these countries are discussed. However, the long-range atmospheric transport of polluted aerosols is not to be neglected.

Keywords: heavy metals; moss; monitoring

*To whom correspondence should be addressed. Oleg Blum, M. M. Gryshko Natl. Botanical Garden, Natl. Acad. Sci. of Ukraine, Timiryazevs'ka 1, 01014 Kyiv, Ukraine; e-mail: blum@voliacable.com

R. N. Hull et al. (eds.), Strategies to Enhance Environmental Security in Transition Countries, 369–385.

1. Introduction

Increasing human activities result in the production of environmental pollutants. Atmospheric deposition of pollutants is in many cases a major source of heavy metals to terrestrial and aquatic ecosystems. Anthropogenic emissions of heavy metals represent potentially hazardous agents for individual elements of the environment and for mankind itself. The increase in the concentration of toxic matters leads to the degradation of flora, as a result of sensitive species elimination. Measurements of the atmospheric supply of metals are therefore of vital importance.

The main sources of pollution in the environment and of the atmosphere, in particular, are the combustion of coal (bituminous and brown coal), ferrous, and nonferrous metallurgy, the chemical industry, as well as usage of lead-containing petrol. The most important elements of industry are power stations, smelting, and chemical factories. As a consequence of these emissions, the high amount of particular matter as dust and volatile ashes is supplied to the environment.

In studies of particulate matter, the coarse and fine particles are distinguished. Coarse particles and volatile ashes are more than 10 μm in diameter. They could quickly precipitate on the surface of the soil and plants and accumulate in the upper layer of soil in the vicinity of the emission site. Fine particles with diameters of 0.1–5 μm form aerosols, and like gases could be spread with the atmospheric masses for long distances.

Some heavy metals, which are emitted into the air from such sources as industrial enterprises and thermoelectric power stations, spread mainly close to the sources of their emission. These areas which are subject to such pollution by heavy metals could range from 10 to 50 km in distance depending on the wind direction and the height of the chimney stacks. As an example of such pollution could be nickel (Ni) and chromium (Cr). Other metals are carried longer distances due to the formation of the gas phase during combustion, leading to the formation of very minute particles, which are easily transported by the wind. This is the case with such metals as arsenic, cadmium, lead, mercury, and zinc.

For investigations of levels of atmospheric heavy metal deposition two approaches are usually used: (1) a technical and (2) a biological one.

Heavy metals were included in Cooperative Programme for Monitoring and Evaluation of the Long-range transmission of Air Pollutants in Europe (EMEP) in 1999. A number of countries have been reporting heavy metals within the EMEP area in connection with different national and international program.

The situation with heavy metal monitoring in air and precipitation in Europe was noticeably improved in recent years. Beginning in 2000, there was established an All-European EMEP network for trace metals, with the first priority elements being Hg, Cd, and Pb and the second priority elements being Cu, Zn, As, Cr and Ni. This network (about ten monitoring sites) provides more monitoring data on concentrations, deposition, and transboundary fluxes of cadmium, lead, and mercury over Europe. Based on these data the Pb, Cd, and Hg transport models, heavy metal critical limits, as well as modeling and mapping of critical loads in terrestrial and aquatic ecosystems are being developed under the Working Group on Effects.

The second approach, based on regional biomonitoring, allows determination of natural (background) levels of heavy metals, and detection of increased concentrations in natural ecosystems as the result of technogenesis.

The use of biological indicators is known to have several advantages over technical methods. The latter requires simultaneous measurements to be made with highly specialized equipment, at many stations, and becomes expensive when research is conducted over large territories and especially in regions difficult to access. On the other hand, bioindication monitoring is cheap, relatively simple, and permits the use of an almost unlimited number of samples for the analysis. In general, both approaches of the direct physicochemical and the biomonitoring are necessary for the realization of a detailed and/or large-scale analysis of the distribution of atmospheric air pollutants.

In the last few decades, plants, animals, fungi, and bacteria have been widely used as bioindicators and biomonitors of air, soil, and water. However, it has been established that bryophytes are among the most effective types of organisms for biomonitoring due to a number of their biological features: species diversity, wide areas of distribution, slow growth, significant absorption ability, and their tendency to accumulate and retain trace elements. As opposed to higher plants, bryophytes lack an advanced root system. Therefore, they can not significantly absorb pollutants, in particular heavy metals, from the substratum and consequently their uptake occurs mainly from the ambient atmosphere. So, mosses provide an effective and cheap method for monitoring of atmospheric pollution by heavy metals.

Mosses have been used extensively for monitoring of heavy metal (As, Cd, Cr, Cu, Fe, Hg, Ni, Pb, V, and Zn) deposition in Europe since the late 1960s (Rühling and Tyler, 1968, 1971; Pakarinen and Tolonen, 1976; Steinnes, 1977). The concept and technique of analyzing deposition of heavy metals in mosses was introduced in the Nordic countries and was first tested on an international scale in 1980 as a joint Danish–Swedish initiative (Rühling and Tyler, 1971; Steinnes, 1980; Gydesen et al., 1983; etc.). Later, from 1985 to 1995, similar moss surveys including an increasing number of European countries were conducted. The *European Survey of Atmospheric Heavy Metal Deposition* was launched in 2001 as a part of the International Cooperative Programme on

Effects of Air Pollution on Natural Vegetation and Crops (ICP Vegetation) and coordinated by ICP Vegetation Coordination Centre at the Centre for Ecology and Hydrology (Bangor, UK) at the United Nations Economic Commission for Europe (UN/ECE).

At present, 90 scientists from 28 European countries take part one way or another in this project. The method of moss monitoring has now become well established and is used on a national basis to evaluate the level of heavy metal pollution in European countries every 5 years. The results of such mosses surveys are reported to the Working Group on Effects of the UN/ECE Convention on Long-Range Transboundary Air Pollution (CLRTAP). Based on the obtained data of heavy metal contents in indicator moss species, the atlas (colored maps with isolines or dot maps and descriptions) of quantitative deposition of the above-mentioned toxic metals from the atmosphere on the territory of the countries participating in the project is composed (Rühling, 1994; Rühling and Steinnes, 1998; Buse et al., 2003).

Thus, the moss technique has become a new tool for environmental studies, especially in industrialized countries, where air pollution studies are of high priority. The extension of the European deposition survey to new areas with natural conditions strongly differing from the territories where the use of the moss technique started has emphasized the need to include new moss species where the original ones do not grow naturally. Evaluation of the specificity of heavy metal retention by various bryophyte species has led to the recommendation of the pleurocarpous, epigeic panboreal *Pleurozium schreberi* and *Hylocomium splendens* and the almost cosmopolitan nemoral *Hypnum cupressiforme* for biomonitoring purposes.

In accordance with a decrease in the ability to accumulate heavy metals, the above-mentioned species can be generally arranged in descending order: *H. cupressiforme* > *H. splendens* > *P. schreberi*.

Other pleurocarpous species (*Pseudoscleropodium purum*, *Abietinella abientina*, *Brachythecium* spp.), and acrocarpous species (*Bryum argenteum*, *Dicaranum* spp., *Polytrichum* spp., *Sphagnum* spp.) are used, too, but considerably less frequently.

The ability for heavy metal sorption strongly depends on morphological and anatomical peculiarities of bryophytes. In general, more is known about the retention efficiency of ions in moss than for particulate matter. As a general rule, a wide variation in metal accumulation from species to species and from habitat to habitat under different microclimatic conditions may take place. Therefore, the determination of intercalibration coefficients is crucial for the future studies of deposition levels and distribution of heavy metals all over Europe.

For the first time, the moss technique was applied in Romania in 1995 to a systematic study of air pollution with heavy metals and other trace elements. It was conducted in several industrialized and urban areas of the Eastern Romanian Carpathians which represent mainly the central regions of Romania. This was done to cover one more "white spot" on the heavy metal atmospheric deposition map of Europe. 84 moss samples (*P. schreberi* and *H. splendens*) were collected (Institute of Atomic Physics, Bucharest) and analyzed by neutron activation analysis (NAA) methods (Joint Institute for Nuclear Research, Dubna, Russia) and by inductively coupled plasma mass spectrometry (ICP-MS) methods (Norwegian University of Science and Technology, Department of Chemistry).

The study was continued by Mocanu R. (University of Iasi) during the 2000–2001 European moss survey in other regions of the southern and western Carpathians. In total 214 samples of mosses were collected and the content of heavy metals in them was determined by the atomic absorption spectrometry method. Moreover, 70 moss samples of *H. cupressiforme* were collected in the Transylvanian Plateau in 1999 within the frame of the project "Atmospheric Deposition of Heavy Metals in Rural and Urban Areas of Romania Studied by the Moss Biomonitoring Technique Employing Nuclear and Related Analytical Techniques". In total, 35 chemical elements including heavy metals were identified by NAA methods (Joint Institute for Nuclear Research, Dubna, Russia) and results of this study were published (Stan et al., 2001). The special investigation of atmospheric deposition of heavy metals around the lead and copper-zinc smelters in Baia Mare using mosses (*P. schreberi*, *P. purum*, and *Rhytidiadelphus squarrosus*) as bioindicators showed that the concentrations of Pb, As, Cu, and Cd are high compared with those observed in other regions of Europe with similar industries. However, the concentrations approached regional background levels at a distance of about 8 km from main source area (Culicov et al., 2002).

2. Fundamentals of the Methodology of Heavy Metal Moss Monitoring

This short protocol is recommended by ICP Vegetation.

2.1. AIMS AND OBJECTIVES

The aims of the heavy metals in moss survey are:

- To characterize qualitatively and quantitatively the regional atmospheric deposition
- To indicate the location of important heavy metal emission sources and the extent of particularly polluted areas

- To produce maps of the regional deposition patterns
- To help to understand the extent of long-range transboundary pollution
- To present retrospective comparisons with similar previous studies

2.2. SAMPLING AND ANALYSES

Two moss species are favoured: *P. schreberi*, and if this species is absent *H. splendens*. However, in some countries it might be necessary to use other species. In that case, the first choice would be *H. cupressiforme*. Other species to be considered are, e.g., *Scleropodium purum* and *Thuidium tamariscinum*. The use of bryophytes other than *Hylocomium* or *Pleurozium* must be preceded by a comparison and calibration of their uptake of heavy metals relative to the main preferred species.

The sampling points should be located at sites representative of nonurban areas. In remote areas the sampling points should be at least 300 m from main roads (highways), villages and industries and at least 100 m away from smaller roads and houses. In more densely populated areas these distances may need to be reduced to, e.g., 100 and 50 m, respectively. In mountainous areas the sampling points should be at an altitude between 500 and 1200 m.

It is recommended to make one composite sample from each sampling point, consisting of 5–10 subsamples, if possible, collected within an area of about $50 \times 50\ m^2$.

Samples should preferably be collected during the period April–October. At least 1.5 moss samples/1000 km^2 should be collected in lowland areas and, it is recommended to have a denser sampling regime in mountain areas.

The metal determination can be performed using various analytical techniques (atomic absorption spectroscopy [AAS], inductively coupled plasma emission spectrometry [ICP-ES], ICP-MS, NAA). Samples of moss should not be subjected to washing. Wet ashing is recommended for the decomposition of organic material. The following elements with mainly anthropogenic and atmospheric origin should be determined: As, Cd, Cr, Cu, Fe, Hg, Ni, Pb, V, and Zn.

3. Comparison of Heavy Metal Concentration in Moss in Romania and Neighboring Countries

Romania is situated in the southeastern part of Central Europe and shares borders with Hungary to the northwest, Yugoslavia to the southwest, Bulgaria to the south, the Black Sea and Ukraine to the southeast, and to the north and the Republic of Moldova to the east. Romania spans a total surface area of 237,500 km^2.

The Carpathian Mountains traverse the center of the country bordered on both sides by foothills and finally the great plains of the outer rim. Forests cover one quarter of the country.

Romania, known for its rich mineral resources, is a highly industrialized country where a great number of metal-processing plants as well as coal-fired power plants are operating. The most important metals are iron, chromium, nickel, aluminium, gold, silver, copper, and zinc; other important elements are arsenic, mercury, vanadium, and rare earth elements. A high concentration of industrial activity is clustered within a limited geographical region in Transylvania. As a result, the environment in the area has reached a state of deep ecological stress. For example, nonferrous metal processing plants pollute the surroundings of Copsa Mica, Zlatna and Baia Mare with heavy metals such as copper lead, tin, and cadmium, with the maximum concentrations exceeding by far the permitted norms (Stan et al., 2001). Thus, lead levels in Copsa Mica factories sometimes reached levels that were 1000 times higher than the allowable international limits.

The heavy metals in mosses surveys 1995–1996 and 2000–2001 were conducted at the same areas of Romania and neighboring countries, from which transboundary transfer of the atmospheric pollutants is possible. Therefore, it is of interest to consider the data on the concentrations of metals in mosses in these countries, which reflect atmospheric heavy metal deposition. The concentration levels of ten elements (As, Cd, Cr, Cu, Fe, Hg, Ni, Pb, V, and Zn) in the indicator moss species sampled in Romania were compared to those from other European countries, in particular, Poland and Slovakia, Hungary, and selected regions of Ukraine (Tables 1 and 2). For comparison, metal concentration in mosses growing at such relatively clean territories as Sweden and Norway were included in the tables also. Most of the metal concentrations from the two latter countries could be considered as background values.

Arsenic (As) – The main source of arsenic emissions is coal combustion, although local mining sources and some industrial processes, such as glass making, produce smaller quantities.

Most arsenic values in mosses studied in Romania were between 0.21 and 17.6 μg/g with a median of 0.10 μg/g (1995) and 0.24 μg/g (2000). But in the Baia Mare region which is affected by the nonferrous metal industry, arsenic concentrations exceed 90 μg/g (Stan et al., 2001; Culicov et al., 2002). Strongly elevated arsenic levels were found also in the northern part of Transylvania (Zagra, Letca, and Magoaja) where there is a long tradition of ore mining. The maximum concentration of As found in mosses was 118 μg/g (Buse et al., 2003). This shows that elevated concentrations of arsenic in Romania were associated substantially with oil refineries and nonferrous smelters.

TABLE 1. Concentration of heavy metals (median, µg/g dw) in indicator mosses in Romania, neighboring countries as well as in Sweden and Norway

Region \ Element		As	Cd	Cr	Cu	Fe	Hg	Ni	Pb	V	Zn
Ukrainian Polissia	1995	0.10	0.18	1.70	6.00	327	0.06	2.64	3.40	1.80	30.0
Ukrainian Carpathians	2000	0.24	0.29	1.50	7.31	313	0.039	2.06	6.80	1.29	29.3
Romania	1995	0.95	0.60	9.10	11.3	1900	–	2.17	25.8	6.40	43.7
	2000	1.56	0.46	8.46	21.5	2510	–	3.35	14.3	7.99	79.5
Poland	1995	–	0.44	1.50	7.60	362	–	1.44	13.6	3.99	43.0
	2000	–	0.36	0.89	8.03	429	–	1.57	9.94	5.84	41.4
Slovakia	1995	0.34	0.26	2.83	17.1	1100	0.09	2.34	8.26	3.71	47.7
	2000	0.71	0.59	6.45	8.76	1560	0.180	3.15	28.3	5.70	55.0
Hungary	1995	–	0.39	3.52	5.27	902	0.03	3.76	10.9	4.13	27.3
	2000	<1.0	0.55	6.40	7.65	1760	<1.0	5.35	15.0	4.20	29.9
Bulgaria	1995	–	0.37	2.29	14.6	1584	–	3.04	18.9	4.88	30.5
	2000	0.21	0.38	2.41	14.5	1410	–	3.33	18.9	4.95	32.6
Sweden	1995	0.14	0.19	0.57	4.54	182	0.064	1.11	6.03	2.15	39.9
	2000	0.16	0.18	0.68	4.36	228	0.017	1.41	4.27	1.31	38.8
Norway	1995	.13	0.13	1.05	5.17	331	0.067	1.62	5.80	2.25	37.7
	2000	0.13	0.09	0.69	4.26	365	0.052	1.11	2.70	1.36	29.4

TABLE 2. Concentration of heavy metals (minimum–maximum, μg/g dw) in indicator mosses in Romania, neighboring countries as well as in Sweden and Norway

Region \ Element		As (min–max)	Cd (min–max)	Cr (min–max)	Cu (min–max)	Fe (min–max)	Hg (min–max)	Ni (min–max)	Pb (min–max)	V (min–max)	Zn (min–max)
Ukrainian Polissia	1995	0.01–2.22	0.08–0.42	0.40–7.80	3.20–24.7	123–1440	0.02–0.15	0.70–6.73	1.30–8.90	0.40–5.90	9.0–128
Ukrainian Carpathians	2000	0.06–0.67	0.10–2.91	0.46–4.38	3.69–48.8	66–1320	0.001–0.114	0.72–7.05	2.26–32.6	0.39–4.32	11.8–107
Romania	1995	0.21–17.6	0.30–8.40	2.60–85.0	5.80–234	48–9960	–	0.44–15.5	13.2–231	2.10–54.2	20.2–413
	2000	0.27–118	0.26–1.03	0.50–51.9	2.21–2420	338–21,300	–	0.26–31.9	6.45–31.5	1.93–31.9	20.1–2940
Poland	1995	–	0.05–2.50	0.10–9.00	3.50–650	87.0–5170	–	0.53–4.72	4.30–86.0	0.20–17.2	19.0–208
	2000	–	0.22–7.17	0.34–10.5	4.53–39.6	216–4240	–	0.72–2.89	3.94–65.6	1.92–16.6	28.3–589
Slovakia	1995	0.02–2.38	0.08–1.38	1.01–22.3	6.60–135	144–8780	0.02–1.33	0.87–16.3	2.33–49.0	1.21–14.8	20.7–227
	2000	0.34–2.21	0.11–1.49	1.10–42.7	3.92–37.1	430–13,700	0.062–3.44	0.70–12.6	9.72–109	1.80–30.3	28.0–179
Hungary	1995	–	0.16–0.97	1.37–5.99	3.26–8.79	488–2290	0.03–0.05	2.15–17.3	6.32–24.5	2.79–8.91	18.9–42.6
	2000	<1.0–<1.0	0.31–1.48	3.00–13.1	3.30–17.6	315–3480	<1.0–<1.0	2.90–25.4	5.00–38.9	2.10–10.1	17.7–114
Bulgaria	1995	–	0.16–4.51	1.00–58.0	6.77–205	542–10,400	–	1.43–45.9	7.14–443	2.70–30.4	18.9–347
	2000	0.08–53.0	0.06–10.6	0.74–53.1	5.34–1860	333–11,600	–	1.49–114	4.55–887	2.02–42.6	12.5–930
Sweden	1995	0.05–3.12	0.01–1.02	0.08–438	1.38–24.8	18.2–2890	0.02–0.50	0.33–9.66	0.30–28.5	0.14–16.4	10.2–206
	2000	0.04–1.14	0.05–0.69	0.11–136	1.69–30.3	12–4270	0.0–0.231	0.46–18.2	0.93–19.4	0.24–12.1	13.6–134
Norway	1995	0.14–2.85	0.01–1.67	0.21–289	2.12–160	86.5–14,900	0.002–0.479	0.41–235	0.88–59.4	0.45–16.5	8.80–603
	2000	0.01–3.43	0.01–2.62	0.13–258	1.74–206	99–11,200	0.022–0.208	0.06–302	0.50–27.7	0.28–22.6	9.71–661

Arsenic concentrations in relatively clean areas such as Polissia and Ukrainian Carpathians ranged between 0.01 and 2.22 μg/g and 0.06 and 0.67μg/g, respectively. The median concentrations were 0.10 μg/g in the Polissia and 0.24 μg/g in the Ukrainian Carpathians area.

Cadmium (Cd) – Cadmium is a trace element present in the earth's crust with an average concentration of 7–10 μg/g, often occurring together with zinc and phosphorus. As a result of wind erosion, soil (ash) particles are carried by the wind and spread throughout the environment. Other emission sources are the phosphate fertilizers used in agriculture.

In urban areas cadmium emissions are often associated with garbage incineration and the combustion of fossil fuels (Mäkinen and Lodenius, 1984). Other emission sources are the mining and metal industries, especially zinc concentrating plants and galvanizing processes. Cadmium is also used in the coating of other metals, in alkaline batteries and pigments, as well as in stabilizers for plastics.

Cadmium is very toxic to all organisms and is a cumulative poison that can be concentrated in nutrient cycles. For instance, some mushrooms accumulate cadmium.

All cadmium values in mosses growing in Romania were between 0.26 μg/g and 8.40 μg/g, with a median of 0.60 μg/g (1995) and 0.46 μg/g (2000). In moss samples from the Baia Mare region, the concentration of Cd exceeds 5 μg/g (Culicov et al., 2002). In another similar moss monitoring study (Stan et al., 2001) with *H. cupressiforme* all Cd concentrations (epithermal neutron activation analysis, ENAA) were between 0.23 μg/g and 55 μg/g. Unfortunately the authors of this study did not give any comment for such high concentrations of cadmium in moss. Nevertheless, in general, the moss Cd concentrations in Romania were two to five fold higher than those in Scandinavia (see Table 1 and 2).

The high cadmium concentration in the Cergau region is mainly due to Zn and Pb smelters. Moreover, the pattern of Cd distribution showed increased concentrations in the Zagra–Letca–Magoaja region, probably a result of ore mining. A high Cd concentration observed in Aiud may be explained by the presence of an oil-fired power station without proper filters (Stan et al., 2001).

Cadmium concentrations in moss in the Ukrainian Polissia and Ukrainian Carpathian areas ranged from 0.08 to 2.91 μg/g. The median concentration was 0.18 μg/g to 0.29 μg/g respectively. The highest values were probably caused by the use of phosphate fertilizers in the Polissia and combustion of fossil fuel in the Ukrainian Carpathians.

Chromium (Cr). Coal and heavy oil combustion are the main sources of increasing chromium deposition in the surrounding ecosystems (Mäkinen, 1983). Chromium is also an important additive used in making stainless steel.

Chromium is an essential micronutrient, but at higher concentrations Cr^{3+} is moderately and Cr^{6+} highly toxic to all organisms.

Chromium concentrations in mosses in Romania were between 0.50 μg/g and 85.0 μg/g with a median of 9.10 μg/g (Survey 1995) and 8.46 μg/g (Survey 2000).

Chromium concentrations in moss in the Ukrainian Polissia and in the Ukrainian Carpathian areas ranged from 0.40 μg/g to 7.80 μg/g. The median concentration was 1.70 μg/g and 1.50 μg/g, respectively. The baseline level in mosses from rural areas evaluated in most countries was less than 2 μg/g. The mean background concentration for the foregoing Ukrainian study areas was 2.75 μg/g.

It is remarkable that very high chromium concentrations in mosses were found in such "background countries" as Sweden and Norway. These local elevated chromium values ranged from 136 μg/g to 438 μg/g, and were mainly connected with emissions from the steel industry and ferrochrome smelters in the above-mentioned countries, and sometimes may be ascribed to soil contamination (Rühling and Steinnes, 1998; Buse et al., 2003).

Copper (Cu). Most copper emissions originate from local sources, i.e., from copper mines and smelters. Smaller amounts (about 6 %) are emitted into the air during the combustion of fossil fuels, as well as from garbage incineration and traffic (Mäkinen and Lodenius, 1984).

Copper is an important micronutrient for all organisms. At higher concentrations, however, it is toxic to vascular plants and very toxic to algae and fungi especially, and is therefore used in fungicides. It is also very toxic to invertebrates. The background level of copper in areas with no local sources is about 4–8 μg/g (Rühling and Steinnes, 1998).

All copper values in mosses in Romania ranged from 3.21 to 2420 μg/g, the median being 11.3 (Survey 1995) and 21.5 (Survey 2000). Copper showed elevated concentrations in the same region as arsenic and has the same source, ore mining and nonferrous metal industry (Stan et al., 2001). The highest levels of Cu were observed in the Baia Mare region.

Very high copper concentrations (1860 μg/g) were observed also in Bulgaria and are connected with copper ore mining near Panagurishte and a copper smelter near the town of Pirdop (Buse et al., 2003). Strong local sources have an effect on copper content in mosses, as also observed in Poland (650 μg/g) and Slovakia (135 μg/g).

In the Ukrainian Polissia and in the Ukrainian Carpathian areas the copper concentration in mosses ranged from 3.2 to 48.8 μg/g. The median concentrations were 6.0 and 7.31 μg/g, respectively.

In Norway, the elevated copper concentrations (160 and 206 μg/g) were the effect of copper–nickel smelters in Kola peninsula in northwest Russia (Buse et al., 2003).

Iron (Fe) – Iron is emitted to the atmosphere mainly from the iron and steel industry and iron mining. Iron is one indicator of human activities because in urban areas its amount in deposition increases (Mäkinen, 1994). It originates, e.g., from coal-fired power plants, the fly ash from which can contain more than 50 mg/g iron (Mäkinen, 1983). The content of iron in mosses does not only reflect the anthropogenic atmospheric deposition, but also the supply of windblown soil particles. The considerable role of the soil factor in Fe accumulation by mosses was indicated by the high Fe–Sc correlation (Stan et al., 2001).

The iron concentration in mosses from rural areas was generally lower than 500 μg/g if the areas were not influenced by soil dust (Rühling and Steinnes, 1998).

Iron is a very important micronutrient for all organisms. However, when accumulated in elevated concentrations in humans it could cause a number of serious illnesses.

In Romania, the maximum concentration of iron in mosses reached 21,300 μg/g. The median concentrations were 1900 and 2510 μg/g in Surveys 1995 and 2000, respectively.

Iron concentrations in moss in the Ukrainian Polissia and in the Ukrainian Carpathian areas ranged from 66 to 1440 μg/g. The median concentrations were 327 and 313 μg/g, respectively.

The median moss concentrations of iron in Norway and Sweden ranged between 182 and 365 μg/g, but a local maximum of Fe concentration in Norway reached 14,900 μg/g.

Lead (Pb) – Most lead emissions are in connection with car traffic and the combustion of leaded fuel. Other lead emission sources are the mining of lead ore, lead smelters, and ferrous metallurgy. During the last few years the increased use of unleaded gasoline has decreased lead emissions in many countries irrespective of the increased traffic density (Rühling et al., 1992).

Lead is a cumulative poison and toxic to all organisms. In spite of the fact that in the natural environment lead is usually tightly bound in the organic top soil layers (humus), in soluble ionic form it is toxic to most organisms. Children seem to be more sensitive then adults (Rühling and Steinnes, 1998).

Very high levels of lead (up to 175 μg/g) in mosses were found in the northeastern part of Romania around the Baia Mare lead smelter (Culicov et al., 2002). All measured Pb concentrations in mosses ranged from 6.45 to 231 μg/g. The median concentrations were 25.8 μg/g (Survey 1995) and 14.3 μg/g (Survey 2000).

Very high lead concentrations (as much as 887 μg/g) were noted for Bulgaria and they are related to old mines and lead–zinc smelting factories (Rühling and Steinnes, 1998; Buse et al., 2003).

Lead concentrations in moss ranged from 1.30 to 8.90 μg/g with a median of 3.40 μg/g in the Ukrainian Polissia and from 2.26 to 32.6 μg/g with the median of 6.80 μg/g in the Ukrainian Carpathian areas. The background concentrations in these areas varied between 3 and 6 μg/g, and was lower than in many other parts of Europe.

Mercury (Hg) – The main sources of mercury emissions are waste incineration, the manufacture of chlorine, nonferrous metal production, and coal combustion.

Mercury and many of its compounds (methyl mercury) are strongly toxic to most living organisms. Humans uptake 15 μg of mercury per day from food and air.

Unfortunately, in European moss surveys 1995/1996 and 2000/2001 mercury was not determined in the samples from Romania.

Among considered countries, Slovakia was the most polluted by mercury. Here, the median concentration was 0.18 μg/g, and the maximum value was 3.44 μg/g, which was ten times higher than mercury concentration in Norway and Sweden.

The mercury concentration in moss in Ukrainian Polissia and Ukrainian Carpathians ranged between 0.001 and 0.15 μg/g.

Nickel (Ni) – The main sources of nickel to the environment are large smelter complexes and the steel industry. Nickel is an important additive in stainless steel. The combustion of heavy oil and coal also increases nickel deposition (Mäkinen, 1983). Nickel is essential to some organisms but also toxic in higher concentrations to most plants, fungi, and animals. It could have allergic and allegedly carcinogenic effects on humans.

Nickel concentrations in mosses in Romania were between 0.26 and 31.9 μg/g with a median of 2.17 μg/g (Survey 1995) and 3.35 μg/g (Survey 2000).

Nickel concentrations in moss in the Ukrainian Polissia and in the Ukrainian Carpathian areas ranged from 0.70 to 7.05 μg/g. The median concentration was 2.64 and 2.06 μg/g, respectively. This is two times as much as background concentration in Sweden and Norway.

Vanadium (V) – Vanadium is mainly emitted from oil combustion and from refineries. Another emission source is the steel industry where vanadium is used as an additive in the manufacture of special steel. On terrain with poor vegetation, soil dust could cause accumulations of this metal in mosses in considerable concentrations. In higher concentrations vanadium is toxic to living organisms.

Vanadium concentrations in mosses in Romania were between 1.93 and 54.2 µg/g with a median of 6.40 (Survey 1995) and 7.99 µg/g (Survey 2000).

Vanadium concentrations in moss in the Ukrainian Polissia and in the Ukrainian Carpathian areas were similar and ranged from 0.40 to 5.90 µg/g and 0.39 to 4.32 µg/g, respectively with medians of 1.80 and 1.29 µg/g, respectively. These values are close to the background concentration observed in Sweden and Norway.

Zinc (Zn) – Zinc is mainly emitted by the metal industry, road transport, waste incineration, and coal combustion. Zinc is an essential micronutrient for all organisms, but at higher concentrations it could be moderately toxic to plants, but only slightly toxic to mammals. The air contains 1–10 $\mu g/m^3$ of zinc.

Zinc concentrations in mosses in Romania were between 20.1 and 2940 µg/g with a median of 43.7 µg/g (Survey 1995) and 79.5 µg/g (Survey 2000). Two regions with high Zn concentration were distinguished in Transylvania. One of them is Cergau, where there is a zinc and lead smelter. The second is Letca–Magoaja–Ileanda, characterized by nonferrous ore mining. The Zn values were between 39 and 2900 µg/g, with the median being 137 µg/g (Stan et al., 2001).

Zinc concentrations in moss in the Ukrainian Polissia and in the Ukrainian Carpathian areas ranged from 9.0 to 128 µg/g with medians of 30.0 and 29.3 µg/g, respectively. These values almost correspond to background concentration observed in Sweden and Norway.

4. Potential Ecological Effects of Heavy Metal and Risk Assessment

The average human life expectancy in Romania is one of the lowest in Europe: 66.56 years for men and 73.17 years for women. The considerable air pollution including heavy metals is likely to be one of the factors responsible for these unfortunate health conditions. In the period after 1989, and especially during the last 3 years, extensive measures have been taken aimed at improving the environmental conditions in Romania. The results shown in Table 1, however, show that there is still a long way to go (Stan et al., 2001).

At elevated levels, in soil and ground water, heavy metals can inhibit plant growth, consequently reduce crop yield, and via the food chain cause harm to human health. In spite of the slow accumulation of heavy metals in the environment, the evaluation of the level of deposition is consequently of vital importance.

Air pollution is a highly localized problem in Romania. While rural areas are often pristine, urban and industrial areas suffer from acid rain and high pollution levels. The state environment report lists 17 urban "hot spots" where air pollution systematically exceeds environmental quality standards. Much of Romania's air pollution stems from its antiquated energy sector. Thermal power plants continue to burn low-efficiency solid fuels (coal) and high-sulfur content heavy fuel (oil).

Another major source of urban air pollution is Romania's transportation sector. Most cars on the road in Romania are old and poorly maintained, running on gasoline that has the highest lead content among Eastern European countries.

As can be seen in data from the moss surveys (Tables 1 and 2), Romania has the highest concentration of cadmium, chromium, and zinc among the European countries participating in the moss survey. It also has a leading position with respect to concentrations of other toxic heavy metals such as arsenic, iron, lead, and vanadium. As mentioned above, the critical areas in Romania with regard to pollution with heavy metals are Copsa Mica, Zlatna and Baia Mare as well as Hunedoara, Calan, and Galati.

Of the neighboring countries to Romania, the moss survey data show Slovakia is the most polluted with heavy metals, in particular by arsenic, cadmium, chromium, lead, and zinc. Among the countries studied (Tables 1 and 2) Hungary takes third place with regard to atmospheric pollution of their territory by such heavy metals as cadmium, chromium, nickel, and arsenic.

Thus, the "Black Triangle" of Romania/Slovakia/Hungry could be distinguished based on the data of biomonitoring of heavy metal pollution of the atmosphere. Transboundary transport of heavy metals across these countries could also be important, particularly for lead.

Among the rest of the countries, in Poland elevated concentrations of cadmium, lead, and zinc were observed, while in Bulgaria comparatively high nickel and lead concentrations were observed.

As to Ukraine, the northern neighbor of Romania, all considered heavy metals found in mosses were considerably lower in concentration in the southwestern regions of the Carpathians compared to the countries of "Black Triangle". In the Ukrainian Carpathians, the sources of heavy metals are industrial and urban areas, such as L'viv, Ivano-Frankivs'k, Uzhgorod, Kalush, and Nadvirna, while Zakarpattia and Chernivtsi regions have no intensive pollution sources. According to our study (Blum et al., 2004) in areas neighboring Romania, the Chernivtsi region there is stronger influence of the crust factor (metals originating from the earth's crust) than of the western transportation of the polluted air masses from Europe. This follows from the principle component analysis of our data and comparison of the percentage contribution to the total dispersion of factor F1: 74.99% in Chernivtsi region against 55.93% in the Zakarpattia region.

In spite of the fact that in Romania heavy metal concentrations exceed by several times those from the neighboring Ukrainian regions, no influence from long-range transport was observed by us. The major reason should be the high relief which divides the regions.

As to the territory of Ukrainian Polissia (the north forested part of Ukraine), it is much less polluted with heavy metals than the majority of areas of Central Europe. Heavy metal concentration in mosses for some relatively clean territories in Norway and Sweden and also in the Ukrainian Polissia served as background levels.

Data from the moss surveys provide information on levels of atmospheric heavy metal deposition and spatial distribution in Romania and neighboring countries. However, further information is needed on the concentrations of heavy metals into environment deposition rates and pathways, and effects on human health and the environment. The results of the moss surveys complement the results of the technical monitoring, which enable authorities to work out legal documents committing countries to reduce pollutant emissions for specific target years and monitoring their success in reducing the impacts of air pollutants on health and the environment.

5. Summary and Conclusions

Analysis of the data of the moss surveys showed that Romania was first among the studied countries with respect to pollution of the atmosphere by toxic heavy metals, in particular by arsenic, cadmium, zinc, lead, chromium, and vanadium. The considerable contribution of emissions of heavy metals into the atmosphere is made by such critical areas as Copsa Mica, Zlatna, and Baia Mare, as well as Hunedoara, Calan, and Galati.

Slovakia takes second place in the pollution of its territory by the above-mentioned heavy metals, while Hungary takes third place due to serious pollution of its territory by such heavy metals as arsenic, cadmium, chromium, and nickel. Thus, the "Black Triangle" of Romania/Slovakia/Hungry could be distinguished, within which transboundary transport of heavy metals could be important, mainly for lead.

Among the rest of the studied countries, in Poland elevated concentrations of cadmium, lead, and zinc were observed, while in Bulgaria comparatively high nickel and lead concentrations were observed.

As to Ukraine, the northern neighbor of Romania, all heavy metals found in mosses were in considerably lower concentrations in the studied southwestern regions of the Carpathians compared to the countries of the "Black Triangle".

References

Blum, O., Culicov, O., Frontasyeva, M.V., 2004, Heavy metal deposition in Ukrainian Carpathians (Zakarpatia and Chernivtsi regions): the regional biomonitoring, in: *Environmental Awareness*, Klump, A., Ansel, W., Klump, G. (eds.), Cuviller Verlag, Göttingen, pp. 249–255.

Buse, A., Norris, D., Harmens, H., Bűker, P., Ashenden, T., Mills, G. (eds.), 2003, Heavy metals in European mosses: 2000/2001 survey, UNECE ICP Vegetation Coordination Centre, CEH Bangor, UK.

Culicov, O., Frontasyeva, M.V., Steinnes, E., Okina, O.S., Santa, Zs., Todoran, R., 2002, Atmospheric deposition of heavy metals around the lead and copper-zinc smelters in Baia Mare, Romania, studied by the moss biomonitoring technique, neutron activation analysis and flame atomic absorption spectrometry, *J. Radioanal. Nucl. Chem.* 254: 109–115.

Gydesen, H., Pilegaard, K., Rasmussen, L., Rühling, Å., 1983, Moss analysis used as a means of surveying the atmospheric heavy metal deposition in Sweden, Denmark and Greenland, Bulletin Statens Naturvardsverk PM 1670.

Mäkinen, A., 1983, Heavy metal and arsenic concentrations of a woodland moss, *Hylocomium splendens* (Hedw.) Br. et Sch., growing around a coal-fires power plant in coastal southern Finland. Project KHM, Teknisk rapport 85: 1–85, The Swedish State Power Board, Välligby.

Mäkinen, A., 1994, Biomonitoring of atmospheric deposition in the Kola Peninsula (Russia) and Finnish Lapland, based on the chemical analysis of mosses, Report 4, Ministry of the Environment, Helsinki, Finland.

Mäkinen, A., Lodenius, 1984, Urban levels of cadmium and mercury in Yelsinki determined from moss-bags and feather mosses, *Ympäristö ja Terveys* 5: 306–317.

Pakarinen, P., 1981, Metal content of ombrotrophic Sphagnum mosses in NW Europe, *Ann. Bot. Fennici*. 18: 281–292.

Pakarinen, P., Tolonen, K., 1976, Regional survey of heavy metals in peat mosses (*Sphagnum*), *Ambio*. 5: 38–40.

Rühling Å., Tyler G. 1968. An ecological approach to the lead problem. *Botaniska Notiser*, 122: 246–342.

Rühling Å., Tyler G. 1971. Regional differences in the deposition of heavy metal over Scandinavia. *J. of Applied Ecol.*, 8: 497–507.

Rühling Å., Brumelis G., Goltsova N., Kvietkus K., Kubin E., Luv S., Magnusson S., Mäkinen A., Sander E., *Steinnes* E., Pilegaard K., Rasmussen L. 1992. Atmospheric heavy metal deposition in Northern Europe 1990. Nord 1992 (12). Nordic Council Ministers, Copenhagen.

Rühling, Å. (ed.), 1994, Atmospheric heavy metal deposition in Europe – estimation based on moss analysis, Nord 1994 (9), Nordic Council Ministers, Copenhagen.

Rühling, Å., Steinnes, E. (eds.), 1998, Atmospheric heavy metal deposition in Europe 1995–1996, Nord 1998 (15), Nordic Council Ministers, Copenhagen.

Stan, O.A., Lucaciu, A., Frontasyeva, M.V., Steinnes, E., 2001, New Results from Air Pollution Studies in Romania, in: *Radionuclides and Heavy Metals in Environment*, Frontasyeva, M.V., Perelygin, V.P., Vater, P. (eds.), NATO Science Series, IV/5, Earth and Environment Sciences, Kluwer Academic, The Netherlands, pp. 179–190.

Steinnes E. 1977. *Atmospheric* deposition of trace elements in Norway studied by means of moss analysis. Institut for Atomenergi, Kjeller Report KR-154. Kjeller, Norway.

Steinnes E. 1980. Atmospheric deposition of heavy metals in Norway studied by the analysis of moss samples using neutron activation analysis and atomic absorption spectrometry. J. *Radioanal. Chem*., 58: 387–391.

Sucharova, J., Suchara, I., 1998, Biomonitoring of the atmospheric deposition of metals and sulphur compounds using moss analyses in the Czech Republic, Results of the international Biomonitoring programme, Research institute for landscape and ornamental gardening Pruhonice, ISBN 80-901916-8-1.

28. POLLUTANT UPTAKE ON AGRICULTURAL LAND: PRACTICAL MODELING

N. GONCHAROVA*, D. BAIRASHEUSKAYA,
V. PUTYRSKAYA
Laboratory of Environmental Monitoring, International Sakharov Environmental University, Minsk, Belarus

Abstract: Problems of ^{137}Cs behavior in the soil–plant system, particularly the main parameters of ^{137}Cs vertical migration through the soil horizons are investigated. The objectives of the work are to study the biological availability of ^{137}Cs on the example of field phytocenosis areas; to analyze the existing models for radionuclide behaviur in soil; as well as to collect experimental data about the "Sudkovo" farm for further research and practical model construction for pollutant uptake and radionuclide behaviur on agricultural lands. A field experiment was carried out on tilled agricultural and forest soddy-podzolic (Podsoluvisols) soils in the south of Belarus affected by the Chernobyl accident. All the data received during the conducted work were added to the experimental data on "Sudkovo" farm, Khoiniki district, Gomel region, collected during the last 20 years from different sources. The present research is directed at solving modern problems connected with methodologies for rehabilitation of the territories polluted with radionuclides. The compilation of a model predicting ^{137}Cs behavior in soil is very important to enable the use of agricultural land and allow improvement of sown areas under conditions of radioactive contamination.

Keywords: transfer factor; vertical migration; soil–plant system

*To whom correspondence should be addressed. N. Goncharova, Laboratory of Environmental Monitoring, International Sakharov Environmental University, Minsk, Belarus; e-mail: nadya_goncharova@yahoo.com

R. N. Hull et al. (eds.), Strategies to Enhance Environmental Security in Transition Countries, 387–399.

1. Introduction

The Chernobyl accident resulted in radioactive contamination covering 23% of Belarus, an area which is populated by 2.2 million people. Agricultural production is conducted on 1.3 million hectares of contaminated land which had ^{137}Cs deposition of 37–1480 kBq/m^2. There are difficulties in producing foodstuff which meets the national permissible levels of ^{137}Cs in some areas.

Radionuclides in the topsoil are potentially available for uptake by plant roots for a long time period. The availability of ^{137}Cs to plants declines significantly due to the process of "aging" and the fixation of ^{137}Cs in the soil. The fraction of mobile ^{137}Cs decreased in different soils on average more than three times. Presently, the ^{137}Cs fixed faction makes up to 87–98% of the total content. Therefore, soil-to-plant transfer coefficients have to be updated periodically to provide the credible forecast of contamination levels of agricultural products.

The contamination level of agricultural products depends on several factors, including the radionuclide deposition rate, soil type, texture and chemical properties, and biological characteristics of plants (Avery, 1996; Bergeijk et al., 1992; Coughtrey et al., 1989).

The main aim of our work was to determine parameters of radionuclide transfer from soil to plants. These parameters are the basis for development of protective measures in the long-term postaccident period to decrease the transfer of ^{137}Cs into the food chain with minimal expenses. In relatively favorable climatic conditions, the level of soil fertility is the most important factor limiting the growth of crop yields with permissible levels of radionuclides.

2. Soil-to-plant transfer of ^{137}Cs in soils of South Belarus

Migration of ^{137}Cs derived from the fall out of the Chernobyl accident is visualized by staining the flow pattern in the soil. Experiments were carried out in a pasture, in a forest soil, and in a cropped soil on the land of Sudkovo farm, Khoiniki district in the south of Belarus. We examined the correlation between the distribution of ^{137}Cs and the presence of preferential flow paths within the upper 50 cm of three different soils:

1. Pasture soil (permanent grassland)
2. Forest soil
3. Field soil (from a farmed area)

During each experiment, 2 plots (1×1 m^2) were irrigated with a *Brilliant Blue FCF* solution to stain the flow paths of water. The stained flow paths and the nonstained matrix were sampled and the activity of ^{137}Cs determined. The plants in the agricultural plots were cut at the 5 cm level aboveground. In the forest plot, the litter was removed and its activity was measured.

2.1. SOIL: PREFERENTIAL FLOW PATH AND MATRIX

Selected chemical and physical properties of the studied soils are presented in Tables 1 and 2.

TABLE 1. Soil characteristics of experimental plots

Plot	Depth(cm)	Soil type	pH[a]
Pasture	0–20	Soddy-podzolic	6.86
	20–40	(Podzoluvisols)	6.85
Forest	0–20	Soddy-podzolic	5.75
	20–40	(Podzoluvisols)	6.65
FIELD	0–20	Soddy-podzolic (Podzoluvisols)	6.90

[a]Measured in a suspension of 2.5 mL/g of soil in deionized water

TABLE 2. Cation exchange capacity of two soddy-podzolic soils of the Sudkovo farm

Plot	Depth (cm)	Exchangeable cations (meq/100 g soil)				
		K^+	Na^+	Mn^{2+}	Ca^{2+}	Fe^{3+}
Pasture	0–20	18.7	2.2	3.7	2.7	2.9
	20–40	16.9	1.6	2.5	2.5	2.7
Forest	0–20	5.6	1.9	3.4	2.4	3.1
	20–40	4.1	1.9	1.5	2.3	2.6

After applying the solution of Brilliant Blue FCF, flow paths stained by the tracer could be distinguished from the unstained zones (matrix). The tracer distribution is shown in Figure 1 for the three different ecosystems.

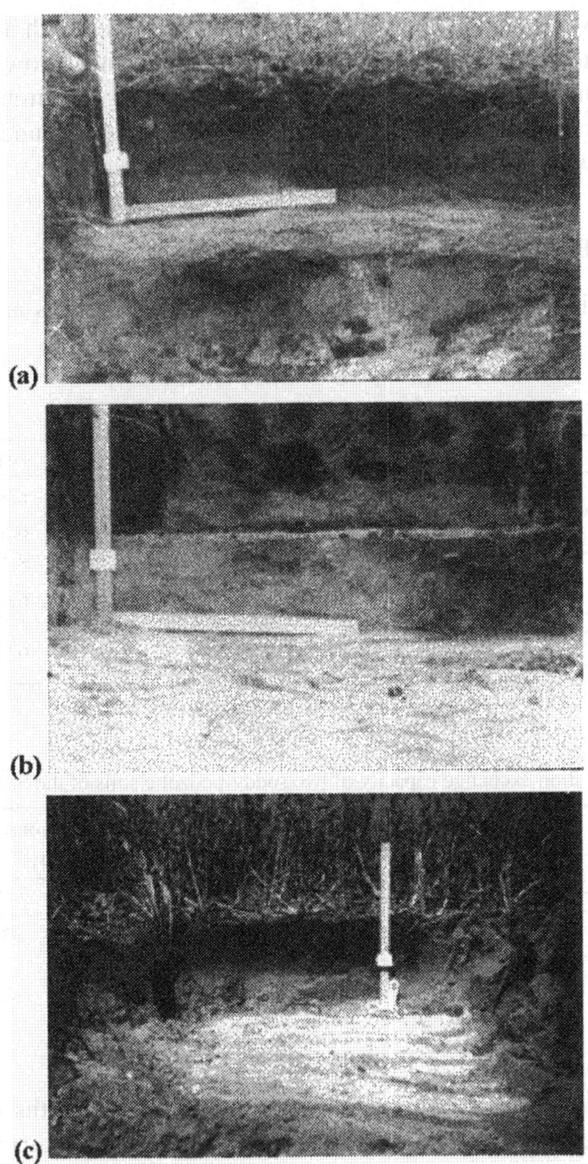

Figure 1. Brilliant Blue FCF distribution over the vertical profile. Vertical profile observation in the pasture (a), forest (b), and agricultural field (c).

In the agricultural soil ^{137}Cs was evenly distributed throughout the tilled layer and, in the upper 0–15 cm, the concentration of ^{137}Cs was higher in the matrix than in the preferential flow paths (Figure 2a).

In the forest soil most of activity remained in the upper 0–5 cm layer with two times higher concentration in the soil matrix than in the preferential flow paths. In the lower layers, the concentration of ^{137}Cs decreased drastically presenting higher values in the soil matrix (Figure 2b).

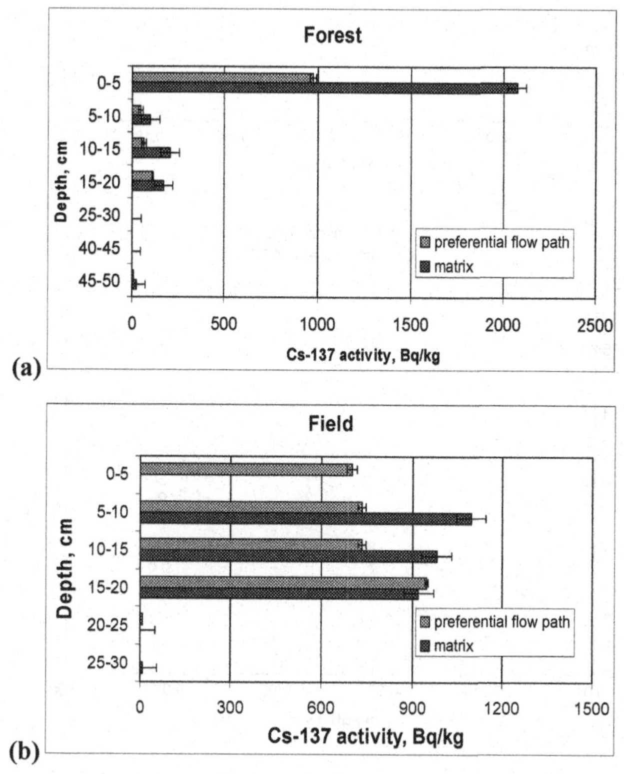

Figure 2. Differences in activities of ^{137}Cs in preferential flow paths and matrix in an agricultural soil (a) and forest soil (b) with depth.

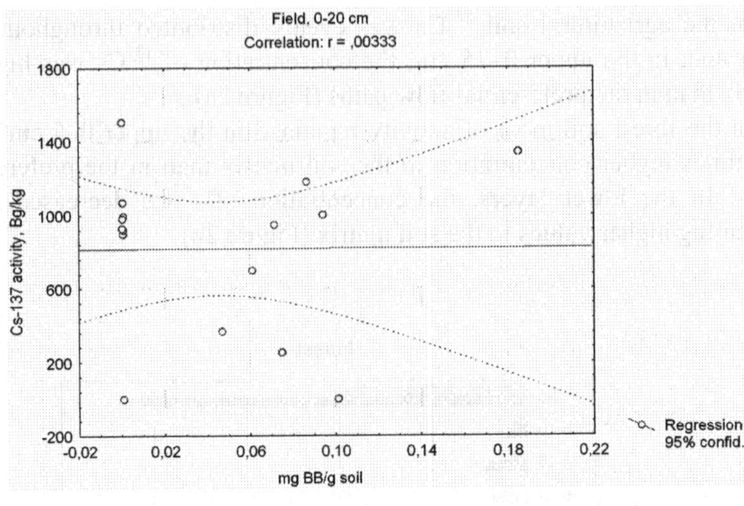

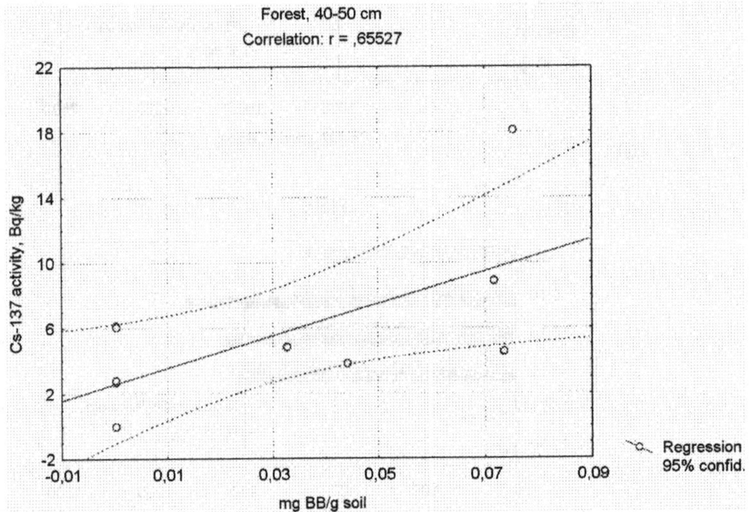

Figure 3. Correlation between brilliant blue FCF concentration and ^{137}Cs activity in the pasture (a) and in the forest (b).

The analysis of the concentration of Brilliant Blue FCF (mg/g of soil) and concentration of ^{137}Cs in soil (Bq/g of soil) in the preferential flow paths displayed no statistically significant correlation (Figure 3).

2.2. ^{137}CS CONCENTRATION IN PLANTS

The results of measurements of ^{137}Cs activity concentration in the root system and in the plant vegetative mass of three studied areas are presented in Table 3. These data show that the highest ^{137}Cs concentrations were observed in grasses (tops and roots) sampled in the forest. Grasses sampled in the pasture showed intermediate ^{137}Cs concentrations, and plants sampled in the agricultural soil had the lowest concentrations.

TABLE 3. The quantitative characteristic of ^{137}Cs accumulation by green mass and roots of different plants grown in the pasture, agricultural field, and in the forest of Sudkovo farm.

Plant sample	^{137}Cs (Bq/kg)
Winter rye (green ears)	3.5 ± 0.02
Winter rye (leaves + stem)	8.7 ± 0.04
Winter rye (roots)	26.1 ± 1.1
Maize (green mass)	36.8 ± 1.4
Maize (roots)	70.7 ± 1.2
Motley grass (green mass)	98.8 ± 12.3
Motley grass (roots)	121.2 ± 17.3
Forest graminae (green mass)	503.8 ± 23.6
Forest graminae (roots)	3149.9 ± 35.6

2.3. TRANSFER FACTOR VALUES

Soil-to-plant transfer factors (TF) were calculated for all experiments. They describe the quantitative characteristics of ^{137}Cs accumulation by plants, which allow comparing the rate of ^{137}Cs entry into the plants of different ecosystems.

We calculated the TF as follows:

$$TF = \frac{C_{\text{plant}}}{C_{\text{soil}}}$$

where C_{plant} is ^{137}Cs activity (Bq/kg) of dry weight plant and C_{soil} is ^{137}Cs activity (Bq/kg) of dry 20 cm depth soil.

TF values calculated for winter rye and motley grass grown in the agricultural field were similar to the expected value given by International Atomic Energy Agency (IAEA) or within the confidence range given therein (Table 4). The TF value calculated for maize was nine-fold higher than the tabulated value given by IAEA. Similarly, the TF value calculated for the forest graminae was higher than the tabulated value, but within the confidence range given by IAEA.

TABLE 4. Comparison between transfer factor (TF) values calculated for experimental data and tabulated data obtained from the International Atomic Energy Agency, Vienna (1994)

Plant species	Calculated value	Tabulated value	95% confidence range
Winter rye	0.02	0.02	0.001–0.01
Maize	0.09	0.01	0.001–0.01
Motley grass	0.11	0.11	0.011–1.1
Forest graminae	0.51	0.11	0.011–1.1

In conclusion, the even distribution of ^{137}Cs with depth in the agricultural soil can be explained by the effect of ploughing which alters the soil structure and creates a homogeneous upper horizon. On the other hand, no clear relation could be shown between structure and radionuclide presence in the Belarussian forest soil. The TFs varied significantly for different plant species which can be due to different uptake of ^{137}Cs by the roots. The higher concentration of ^{137}Cs in grasses grown in the forest can be due to their shallow root systems foraging in the soil with high ^{137}Cs content and to the higher availability of ^{137}Cs in this soil where it is weakly bound onto organic matter compounds (Rigol et al., 2002).

3. Review of the Database

The database of soil-to-plant TFs for radiocaesium has been compiled for Sudkovo farm, Hoiniki district, Gomel region, South Belarus. It contains auxiliary data on important soil characteristics. The database is subdivided into several soil–crop combinations, covering three soil types and six plant groups (Table 5).

TABLE 5. Summary of data recorded in the Sudkovo farm database

Parameter	Classification
Soil type	Podzoluvisol, Gleysol, Histosol
Plant species	Natural field grass, natural herbage grass, legume crops, forage crops, industrial crops, woody plants
Soil characteristics	pH, organic matter content, cation exchange capacity (CEC), exchangeable K, N, and CA[a]
TF	Calculated values

[a]Only for 1992

The database includes information on the following soil and plant characteristics: year of experiment (1988–2001), pH, soil type, TF, plant species, humus content (%), soil radioactivity (Bq/kg), nutrient components (mg/100g soil), plant radioactivity (Bq/kg), ^{137}Cs activity in soil layers (0–45 cm).

In Figures 4–6, the results of GIS design development and data visualization are presented.

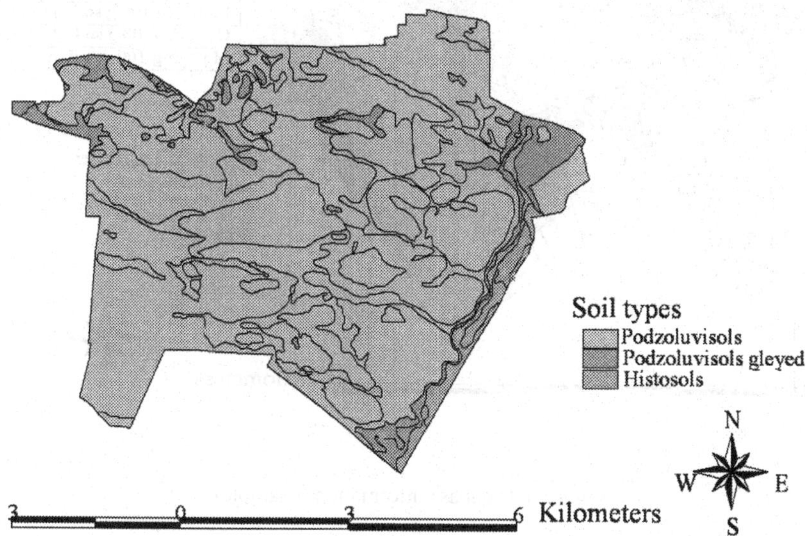

Figure 4. Soil types of Sudkovo farm, Hoiniki district.

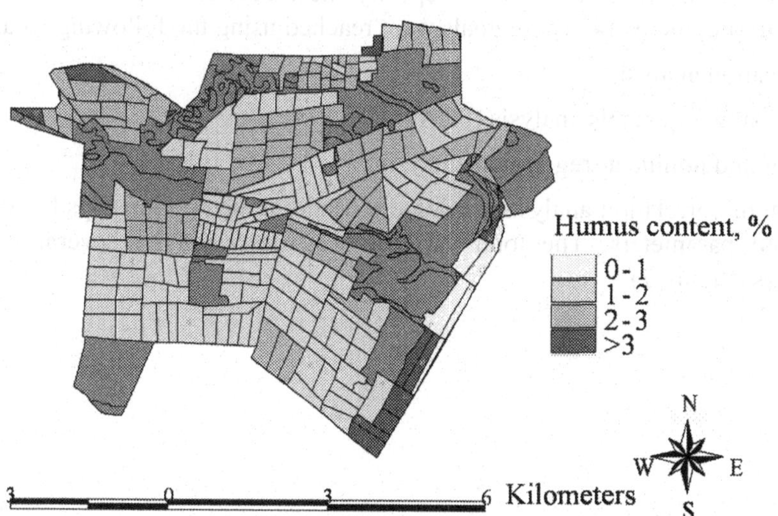

Figure 5. Humus content of Sudkovo farm, Hoiniki district.

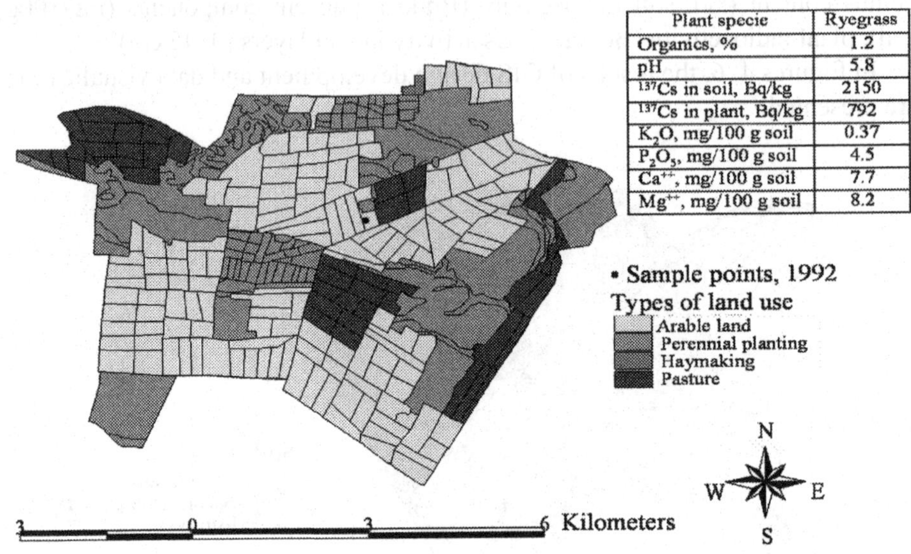

Plant specie	Ryegrass
Organics, %	1.2
pH	5.8
^{137}Cs in soil, Bq/kg	2150
^{137}Cs in plant, Bq/kg	792
K_2O, mg/100 g soil	0.37
P_2O_5, mg/100 g soil	4.5
Ca^{++}, mg/100 g soil	7.7
Mg^{++}, mg/100 g soil	8.2

Figure 6. Database information (example).

Statistical analysis was used to find the dependence of TF on soil characteristics as well as the variables that explain the main variance of the original data, to be able to apply an equation which describes variable contribution to the TF, and to try to specify the model relying on the results obtained during the work. These goals were reached using the following means:

- Correlation analysis
- Principal components analysis (PCA)
- Linear and nonlinear regression analysis

The result of correlation analysis was used to find out the relationships between all studied parameters. The total variance was described by several main parameters (Figure 7).

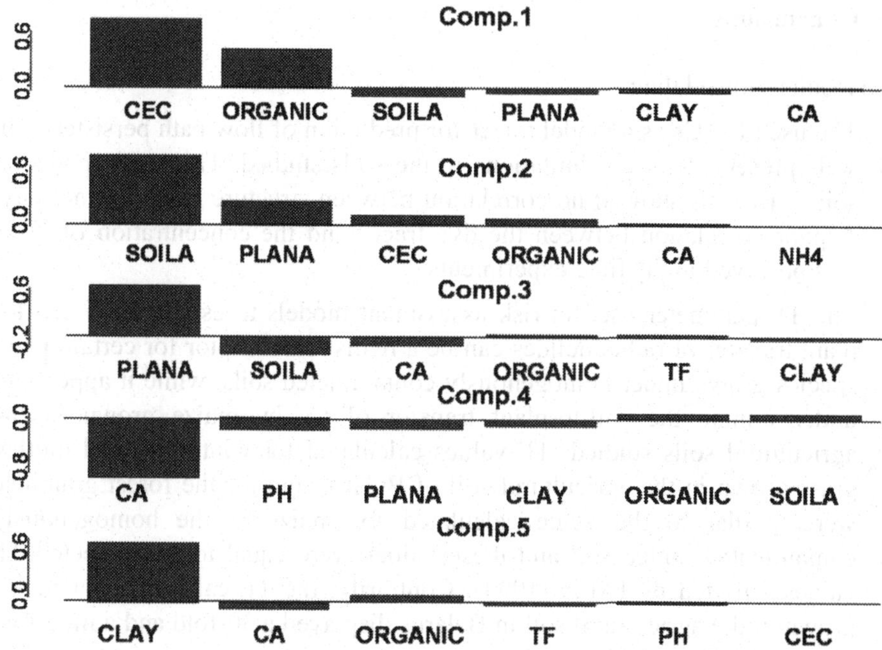

Figure 7. Input of different variables in the components (1992).

This allowed us to describe the behavior in the whole system with the help of one formula. Taking into account that after preliminary analysis the cation exchange capacity (CEC) and organic content were the main parameters influencing the TF values, we proposed the following equation for the calculation of the TF:

$$TF = A_0\ \text{CEC}\cdot\text{organics} + A_1,$$

where A_0 and A_1 are optimized free parameters (Table 6).

TABLE 6. Free parameters for the transfer factor calculation

Free parameter	Calculated value	Standard deviation
A_0	–0.019	0.014
A_1	0.205	0.016

4. Conclusions

The results showed that:

1. The use of ^{137}Cs as a model tracer for prediction of flow path persistence in soils presented strong limitations in the soils studied. The soddy-podzolic soil in Belarus showed no correlation between structure and ^{137}Cs activity. A poor correlation between the dye tracer and the concentration of ^{137}Cs was observed in the field experiments.
2. The TF parameter used in risk assessment models to estimate the soil-to-plant transfer of radionuclides can be a realistic descriptor for certain plant species grown under homogenously contaminated soils, while it appears to underestimate the soil-to-plant transfer of Cs by maize grown in the agricultural soils studied. TF values calculated for winter rye and motley grass grown in the agricultural soil of Belarus and for the forest graminae were similar to the value calculated for maize in the homogenously contaminated potted soil and these values were equal to the expected TF values reported by IAEA (1994). Contrarily, the TF calculated for maize grown in the agricultural soil in Belarus displayed a 30-fold and a nine-fold higher value, respectively.
3. The analysis of the database obtained from data collected in the Gomel region, south of Belarus, indicated that CEC and soil organic content were the main factors influencing the TF. Based on these results, a modified equation for the calculation of the TF is proposed.

Acknowledgments

The present work was carried out in collaboration with PLANT Nutrition Group, Institute of Plant Science, ETH Zurich, Switzerland and supported by a Swiss National Science Foundation (SCOPES) grant.

The authors express their hearty thanks to the head of the Ecological Laboratory of the Fachhochschule Ravensburg-Weingarten, Germany for support in measurements of soil and plant samples.

References

Albrecht, A., Schultze, U., Liedgens, M., Fluhler, H., Frossard, 2002, Incorporation soil structure and root distribution into plant uptake models for radionuclides: toward a more physically based transfer model, *J. Environ. Radioact.* 59: 329–350.

Avery, S.V., 1996, Fate of caesium in the environment: distribution between the abiotic and biotic components of aquatic and terrestrial ecosystems, *J. Environ. Radioact.* 30 (2): 139–171.

Bundt, M., Albrecht, A., Froidevaux, P., Blaser, P., Flühler, H., 2000, Impact of preferential flow on radionuclide distribution in soil, *Environ. Sci. Technol.* 34: 3895–3899.

Coughtrey, P. J., Kirton, J. A., Mitchell, N. G., Morris, C. 1989. Transfer of radioactive caesium from soil to vegetation and comparison with potassium in upland grasslands. Environmental Pollution 62: 281-315

IAEA, 1994, Handbook of parameter values for the prediction of radionuclide transfer in temperate environments, International Atomic Energy Agency, Vienna, Austria.

Rigol, A., Vidal, M., Rauret, G., 2002, An overview of the effect of organic matter on soil radiocaesium interaction: implications in root uptake, *J. Environ. Radioact.* 58: 191–216.

van Bergeijk KE, Noordijk H, Lembrechts J, Frissel MJ. 1992. Influence of pH, soil type and soil organic matter content on soil-to-plant transfer of radiocesium and -strontium as analyzed by a nonparametric method. J Environ Radioactivity 15: 265-276.

29. PLANT–MICROBE SYMBIOSIS FOR BIOREMEDIATION OF HEAVY METAL CONTAMINATED SOIL: PERSPECTIVES FOR BELARUS

ALENA LOPAREVA*, NADEZHDA GONCHAROVA
Laboratory of Environmental Monitoring, International Sakharov Environmental University, Minsk, Belarus

Abstract: Problems of heavy metal and radionuclide contamination are very important for Belarus. About 23% of its territory is contaminated with radionuclides, some areas are heavily contaminated with Cu, Zn, Pb, and Cd. Heavy metals can be carcinogenic and mutagenic even at low doses, are not biodegradable and are easily accumulated by live organisms. Phytoremediation and bioremediation technologies are just beginning to be developed in Belarus. Some research studies in the field of phytostabilization and phytoextraction of heavy metals in soil are realized. However, mechanisms of heavy metal accumulation and hyperaccumulation for cleanup of contaminated soils need more consideration. The problem of hyperaccumulator plants consists of their low biomass and slow growth. So, different ways to improve heavy metal phytoremediation should be considered. Bacteria play an important role in soil–plant interaction and their influence on heavy metal contamination attracted attention. Plant-assisted bioremediation involves remediation of contaminated soils with both plants and microorganisms associated to their rhizosphere. Alfalfa with its symbiotic bacteria was successfully applied to the final step of remediation of organically polluted areas. But the mechanisms and the role of bacteria in heavy metal accumulation by plants need further research. We present our results concerning the role of bacterial polysaccharides in metal accumulation by plants. It is shown that metal transfer into solution from elemental form depends on the exopolysaccharide (EPS) concentration in solution and glucuronic acid content in EPS. Bacterial polysaccharides do influence zinc uptake by the hyperaccumulator plant *Thlaspi caerulescens*. Possible perspectives for plant–bacterial remediation systems are discussed.

*To whom correspondence should be addressed. Lopareva A. Laboratory of Environmental Monitoring, International Sakharov Environmental University, Minsk, Belarus; e-mails: hotelinka@tut.by; lopareva@gmail. com

R. N. Hull et al. (eds.), Strategies to Enhance Environmental Security in Transition Countries, 401–414.

Keywords: phytoremediation; Cd; Zn; Ni; *Thlaspi caerulescens*; Rhizobiaceae

1. Introduction

Problems of heavy metal and radionuclide contamination are very important for Belarus. About 23% of its territory is contaminated with radionuclides, some areas are heavily contaminated with Cu, Zn, Pb, and Cd. Heavy metals can be carcinogenic and mutagenic even at low doses, are not biodegradable and are easily accumulated by live organisms. Soils in Belarus are mostly contaminated by Pb: concentration ranges between 109 and 370 mg/kg soil dry weight, Zn: 21 mg/kg, Cd: 0.4 and 3.5 mg/kg, Cu: 86 and 138 mg/kg, sulfates and other pollutants (Lukashev and Okun, 1996).

Conventional remediation methods such as soil excavation followed by coagulation–filtration or ion exchange are expensive and disruptive to ecosystems. *In situ* bioremediation and phytoremediation attract attention as low-cost and effective methods for restoration and remediation under many site conditions. Phytoremediation, the use of hyperaccumulator plants to extract heavy metals from contaminated soils, is one prospective method, gaining much attention recently. One of the possible strategies is phytoextraction of heavy metals by hyperaccumulator plants. Phytoextraction is the use of plants to remove toxic elements from contaminated environments. The metal is extracted from soil and translocated to harvestable parts of the plant (Baker et al., 1994).

The practical utility of many hyperaccumulators for phytoremediation may be limited, because many of these species, including *T. caerulescens* J et C Presl, are slow growing and produce little shoot biomass, severely constraining their potential for large-scale decontamination of polluted soils (Lasat et al., 2000). The effectiveness of phytoextraction depends on the capacity of hyperaccumulator plants to grow, to develop their root system, and to accumulate available metals into vegetable parts of plant (Schwartz et al., 1999). Increasing metal phytoavailability may be beneficial for phytoextraction and recovery using (hyper) accumulating plants, whereas an essential reduction in toxic metal phytoavailability (phytostabilization) may be promising for planting agricultural crops.

However, a lack of understanding of the mechanisms involved in heavy metal tolerance and hyperaccumulation prevents the optimization and wide use of the technique. It is therefore a research priority to gain basic information on sink capacities and transport mechanisms of heavy metals into the cells of hyperaccumulator species. *Thlaspi caerulescens*, a hyperaccumulator of Zn, Cd, and Ni (depending on the population tested) has been recognized as an interesting model for studying hyperaccumulation (Assuncao et al., 2003).

The exudates of bacteria in the plant rhizosphere play an important role in plant–microbial interactions. Soil microorganisms are constantly challenged by environmental fluctuations such as nutrient limitation, water shortage, exposure to elevated temperature, acidity, heavy metals, high osmolarity, or high salinity. To adapt to these adverse conditions, microbes have evolved mechanisms to monitor the environment and to alter gene expression patterns, enzyme activity, and transport proteins that enable them to cope with the new environmental conditions. Soils may become contaminated with metals from a variety of anthropogenic sources such as smelters, mining, power stations, industry, and application of metal pesticides, fertilizers, and sewage sludge to land (McGrath et al., 1995).

Mechanisms of heavy metal tolerance in bacteria are diverse and may involve energy-dependent efflux of the metal, precipitation of the metals as insoluble salts, alteration in membrane permeability to the metal, immobilization of the metal within the cell wall, production of chelating agents, such as metallothioneins, phytochelatines, and polysaccahrides (Mejáre and Bülow, 2001).

Microbial-driven processes are essential also for mineral weathering (Banfield et al., 1999). A complex role is played by microbial polysaccharides and proteinaceous structures, which can either inhibit mineral dissolution under some conditions by forming biofilms or promote chemical weathering by producing extracellular polymers of an acidic nature and increase metal bioavailability and phytoavailability and thus affecting the fertility of soils. The role of rhizobium polysaccharides in plant–host interaction is quite well studied, but their capacity to absorb heavy metals was investigated only recently (Mamaril et al., 1989).

Despite recent advances (Assuncao et al., 2003) the mechanisms underlying metal hyperaccumulation are still largely unknown and the role of soil bacteria on the behavior of heavy metals in soil and the mechanisms of metal accumulation by plants needs more deep understanding.

In this study, we investigated the physiological and molecular basis for plant heavy metal hyperaccumulation through an investigation of Cd, Zn, and Ni metal dissolution from elemental form into solution in the presence of rhizobium exopolysaccharides (EPS), and Cd and Zn transport and accumulation in shoots and roots of the metal hyperaccumulator plant *T. caerulescens* relative to the uronic acid content in EPS.

2. Materials and Methods

We previously isolated mutant Rhizobiaceae strains producing new EPS with promising properties from *Medicago* plants, which were exposed to chronic irradiation in the Chernobyl area (Lopareva et al., 2003).

For this experiment, four strains were chosen, producing EPS with different structures. Strains Str 1 and Str 5 were isolated from *Medicago sativa* plants grown on a radionuclide-contaminated area, where they probably acquired resistance to environmental pollution. A collection of strains with known properties were used as controls: *Sinorhizobium meliloti* M5N1, producing EPS succinoglucan and *S. meliloti* M5N1 CS, producing glucuronan from Laboratoire des Polysaccharides Microbiennes et Vegetaux, LPMV, Amiens, France (Table 1).

TABLE 1. Characteristics of chosen exopolysaccharides

EPS	pH	Uronic acids (%)	Glucose:galactose:glucuronic acid:fucose	Pyruvat:acetate: suucinate
Succinoglycan	7.8 ± 0.2	0	7:1:0:0	1:1:1
EPS Str 1	6.5 ± 0.2	36	1,5:1:2:1	1:2:0
EPS Str 5	5.0 ± 0.1	44	1,4:1:2:0	3:2:0
Glucuronan	5.4 ± 0.1	100	0:0:1:0	0:0:0

EPS were obtained in a fermentor, precipitated by three volumes of ethanol and lyophilized.

Pure Cd^0 and Zn^0 dissolution experiments in water and a nutritive solution of Arnon and Hoagland were conducted in 100 mL plastic pots over a period of 24 h at 28°C with manual shaking three times per day. EPS were added at a concentration of 0.1 g/L. Metal concentrations were Cd^0 at 280 mg/L (Prolabo, Cd shots, 99.5%), Zn^0 at 200 mg/L. Ni was added in Ni-goethite form at 200 mg/L.

Experiments with plants were conducted over 5 weeks in hydroponics pots in sterile conditions. Glucuronan and EPS Str 5 were injected at concentrations of 0.1 g/L in *T. caerulescens* J&C Presl (Brassicaceae) rhizosphere on 90 mL odified media Hoagland and Arnon (1938). The experiment was conducted in hydroponics in a phytotron under controlled standard conditions. Plants were grown in growth chambers under conditions of 120 μmol/ m^2/s white light, 16 h photoperiod, 25°C day, 19°C night, and 60% relative humidity. Metals were added in elemental forms as previously described for dissolution experiments.

Morphological characterization of the plants was performed after scanning plant roots using special equipment (WinRhizo scan system, Regent Instruments Image Analysis system) (Allison et al., 1991). The total Cd, Zn, and Ni content was analyzed after sulfuric acid mineralization by inductively coupled plasma mass spectrometer (ICP-MS) methods. Statistical analyses of the metal accumulation data were conducted using Student's *t*-test and analysis of variance (ANOVA) followed by Student–Nueman–Keuls posthoc analysis (Statistika 6.0). Samples sizes for *T. caerulescens* accessions were $n = 5$; for metal dissolution experiments $n = 3$.

3. Metal Dissolution in the Presence of Rhizobial Polysaccahrides

Experiments to determine the effect of EPS on metal dissolution in free form (Zn^0, Cd^0) showed that glucuronan, containing 100% glucuronic acid, significantly stimulated metal dissolution compared to the control (Table 2).

TABLE 2. EPS influence on elemental metals dissolution in water solution

EPS	Zn^{2+} (μg/L)	Cd^{2+} (μg/L)
Control	416.9 ± 47.7	53.48 ± 21.17
Succinoglycan	92.4 ± 6.9	2.34 ± 0.69
EPS Str 5	378.6 ± 54.6	87.25 ± 29.38
Glucuronan	6402.4 ± 220.4	255.7 ± 1.87

This effect is more significant in distilled water than in nutritive media for plants. Glucuronan increased the concentration of Cd ions in water solution by 4.8 times ($P < 0.05$) and Zn by 15 times ($P < 0.01$) compared to the control. Str 5 containing 50% glucuronic acid was not as efficient in Cd dissolution and increased the Cd ion concentration in solution only 1.6 times compared to the control. When using Succinoglycan, a polysaccharide, which does not contain glucuronic acid, Cd, and Zn concentration in solution was 23 ($P < 0.01$) and 4 ($P < 0.05$) times less, respectively. These results can be explained by the differences in EPS structure, as Succinoglycan did not stimulate metal dissolution and chelated ions from solution.

The analysis of Cd and Zn ion concentrations in solution and uronic acid concentration in EPS content (EPS concentration 0.1 g/L) showed significant correlation (Figure 1).

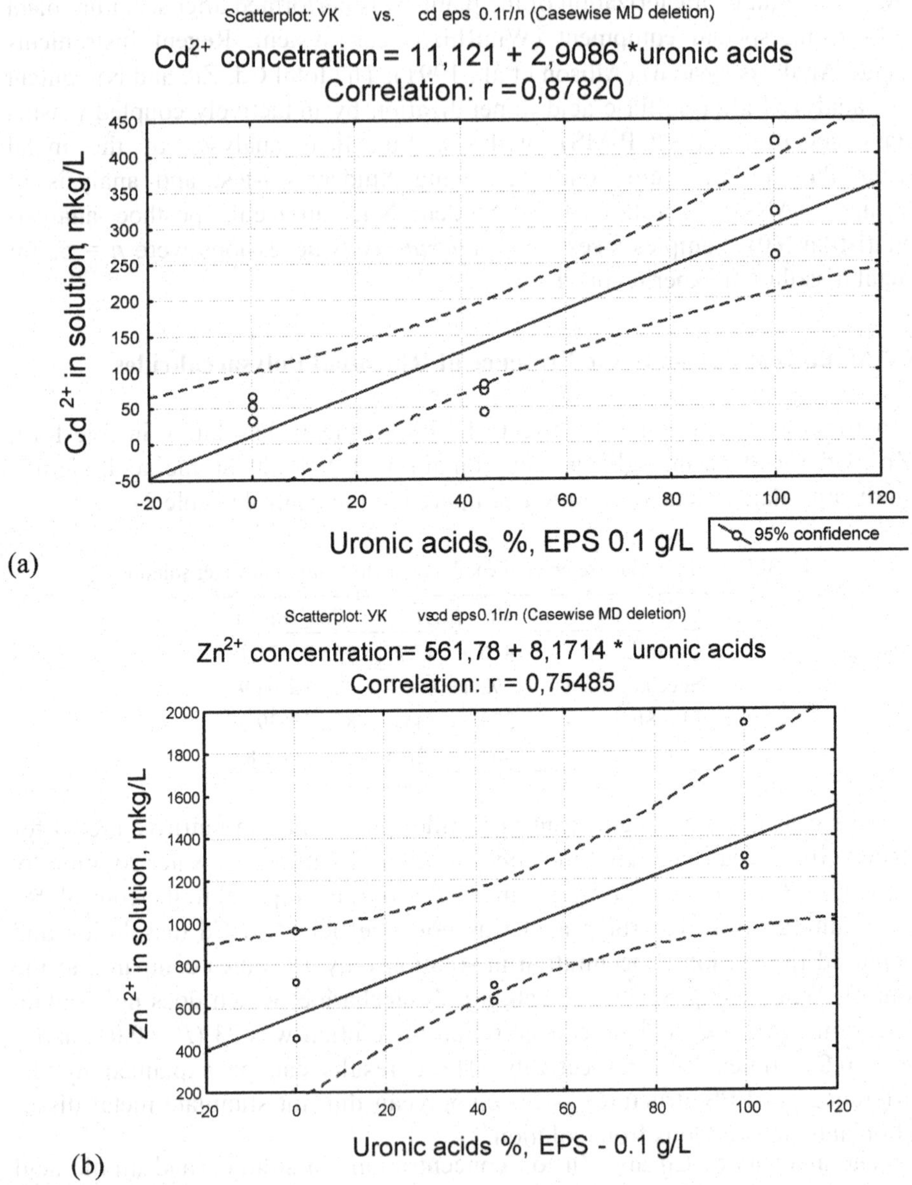

Scatterplot: УК vs. cd eps 0.1г/л (Casewise MD deletion)
Cd2+ concetration = 11,121 + 2,9086 *uronic acids
Correlation: r =0,87820
Cd 2+ in solution, mkg/L
Uronic acids, %, EPS 0.1 g/L
95% confidence
(a)
Scatterplot: УК vscd eps0.1г/л (Casewise MD deletion)
Zn2+ concentration= 561,78 + 8,1714 * uronic acids
Correlation: r = 0,75485
Zn 2+ in solution, mkg/L
Uronic acids %, EPS - 0.1 g/L
(b)

Figure 1. Correlation between uronic acid concentration in EPS and Cd (a) and Zn (b) ion concentration in solution.

Our results are in accordance with a study by Welch et al. (1999), which demonstrated that, like low-molecular weight organic ligands, the high-molecular weight acid polysaccharides can also affect mineral weathering

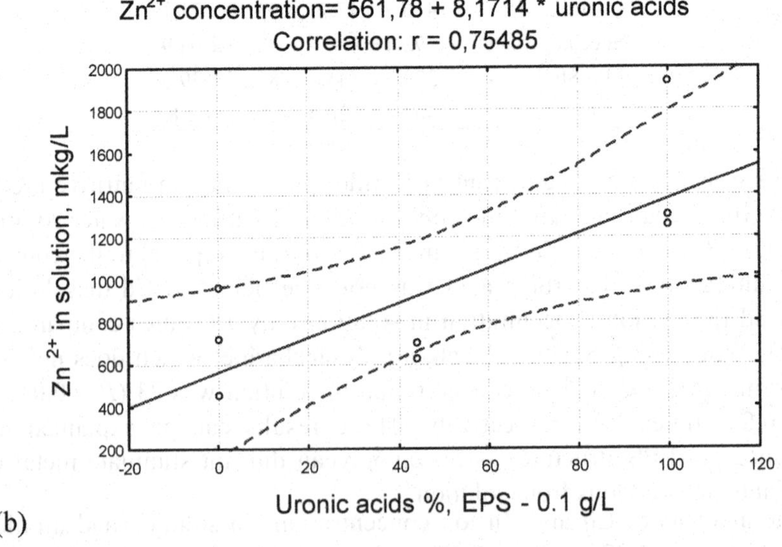

reactions. Microbial extracellular polymers can impact metal bioavailability by several mechanisms. Microbial capsules or slimes are composed of 90–99% water, and serve as antidesiccation protection for the cell. Capsules also help retain water at the mineral surface, increasing the time available for mineral hydrolysis reactions while maintaining diffusion pathways as water potential decreases (Chenu and Roberson, 1996). In addition to these indirect effects, microbial EPS can directly affect mineral weathering reactions. The compounds often contain carboxylic acid or other functional groups that complex with mineral ions, lowering the solution saturation state, thereby enhancing dissolution.

Our results show that metal transfer from elemental to ion form in solution strongly depends on the EPS concentration in solution and uronic acid content in the polysaccharide. Also the effectiveness of the polysaccharide depends on its polymer structure and composition, the orientation of complexing functional groups, dissociation, and metal–ligand stability constants of functional groups, as well as solution composition (Barker et al., 1998).

EPS interaction with Ni was studied on four polymers: glucuronan (100% glucuronic acid), Str 1 and Str 5 (about 50% glucuronic acid), and succinoglycan (0% glucuronic acid). Ni was added in goethite form (alpha-FeOOH), where Ni can replace Fe^{3+} ions, and is in a form not readily available to plants. Only about 0.25% of the Ni in goethite transferred in *T. caerulescens* plants (Echevarria et al., 1998).

Results are shown in Table 3. The presence of glucuronan increased the nickel concentration in water by 1.4 times compared to the control ($P < 0.01$).

EPS Str 1 and Str 5, containing 36% and 44% of glucuronic acid, respectively, increased the Ni^{2+} concentration 1.1 times greater than the control, while succinoglycan did not significantly change the nickel ion concentration in solution compared to the control.

TABLE 3. EPS influence on Ni^{2+} transfer from goethite into water solution

EPS	Ni concentration (μg/L)
Control	160.15 ± 2.03
Succinoglycan	157.87 ± 12.28
EPS Str 1	180.45 ± 1.45
EPS Str 5	172.45 ± 3.54
Glucuronan	216.31 ± 2.36

Correlation analysis showed that Ni^{2+} transfer from bound Ni in goethite into water solution depends on uronic acid content in EPS (Figure 2).

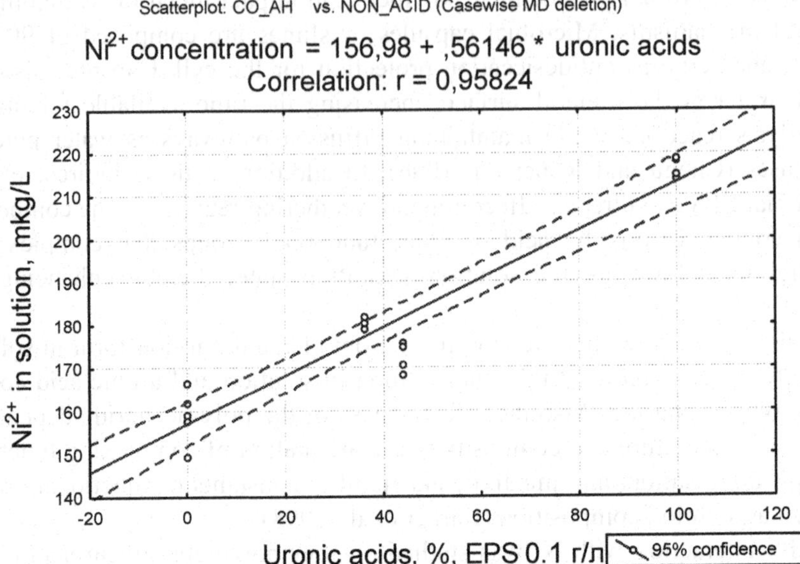

Figure 2. Correlation between nickel ion transfer into solution from goethite form and uronic acid content in polysaccharides.

The studied polysaccharides can be arranged in descending order by their affect on Ni desorption from goethite: glucuronan > Str 5 > Str 1 > succinoglycan.

The influence of polysaccharide concentration in solution on Cd^0 and Zn^0 dissolution was studied, for EPS concentrations: 0.01; 0.1; and 1.0 g/L. Results are presented in Figure 3.

Correlation analysis of Cd and Zn ions dissolved in solution from their elemental form, and EPS concentration showed that there is a strong correlation between the glucuronan concentration and Cd ion ($r_{\text{Spirmen}} = 0.95$, $P < 0.01$) and Zn ion ($r_{\text{Spirmen}} = 0.94$, $P < 0.001$) concentrations in solution. For EPS Str 5 there was not a significant correlation in the case of Cd ($r_{\text{Spirmen}} = 0.16$, $P < 0.01$), and a negative correlation in the case of Zn ($r_{\text{Spirmen}} = -0.88$; $P < 0.001$). The fact that Zn decreased with an EPS Str 5 increase in solution can be explained by differences in polymer structure and composition from glucuronan, as polymer Str 5 was more viscous and chelated metal ions from solution.

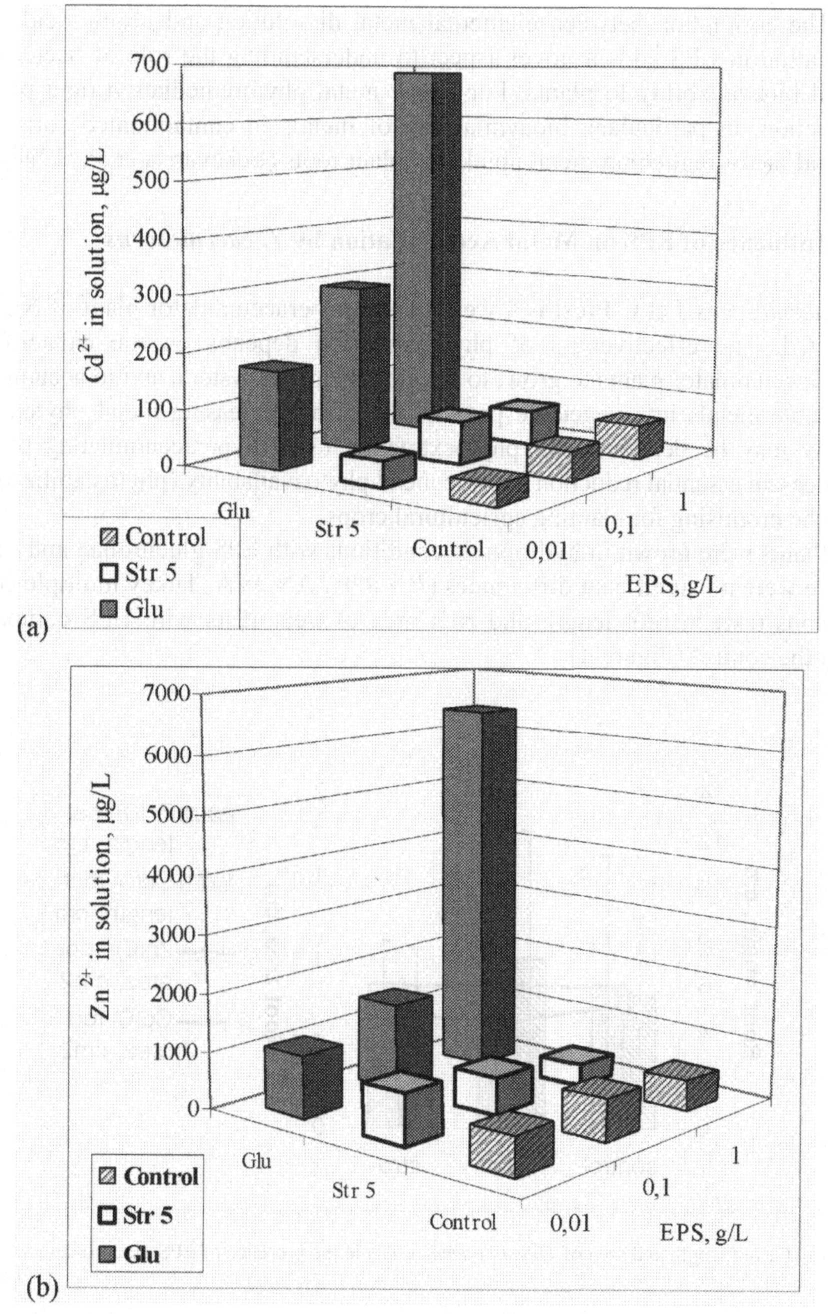

Figure 3. Elemental Cd° (a) and Zn° (b) dissolution in water solution, related to EPS concentration.

The correlation between elemental metal dissolution and uronic acid concentration in EPS adds a novel aspect to understanding the role of bacteria in metal bioavailability to plants. For heavy metal phytoremediation (and phytoextraction, in particular), bioavailability of metals in contaminated soils is a crucial factor regulating metal uptake by plant roots (Echevarria et al., 1998).

4. Influence of EPS on Metal Accumulation by *T. caerulescens*

T. caerulescens J et C Presl is a well-known hyperaccumulator plant of Ni, Zn, and Cd. The effectiveness of phytoextraction depends on the capacity of hyperaccumulator plants to grow, to develop their root system, and to accumulate available metals into vegetable parts of the plant. Increasing metal phytoavailability may be beneficial for phytoextraction using hyperaccumulating plants, whereas an essential reduction in toxic metal phytoavailability (phytostabilization) may be promising for planting agricultural crops.

Plants were grown in hydroponic conditions with EPS glucuronan and Str 5. There were no significant differences ($P > 0.05$, ANOVA, Tukey multiple comparisons test) in root length and root area of treatments with EPS compared with the control (Figure 4).

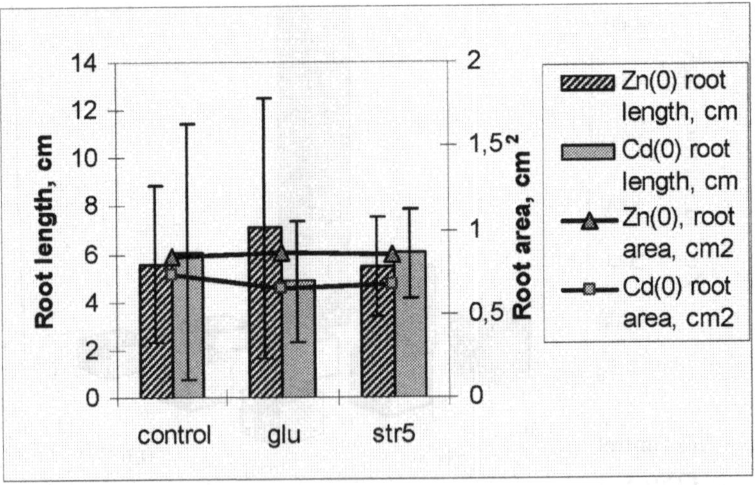

Figure 4. Root length and area of *Thlaspi caerulescens* in the presence of EPS glucuronan and Str 5.

Treatment of plants with glucuronan and EPS Str 5 stimulated Cd dissolution from the pure metal form (Cd^0) and its concentration in plant rhizosphere and roots. After 5 weeks of treatment, the Cd concentration was 17–20 mg/L in the presence of Str 5 and glucuronane, which is seven times greater than the control (2.8 mg/L). No visible toxic effects to plants were found. Our results are in accordance with Pr. Morel and colleagues who demonstrated that the concentration of Cd 24 mg/L is not toxic for *T. caerulescens* plants (Schwartz et al., 1999).

Despite the fact that the presence of polysaccharide-stimulated dissolution of pure metal and the Cd concentration in solution was high, this liberated Cd did not transfer to shoots (Figure 5).

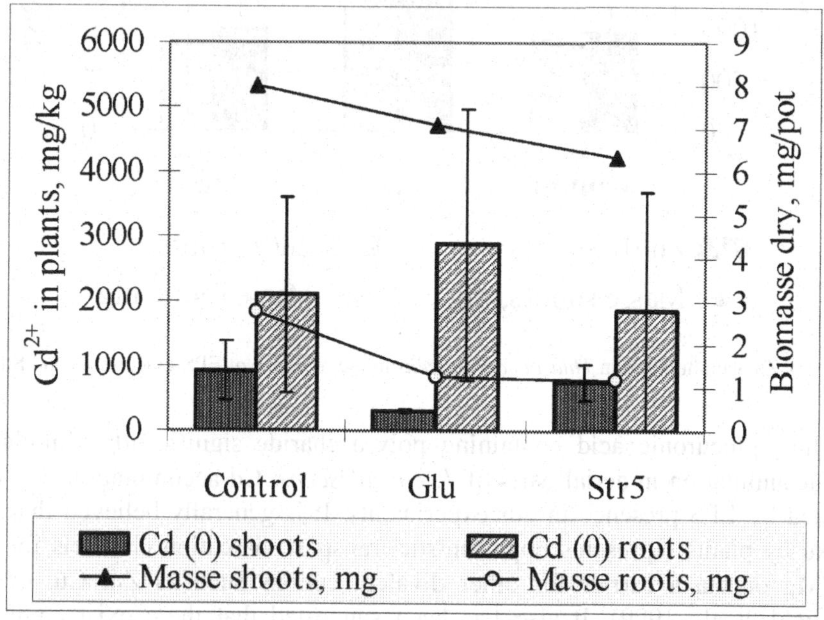

Figure 5. Cd accumulation in *Thlaspi caerulescens* in the presence of EPS Glucuronan and Str 5.

The Cd concentration in shoots treated with EPS Str 5 did not change and was three times less ($P = 0.005$) with glucuronan, when compared to control. But, the Cd concentration in roots was slightly (1.6 times) greater than in the control case.

Glucuronan presence in the *T. caerulescens* rhizosphere significantly stimulated Zn accumulation in shoots, at 1.7 times ($P < 0.01$) compared to the control (Zn^0 in nutritive media without EPS) (Figure 6).

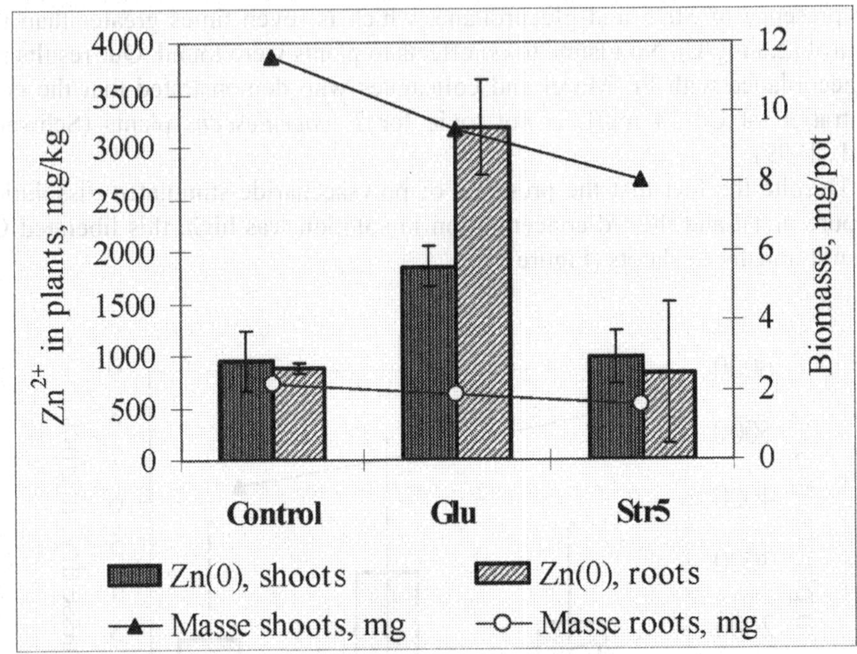

Figure 6. Zn accumulation in *Thlaspi caerulescens* in the presence of EPS glucuronan and Str 5.

Thus, glucuronic acid containing polysaccharide significantly stimulated zinc accumulation in aerial parts of *T. caerulescens*. Cd accumulation was not affected by EPS presence in our experiments. It is generally believed that Cd uptake by plants represents opportunistic transport via cation channels for Ca and Mg or via a carrier for other divalent cations such as Zn, Cu, or Fe (Banfield et al., 1999). It also has been suggested that there exist common mechanisms of absorption and transport of Zn and Cd in *T. caerulescens* because Cd and Zn have a similar electronic structure (Baker et al., 1994). There is evidence that the Zn transporter, *ZNT*1, recently cloned from the Prayon ecotype of *T. caerulescens* also can mediate transport of Cd. Zn was found in some cases to depress Cd uptake, indicating some kind of interaction between these metals (Lombi et al., 2000). However, because of differences in Cd uptake found between populations, it seems now that there might be differences between Cd and Zn for uptake and accumulation by *T. caerulescens* (Lombi et al., 2000; Assuncao, 2001, 2003). Our data confirms that there can exist various mechanisms as accumulation of Zn and Cd by *T. caerulescens* differed in the presence of glucuronic acid-containing polysaccharides.

5. Conclusion

The microbial soil community can alter the soil composition and properties in the course of metabolic processes, in particular by biodegrading and mineralizing organic matter. Thus, heavy metal phytoavailability can be affected by soil microbiota.

Production of organic acids by soil fungi and bacteria, including rhizobacteria, may promote solubilization, mobility, and bioavailability of metals and accompanying anions (e.g., phosphate) by lowering the pH and supplying metal-complexing organic acid ligands. An important role is played by bacterial exudates, in particular exogenous polysaccharides (Welch et al., 1999). This is confirmed by the correlation of heavy metal dissolution and uronic acid content in EPS shown in this work. Increasing metal phytoavailability may be beneficial for phytoextraction and recovery using hyperaccumulating plants, whereas an essential reduction in toxic metal phytoavailability (phytostabilization) may be promising for planting agricultural crops.

The effect of EPS influence on heavy metal sorption by plants and EPS influence on metal dissolution can be one of the possible explanations of increased accumulation of different pollutants by leguminous plants. Plant-associated symbiotic bacteria (*Rhizobium*) can produce polysaccharides with varied structure and monosaccharide composition, including uronic acid, which can affect the metal form in soil and its availability to plants. When planting leguminous crops on polluted areas it is necessary to consider EPS chemical structure produced by leguminous symbiotic bacteria.

The results show the role of rhizobium EPS in heavy metal accumulation by plants. However, field experiments with hyperaccumulator plants and leguminous plants infected by rhizobium strains, which produce EPS with different glucuronic content, remain to be conducted. These approaches are to be investigated in the near future.

Acknowledgments

The present work was carried out in collaboration with Laboratoire Sols et Environnement (LSE), ENSAIA/INPL INRA, Nancy, France and supported by NATO basic fellowship "Effects of cultivation in association of heavy metal hyperaccumulator plants and leguminous plants in contaminated soils". The authors are grateful to professors Josiane and Bernard Courtois for EPS (glucuronan, succinoglycan property of LPMV-CNRS FRE 2779, IUT Amiens) used in this study.

References

Allison, J.D., Brown, D.S., Novo-Gradac, K.J., 1991, MINTEQA2/PRODEFA2, A geochemical assessment model for environmental systems: Version 3.0 Users manual, EPA 600/3-91/021, US Environmental Protection Agency, Athens, GA.

Assuncao, A.G.L., Martins, P.D., De Folter, S., Vooijs, R., Schat, H., Aarts, M.G.M., 2001, Elevated expression of metal transporter genes in three accessions of the metal hyper-accumulator *Thlaspi caerulescens*, *Plant Cell Environ.* 24: 217–226.

Assuncao, A.G.L., Schat, H., Aarts, M.G.M., 2003, *Thlaspi caerulescens*, an attractive model species to study heavy metal hyperaccumulation in plants, *New Phytol.* 159: 351–360.

Baker, A.J.M., Reeves, R.D., Hajar, A.S.M., 1994, Heavy metal accumulation and tolerance in British populations of the metallophyte *Thlaspi caerulescens* J & C Presl (Brassicaceae), *New Phytol.* 127: 61–68.

Banfield, J.F., Barker, W.W., Welch, S.A., Taunton, A., 1999, Biological impact on mineral dissolution: application of the lichen model to understanding mineral weathering in the rhizosphere, *Proc. Natl. Acad. Sci. USA* 96: 3404–3411.

Barker, W.W., Welch, S.A., Banfield, J.F., 1998, Experimental observations of the effects of microorganisms on silicate mineral weathering, *Am. Mineralogist* 83: 1551–1563.

Chenu, C., Roberson, E.B., 1996, Diffusion of glucose in microbial extracellular polysaccharide as affected by soil water potential, *Soil Biol. Biochem.* 28: 877–884.

Echevarria, G., Morel, J.L., Fardeau, J.C., Leclerc-Cessac, E., 1998, Assessment of phyto-availability of nickel in soils, *J. Environ. Qual.* 27(Series D): 1064–1070.

Lasat, M.M., Pence, N.S., Garvin, D.F., Ebbs, S.D., Kochian, L.V., 2000, Molecular physiology of zinc, transport in the Zn hyperaccumulator *Thlaspi caerulescens*, *J. Exp. Bot.* 51: 71–79.

Lombi, E., Zhao, F.J., Dunham, S.J., McGrath, S.P., 2000, Cadmium accumulation in populations of *Thlaspi caerulescens* and *Thlaspi goesingense*, *New Phytol.* 145: 11–20.

Lopareva, A., Pawlicki-Julian, N., Goncharova, N., Courtois, B., Courtois, J., 2003, Characterization of EPS from Rhizobium of plants grown on the area contaminated with radionuclides. *Proceedings of Working Group 3, Soil–Plant Relationships*, XXXIII Annual Meeting of ESNA, Viterbo, Italia, pp. 42–46.

Lukashev, V.K., Okun, L.V., 1996, Heavy metal contamination of Minsk, Mn. 80p.

Mamaril, J.C., Paner, E.T., Capuno, V.T., Trinidad, L.C., 1989, Reduction of heavy metal concentration in liquids by Rhizobium polysaccharides, *Asean J. Sci. Technol. Dev.* 6(2): 75–86.

McGrath, S.P., Chaudri, A.M., Giller, K.E., 1995, Long-term effects of metals in sewage sludge on soils, microorganisms and plants, *J. Ind. Microbiol. Biotechnol.* 14: 94–104.

Mejáre, M., Bülow, L., 2001, Metal-binding proteins and peptides in bioremediation and phytoremediation of heavy metals, *Trends Biotechnol.* 19(2): 67–73.

Schwartz, C., Morel, J.L., Saumier, S., Withing, S.N., Baker, A.J.M., 1999, Root structures of the Zn-hyperaccumulator seedling *Thlaspi caerulescens* have affected by metal origin, content and localization in soil, *Seedling Soil* 208: 103–115.

Welch, S.A., Barker, W.W., Banfield, J.F., 1999, Microbial extracellular polysaccharides and plagioclase dissolution, *Geochim. Cosmochim. Acta* 63: 1405–1419.

PART V

SUMMARY CHAPTER

30. ENVIRONMENTAL SECURITY IN TRANSITION COUNTRIES: KNOWLEDGE GAPS, HURDLES, AND EFFECTIVE STRATEGIES TO ADDRESS THEM

SUSAN ALLEN-GIL[*]
Environmental Studies Program, Ithaca College, Ithaca, NY 14850, USA

OLENA BORYSOVA
Kharkiv National Academy of Municipal Economy, Dept. of Environmental Engineering and Management, 12 Revolutzii str Kharkiv 61002 Ukraine

Abstract: Achieving environmental security, which addresses the negative impact of human activities on the environment, direct and indirect effects of environmental scarcity and degradation, and the insecurity individuals and groups experience due to environmental change, is a critically important issue for transition countries confronting postindustrial pollution and/or resource scarcity. For some countries, knowledge gaps in how to apply scientific information to public policy present a formidable obstacle. In other cases, the rapid rate of political and economic change hinders the ability of jurisdictions to tackle complex environmental problems on large spatial or temporal scales. Also, the lack of a strong economy and funding limits the progress in remediating postindustrial contaminated sites. Effective strategies to overcome these knowledge gaps and implementation hurdles include using risk assessment to identify priority issues and/or sites, and to use entry into the European Union (EU) as a lever to implement and enforce legislation that will ensure environmental security.

Keywords: challenges; strategies; education; policy; management

[*]To whom correspondence should be addressed. Susan Allen-Gil. Environmental Studies Program, Ithaca College, Ithaca, NY 14850 USA

R. N. Hull et al. (eds.), Strategies to Enhance Environmental Security in Transition Countries, 417–424.

1. Introduction

As regionalization through the expansion of the European Union (EU) and globalization through international trade and tourism expand, environmental concerns tend to become increasingly complex and cover large spatial and temporal scales. Since the preindustrial times, environmental concerns have grown from localized pollution issues (such as cholera in drinking water and industrial urban smog) to global concerns regarding ozone depletion, climate change, and loss of biodiversity. The sheer number of environmental problems, the ways in which they interact with each other, and the potential for global ecological collapse makes environmental security a priority for all nations. The goal of environmental security is to protect people from the immediate and long-term ravages of nature, human-induced threats to nature, and deterioration of the natural environment (Barbu et al., 2007, this volume). It encompasses concerns about the negative impact of human activities on the environment, direct and indirect effects of environmental scarcity and degradation, and the insecurity individuals and groups experience due to environmental change such as water scarcity, air pollution, and global warming (Barbu et al., 2007, this volume). The importance of environmental security is evident in the joint development among the United Nations (UN), North Atlantic Treaty Organization (NATO), and other regional consortiums of the Environment and Security Initiative (EnvSec), which focuses on assessment of environmental and security risks, and capacity building, and institutional development to strengthen environmental cooperation as well as integration of environmental and security concerns and priorities in international and national policy making (Ozunu et al., 2007, this volume).

Issues of environmental security are particularly acute in transition countries where there is a legacy of severe industrial pollution almost unparalleled anywhere in world, a very rapid rate of cultural and economic change, and often insufficient financial resources and managerial expertise to adequately address them. For example, the Copsa Mica region of Romania is heavily polluted from previous lead smelting and carbon-processing operations. The spatial sale of contamination is so large that soil cleanup is economically prohibitive. The loss of other economic opportunities in the area has raised local unemployment to 60%, resulting in community support for reopening the smelter under new management (Gurzau, 2006, Sibiu Romania personal communication). This is just one example of the processes and challenges taking place from Eastern Germany to the Ural Mountains.

The participants of the Advanced Research Workshop were divided into two Working Groups (see Appendix A). The first group summarized the challenges facing transition countries, as well as some strategies to address these challenges. The second group proposed a path forward to apply these strategies and methods in a case study.

2. Knowledge Gaps and Hurdles

Achieving environmental security in transition countries is exceedingly challenging at present, in part because of the knowledge gaps and hurdles to effective practices and strategies.

The largest knowledge gap is in how to approach complex environmental problems. Soviet bloc scientists have strong technical knowledge in hard sciences, but translating that knowledge into public policy is not easy. The education system under the former Communist bloc emphasized natural history and detailed investigations of specific biological, chemical, and physical processes with insufficient attention to the interactions among them. Additionally, tradition of interdisciplinary education and research that looks into relations between ecological and social processes is younger than in the Western countries. Problem-solving and managerial skills were also never high on the university education agenda. There have been two significant consequences: (1) there is a disconnect between the basic science and the ability to apply it to solve complex, large-scale problems (2) policy makers are typically ill-equipped to approach concerns from an interdisciplinary perspective. This will require a continued education of scientists and policy makers to bridge the basic and applied sciences, and a retooling of environmental curriculum in higher education in transition and countries of the former Communist bloc toward a more interdisciplinary and applied approach.

The biggest hurdles to environmental security are great inefficiency and ineffectiveness of responsible structures resulting from little cooperation among different levels of government, when managers and rules are in constant flux because of the rate of political and economic change.

Lack of a strong economy and funding is clearly one the greatest limitations to cleaning up postindustrial contaminated sites. Even with significant international assistance, cleanup efforts have been inadequate due to the enormity of previous pollution, especially in areas of the "Black Triangle" (Malina, 2007, this volume) and due to the very character of environmental problems that cannot be solved overnight and therefore pose a threat to future generations. Another problem in Romania and other transition countries is that international organizations are not very effective because they usually are not familiar with business and management policies within the country and lack coordination among themselves.

Lack of a strong economy also hinders moving toward environmental security because poverty and social unrest can result in continued environmental damage out of individual desperation. Without concurrent social and economic programs, environmental concerns cannot be expected to improve and environmental security will remain elusive. It is difficult (and probably unfair) and unreasonable to expect people who spent more then 70% of their budget on food (statistics true for more than half of the population of transitional countries) to have environmental quality as their first priority. The consequence is that the level of awareness and recognition of environmental values in transition countries is relatively low, and this points to the crucial role of education once again.

3. Effective Strategies

While it is not possible to correct over 50 years of industrial pollution in a matter of a few years, there are some strategies to effectively remediate polluted sites, minimize future environmental concerns, and move toward greater environmental security.

Given the abundance and severity of polluted sites, it is important to adopt a general method to approaching environmental problems. One effective strategy to approach environmental concerns and identify local, regional, or international priorities is risk assessment. Risk assessment is a step-by-step process to compare the relative risks of taking various actions with respect to cleaning up contaminated areas (Swanson, 2007, this volume). Risk assessments can be conducted on various levels, from very broad-scale investigations comparing different sites to very detailed species-specific analyses within a particular site (Semenzin et al., 2007, this volume; Hull et al., 2007, this volume). This approach can be very powerful where finances to restore an area to pristine conditions are not available, and prioritization of actions for human health or ecological protection must be identified. The next step in an effective strategy is to rely on a combination of national and international expertise, if the appropriate academicians can be identified to provide specific cleanup advice at minimal cost. Educational institution and volunteer organizations can also assist in providing maximum cost efficiency.

While cleaning up polluted sites to prevent ongoing or additional contamination, it is also imperative that future activities are conducted in such a way that environmental concerns are minimized. To this end, one effective strategy is to use entry into the EU as the incentive and the EU mandates as the model for national environmental management and legislation. Before transition countries can apply to join the EU, they must demonstrate that they will be able to conduct business and ensure environmental protection at EU standards. This has prompted many countries, such as Romania and Bulgaria to adopt more stringent environmental legislation. In countries such as Russia or Moldova, where EU membership is not being pursued, enforcement of environmental legislation is typically insufficient.

Adopting best EU practices could be very efficient, especially when it comes to managerial and decision-making processes where changes could be introduced (given there is a political will) through awareness raising, training, and information dissemination. Such changes may not require huge technical resources but may have impacts far more significant for environmental security than any actual cleanup activities. One example of a proactive approach toward environmental problems is strategic environmental assessment (SEA) that aims to streamline and elevate environmental issues in the decision-making process. SEA is the formalized, systematic, and comprehensive process of evaluating the environmental impacts of a policy, plan, or program and its alternatives, including the preparation of a written report on the findings of that evaluation, and using the findings in publicly accountable decision making (Therievel et al., 1992). SEA has gained recently a reputation as an effective and efficient sustainability development tool, and its introduction in the planning and decision-making context of transition countries might prove very beneficial (Borysova, 2007, this volume).

Application and entry to the EU usually results in accelerated national and international economic activity. Although it is unlikely that investors will be in the position to provide enough funding to remediate formerly polluted sites, some of the profits from foreign investment should be used to develop standards for environmental policies and provide local, community-based environmental programs. Funding for environmental security must rely partially on the private sector, as national and international funding has been insufficient to date. In some countries, however, private business can, and does, play a vital role in addressing environmental concerns.

4. Conclusions

The knowledge and approaches described in this chapter, and discussed at the Advance Research Workshop can and should be tested. We propose doing so using a case study approach. A good case study would:

- Focus on a site with a broad range of issues (e.g., several contaminated media) or multiple stressors (e.g., radiation and metals)
- Address transboundary issues
- Be conducted in a transparent manner, and involve stakeholders (including government and universities)
- Allow for practical implementation or practical results
- Develop methods suitable for transition countries
- Stimulate networking among scientists of transition countries

- Disseminate the results widely (media, or government, or wider scientific community), with effective communication methods according to the audience (e.g., webpage, public meetings)
- Leave a legacy (e.g., knowledge and equipment)

Remediating heavily contaminated sites and minimizing future environmental impact are important, but will not in and of themselves provide true environmental security. This security can only be provided when the mind-sets of the people and the politicians include ecological and environmental health as a priority. This will require an aggressive education campaign in which environmental literacy is embedded in public education, environmental concerns are brought to the attention of the public and government, and managerial and decision-making skills are reinforced. It will also be critical that public citizens are included as active stakeholders in environmental policy at the local, regional, and national levels.

References

Barbu, C.-H., Sand, C., Oprean, L., 2007, Introduction to environmental security, in: *Strategies to Enhance Environmental Security in Transition Countries*, Hull, R., Barbu, C.-H., Goncharova, N. (eds.), NATO Security through Science Series C, Environmental Security, Springer, Dordrecht.

Borysova, O., 2007, Potential of strategic environmental assessment as an instrument for sustainable development of the countries in transition, in: *Strategies to Enhance Environmental Security in Transition Countries*, Hull, R., Barbu, C.-H., Goncharova, N. (eds.), NATO Security through Science Series C, Environmental Security, Springer, Dordrecht.

Hull, R.N., Barbu, C.-H., Hilts, S.R., 2007, Which human and terrestrial ecological receptors are most at risk from smelter emissions? In: *Strategies to Enhance Environmental Security in Transition Countries*, Hull, R., Barbu, C.-H., Goncharova, N. (eds.), NATO Security through Science Series C, Environmental Security, Springer, Dordrecht.

Malina, G., 2007, Integrated management strategies for environmental risk reduction at post-industrial areas, in: *Strategies to Enhance Environmental Security in Transition Countries*, Hull, R., Barbu, C.-H., Goncharova, N. (eds.), NATO Security through Science Series C, Environmental Security, Springer, Dordrecht.

Ozunu, A., Cadar, D., Petrescu, D.C., 2007, Environmental security – priority issues and challenges in Romania and European transition countries, in: *Strategies to Enhance Environmental Security in Transition Countries*, Hull, R., Barbu, C.-H., Goncharova, N. (eds.), NATO Security through Science Series C, Environmental Security, Springer, Dordrecht.

Semenzin, E., Critto, A., Marcomini, A., Rutgers, M., 2007, DSS-ERAMANIA: a decision support system for site-specific ecological risk assessment of contaminated sites, in: Hull, R., Barbu, C.-H., Goncharova, N. (eds.), *Strategies to Enhance Environmental Security in Transition Countries*, NATO Security through Science Series C, Environmental Security, Springer, Dordrecht.

Swanson, S., 2007, A Process for focusing clean-up actions at contaminated sites: lessons learned from remote northern sites in Canada, in: *Strategies to Enhance Environmental Security in Transition Countries*, Hull, R., Barbu, C.-H., Goncharova, N. (eds.), NATO Security through Science Series C, Environmental Security, Springer, Dordrecht.

Therievel, R., Wilson, E., Thompson, S., Heaney, D., Pritchard, D., 1992, *Strategic Environmental Assessment*, Earthscan, London.

Appendix A: Working Group participants

Group 1

Susan Allen-Gil – USA
Oleg Blum – Ukraine
Olena Borysova – Ukraine
Doina Botzan – Romania
Zdenek Filip – Czech Republic
Igor Gishkeluk – Belarus
Clive Lipchin – Israel
Grzegorz Malina – Poland
Sergey Melnov – Belarus
Alexandra Minis – Moldova
Jorg Perner – Germany
Ioana Tanasescu – Romania
Gagik Torosyan – Armenia

Group 2

Shmuel Brenner – Israel
Stanislav Geras'kin – Russia
Nadya Goncharova – Belarus
Plamen Gramatikov – Bulgaria
Inga Grdzelishvili – Republic of Georgia
Eduard Interwies – Germany
Arina Lahina – Moldova
Mihaela Lazarescu – Romania
Alena Lopareva – Belarus
Waldemar Nawrocki – Poland
Elena Semenzin – Italy
Stella Swanson – Canada
Valentina Tasoti – Romania

INDEX

A

B

C

D

H

I

L

M

N

P

R

S

T

V

W

X

Z